Receptor-Based Drug Design

DRUGS AND THE PHARMACEUTICAL SCIENCES

A Series of Textbooks and Monographs

1. Pharmacokinetics, *Milo Gibaldi and Donald Perrier*
2. Good Manufacturing Practices for Pharmaceuticals: A Plan for Total Quality Control, *Sidney H. Willig, Murray M. Tuckerman, and William S. Hitchings IV*
3. Microencapsulation, *edited by J. R. Nixon*
4. Drug Metabolism: Chemical and Biochemical Aspects, *Bernard Testa and Peter Jenner*
5. New Drugs: Discovery and Development, *edited by Alan A. Rubin*
6. Sustained and Controlled Release Drug Delivery Systems, *edited by Joseph R. Robinson*
7. Modern Pharmaceutics, *edited by Gilbert S. Banker and Christopher T. Rhodes*
8. Prescription Drugs in Short Supply: Case Histories, *Michael A. Schwartz*
9. Activated Charcoal: Antidotal and Other Medical Uses, *David O. Cooney*
10. Concepts in Drug Metabolism (in two parts), *edited by Peter Jenner and Bernard Testa*
11. Pharmaceutical Analysis: Modern Methods (in two parts), *edited by James W. Munson*
12. Techniques of Solubilization of Drugs, *edited by Samuel H. Yalkowsky*
13. Orphan Drugs, *edited by Fred E. Karch*
14. Novel Drug Delivery Systems: Fundamentals, Developmental Concepts, Biomedical Assessments, *Yie W. Chien*
15. Pharmacokinetics: Second Edition, Revised and Expanded, *Milo Gibaldi and Donald Perrier*
16. Good Manufacturing Practices for Pharmaceuticals: A Plan for Total Quality Control, Second Edition, Revised and Expanded, *Sidney H. Willig, Murray M. Tuckerman, and William S. Hitchings IV*
17. Formulation of Veterinary Dosage Forms, *edited by Jack Blodinger*
18. Dermatological Formulations: Percutaneous Absorption, *Brian W. Barry*
19. The Clinical Research Process in the Pharmaceutical Industry, *edited by Gary M. Matoren*
20. Microencapsulation and Related Drug Processes, *Patrick B. Deasy*
21. Drugs and Nutrients: The Interactive Effects, *edited by Daphne A. Roe and T. Colin Campbell*

22. Biotechnology of Industrial Antibiotics, *Erick J. Vandamme*
23. Pharmaceutical Process Validation, *edited by Bernard T. Loftus and Robert A. Nash*
24. Anticancer and Interferon Agents: Synthesis and Properties, *edited by Raphael M. Ottenbrite and George B. Butler*
25. Pharmaceutical Statistics: Practical and Clinical Applications, *Sanford Bolton*
26. Drug Dynamics for Analytical, Clinical, and Biological Chemists, *Benjamin J. Gudzinowicz, Burrows T. Younkin, Jr., and Michael J. Gudzinowicz*
27. Modern Analysis of Antibiotics, *edited by Adjoran Aszalos*
28. Solubility and Related Properties, *Kenneth C. James*
29. Controlled Drug Delivery: Fundamentals and Applications, Second Edition, Revised and Expanded, *edited by Joseph R. Robinson and Vincent H. Lee*
30. New Drug Approval Process: Clinical and Regulatory Management, *edited by Richard A. Guarino*
31. Transdermal Controlled Systemic Medications, *edited by Yie W. Chien*
32. Drug Delivery Devices: Fundamentals and Applications, *edited by Praveen Tyle*
33. Pharmacokinetics: Regulatory • Industrial • Academic Perspectives, *edited by Peter G. Welling and Francis L. S. Tse*
34. Clinical Drug Trials and Tribulations, *edited by Allen E. Cato*
35. Transdermal Drug Delivery: Developmental Issues and Research Initiatives, *edited by Jonathan Hadgraft and Richard H. Guy*
36. Aqueous Polymeric Coatings for Pharmaceutical Dosage Forms, *edited by James W. McGinity*
37. Pharmaceutical Pelletization Technology, *edited by Isaac Ghebre-Sellassie*
38. Good Laboratory Practice Regulations, *edited by Allen F. Hirsch*
39. Nasal Systemic Drug Delivery, *Yie W. Chien, Kenneth S. E. Su, and Shyi-Feu Chang*
40. Modern Pharmaceutics: Second Edition, Revised and Expanded, *edited by Gilbert S. Banker and Christopher T. Rhodes*
41. Specialized Drug Delivery Systems: Manufacturing and Production Technology, *edited by Praveen Tyle*
42. Topical Drug Delivery Formulations, *edited by David W. Osborne and Anton H. Amann*
43. Drug Stability: Principles and Practices, *Jens T. Carstensen*
44. Pharmaceutical Statistics: Practical and Clinical Applications, Second Edition, Revised and Expanded, *Sanford Bolton*
45. Biodegradable Polymers as Drug Delivery Systems, *edited by Mark Chasin and Robert Langer*
46. Preclinical Drug Disposition: A Laboratory Handbook, *Francis L. S. Tse and James J. Jaffe*

47. HPLC in the Pharmaceutical Industry, *edited by Godwin W. Fong and Stanley K. Lam*
48. Pharmaceutical Bioequivalence, *edited by Peter G. Welling, Francis L. S. Tse, and Shrikant V. Dinghe*
49. Pharmaceutical Dissolution Testing, *Umesh V. Banakar*
50. Novel Drug Delivery Systems: Second Edition, Revised and Expanded, *Yie W. Chien*
51. Managing the Clinical Drug Development Process, *David M. Cocchetto and Ronald V. Nardi*
52. Good Manufacturing Practices for Pharmaceuticals: A Plan for Total Quality Control, Third Edition, *edited by Sidney H. Willig and James R. Stoker*
53. Prodrugs: Topical and Ocular Drug Delivery, *edited by Kenneth B. Sloan*
54. Pharmaceutical Inhalation Aerosol Technology, *edited by Anthony J. Hickey*
55. Radiopharmaceuticals: Chemistry and Pharmacology, *edited by Adrian D. Nunn*
56. New Drug Approval Process: Second Edition, Revised and Expanded, *edited by Richard A. Guarino*
57. Pharmaceutical Process Validation: Second Edition, Revised and Expanded, *edited by Ira R. Berry and Robert A. Nash*
58. Ophthalmic Drug Delivery Systems, *edited by Ashim K. Mitra*
59. Pharmaceutical Skin Penetration Enhancement, *edited by Kenneth A. Walters and Jonathan Hadgraft*
60. Colonic Drug Absorption and Metabolism, *edited by Peter R. Bieck*
61. Pharmaceutical Particulate Carriers: Therapeutic Applications, *edited by Alain Rolland*
62. Drug Permeation Enhancement: Theory and Applications, *edited by Dean S. Hsieh*
63. Glycopeptide Antibiotics, *edited by Ramakrishnan Nagarajan*
64. Achieving Sterility in Medical and Pharmaceutical Products, *Nigel A. Halls*
65. Multiparticulate Oral Drug Delivery, *edited by Isaac Ghebre-Sellassie*
66. Colloidal Drug Delivery Systems, *edited by Jörg Kreuter*
67. Pharmacokinetics: Regulatory • Industrial • Academic Perspectives, Second Edition, *edited by Peter G. Welling and Francis L. S. Tse*
68. Drug Stability: Principles and Practices, Second Edition, Revised and Expanded, *Jens T. Carstensen*
69. Good Laboratory Practice Regulations: Second Edition, Revised and Expanded, *edited by Sandy Weinberg*
70. Physical Characterization of Pharmaceutical Solids, *edited by Harry G. Brittain*
71. Pharmaceutical Powder Compaction Technology, *edited by Göran Alderborn and Christer Nyström*

72. Modern Pharmaceutics: Third Edition, Revised and Expanded, *edited by Gilbert S. Banker and Christopher T. Rhodes*

73. Microencapsulation: Methods and Industrial Applications, *edited by Simon Benita*

74. Oral Mucosal Drug Delivery, *edited by Michael J. Rathbone*

75. Clinical Research in Pharmaceutical Development, *edited by Barry Bleidt and Michael Montagne*

76. The Drug Development Process: Increasing Efficiency and Cost Effectiveness, *edited by Peter G. Welling, Louis Lasagna, and Umesh V. Banakar*

77. Microparticulate Systems for the Delivery of Proteins and Vaccines, *edited by Smadar Cohen and Howard Bernstein*

78. Good Manufacturing Practices for Pharmaceuticals: A Plan for Total Quality Control, Fourth Edition, Revised and Expanded, *Sidney H. Willig and James R. Stoker*

79. Aqueous Polymeric Coatings for Pharmaceutical Dosage Forms: Second Edition, Revised and Expanded, *edited by James W. McGinity*

80. Pharmaceutical Statistics: Practical and Clinical Applications, Third Edition, *Sanford Bolton*

81. Handbook of Pharmaceutical Granulation Technology, *edited by Dilip M. Parikh*

82. Biotechnology of Antibiotics: Second Edition, Revised and Expanded, *edited by William R. Strohl*

83. Mechanisms of Transdermal Drug Delivery, *edited by Russell O. Potts and Richard H. Guy*

84. Pharmaceutical Enzymes, *edited by Albert Lauwers and Simon Scharpé*

85. Development of Biopharmaceutical Parenteral Dosage Forms, *edited by John A. Bontempo*

86. Pharmaceutical Project Management, *edited by Tony Kennedy*

87. Drug Products for Clinical Trials: An International Guide to Formulation • Production • Quality Control, *edited by Donald C. Monkhouse and Christopher T. Rhodes*

88. Development and Formulation of Veterinary Dosage Forms: Second Edition, Revised and Expanded, *edited by Gregory E. Hardee and J. Desmond Baggot*

89. Receptor-Based Drug Design, *edited by Paul Leff*

ADDITIONAL VOLUMES IN PREPARATION

Dermal Absorption and Toxicity Assessment, *edited by Michael S. Roberts and Kenneth A. Walters*

Automation and Validation of Information in Pharmaceutical Processing, *edited by Joseph F. deSpautz*

Receptor-Based Drug Design

edited by
Paul Leff
Astra Charnwood
Loughborough, Leicestershire, England

CRC Press
Taylor & Francis Group
Boca Raton London New York

CRC Press is an imprint of the
Taylor & Francis Group, an **informa** business

CRC Press
Taylor & Francis Group
6000 Broken Sound Parkway NW, Suite 300
Boca Raton, FL 33487-2742

First issued in paperback 2019

© 1998 by Taylor Francis Group, LLC
CRC Press is an imprint of Taylor & Francis Group, an Informa business

No claim to original U.S. Government works

ISBN-13: 978-0-8247-0162-8 (hbk)
ISBN-13: 978-0-367-40052-1 (pbk)

Library of Congress Cataloging-in-Publication Data

Receptor-based drug design / edited by Paul Leff.

p.   cm. -- (Drugs and the pharmaceutical sciences ; v. 89)
Includes bibliographical references and index.
ISBN 0-8247-0162-3 (alk. paper)
1. Drug receptors.   2. Drugs--Design.   3. G proteins.   4. Ion channels.   I. Leff, Paul.   II. Series.
[DNLM: 1. Drug Design.   2. Receptors, Cell Surface--physiology.
3. Receptors, Cell Surface--metabolism.   4. G-Proteins--physiology.
5. Ligands.   WI DR893B v.89 1998 / QV 744 R295 1998]
RM301.41.R39   1998
615' . 19--dc21
DNLM/DLC
for Library of Congress   98-13320
                                                                                CIP

**Visit the Taylor & Francis Web site at
http://www.taylorandfrancis.com**

**and the CRC Press Web site at
http://www.crcpress.com**

Preface

Receptor-based drug design is an enormous subject and one which, in principle, extends well beyond the scope of a single volume. The major challenge in assembling the chapters of this book, therefore, was to convey the essentials of the subject while confining the work to a practical size. Inevitably, this meant that hard choices had to be made about the material to be covered. The first of these was to restrict the receptors in question to those described by Sir James Black in his introductory chapter as *hormone receptors*, namely, those receptors that subserve the physiological effects of chemical messengers in the body. Then, I decided to concentrate on the two major superfamilies of receptors, G-protein-coupled receptors and ligand-gated ion channels, since, in my view, they best exemplify the principles of pharmacological analysis and classification on which rational drug design is based. The book begins with an exposition of these principles and descriptions of the experimental and theoretical methods used to characterize interactions between ligands and receptors. The pharmacological information that these methods provide establishes the basis for approaching and solving therapeutic problems; the numbers they provide are the basic material used by medicinal chemists in the design process. A range of examples are then presented that illustrate how these approaches have been, and continue to be, successfully employed to invent new medicines. Finally, some indications are given of the ways in which theoretical and technological advances are impacting on the drug-design process. It is hoped that, while far from comprehensive, this volume will provide the reader with a guide to the thinking and analytical processes with which the design of receptor-based drugs is approached, and act as a stimulus to challenge, and to participate in improving, the practice of design in the future.

I am deeply indebted to all the contributors for giving of their time, energy, and intellects to produce this book. Also, I must thank Jacky Wilkinson, my secretary, for taking on the considerable administrative burden associated with this endeavor, for sharing the editorial tasks, and for ensuring the successful completion of the project.

Paul Leff

Contents

Preface   iii
Contributors   ix

Introduction   1
James Black

Part One   PRINCIPLES AND MEASUREMENT OF
              DRUG-RECEPTOR INTERACTIONS

1. The Characterization and Classification of Receptors   7
    Patrick P. A. Humphrey

2. Functional Methods for Quantifying Agonists
    and Antagonists   25
    Iain G. Dougall

3. Quantification of Receptor Interactions Using Binding
    Methods   49
    Sebastian Lazareno

4. Drug Analysis Based on Signaling Responses to
    G-Protein-Coupled Receptors   79
    T. Kendall Harden, José L. Boyer, and Robert W. Dougherty

5. The Use of Electrophysiology to Improve Understanding of
    Drug-Receptor Interactions   107
    John G. Connolly and Charles Kennedy

Part Two   RECEPTOR-BASED DRUGS IN MEDICINE

6. **Selective Adrenergic and Glucocorticoid Treatments for Asthma**   135
   David Jack

7. **Serotonin 5-HT$_{1B/D}$ Receptor Agonists**   173
   Graeme R. Martin

8. **Histamine H$_2$-Receptor Antagonists**   195
   Michael E. Parsons

9. **Angiotensin Antagonists**   207
   Mark J. Robertson

Part Three   EXPERIMENTAL AND EMERGING DRUGS

10. **α-Adrenoceptor Agonists and Antagonists**   231
    J. Paul Hieble and Robert R. Ruffolo, Jr.

11. **P$_{2U}$-Purinoceptor Agonists: Emerging Therapy for Pulmonary Diseases**   253
    Eduardo R. Lazarowski, William C. Watt, T. Kendall Harden, and Richard C. Boucher

12. **Muscarinic Receptor Antagonists: Pharmacological and Therapeutic Utility**   273
    Richard M. Eglen

13. **NMDA Receptor Antagonists**   297
    John A. Kemp and James N. C. Kew

Part Four   NEW TRENDS IN DRUG DESIGN

14. **Computational Chemistry in Receptor-Based Drug Design**   323
    Ad P. IJzerman, Eleonora M. van der Wenden, and Wilma Kuipers

Contents

15. Recombinant G-Protein-Coupled Receptors as Screening Tools for Drug Discovery 345
A. Donny Strosberg and Prabhavathi B. Fernandes

16. Inverse Agonists and G-Protein-Coupled Receptors 363
Richard A. Bond and Michel Bouvier

Index 379

Contributors

James Black   Professor Emeritus, Department of Analytical Pharmacology, King's College London, London, England

Richard A. Bond, Ph.D.   Assistant Professor, Department of Pharmacological and Pharmaceutical Sciences, University of Houston, Houston, Texas

Richard C. Boucher, M.D.   Professor, Department of Medicine, University of North Carolina at Chapel Hill, Chapel Hill, North Carolina

Michel Bouvier, Ph.D.   Department of Biochemistry, University of Montreal, Montreal, Quebec, Canada

José L. Boyer, Ph.D.   Research Assistant Professor, Department of Pharmacology, University of North Carolina at Chapel Hill, Chapel Hill, North Carolina

John G. Connolly, Ph.D.   Department of Physiology and Pharmacology, University of Strathclyde, Glasgow, Scotland

Iain G. Dougall, B.Sc.   Principal Research Pharmacologist, Department of Pharmacology, Astra Charnwood, Loughborough, Leicestershire, England

Robert W. Dougherty, Ph.D.   Senior Research Biologist, Inspire Pharmaceuticals, Inc., Durham, North Carolina

Richard M. Eglen, Ph.D.   Vice President and Director, Center for Biological Research, Roche Bioscience, Palo Alto, California

Prabhavathi B. Fernandes, Ph.D.   Chief Executive Officer, Small Molecule Therapeutics, Inc., Monmouth Junction, New Jersey

T. Kendall Harden, Ph.D.   Professor, Department of Pharmacology, University of North Carolina at Chapel Hill, Chapel Hill, North Carolina

J. Paul Hieble, Ph.D.   Director, Department of Cardiovascular Pharmacology, SmithKline Beecham Pharmaceuticals, King of Prussia, Pennsylvania

Patrick P. A. Humphrey   Professor of Applied Pharmacology, Department of Pharmacology, University of Cambridge, and Glaxo Institute of Applied Pharmacology, Cambridge, England

Ad P. IJzerman, Ph.D.   Associate Professor, Division of Medicinal Chemistry, Leiden/Amsterdam Center for Drug Research, Leiden, The Netherlands

David Jack   Former Research and Development Director, Glaxo Holdings plc, Wheathampstead, Hertfordshire, England

John A. Kemp, Ph.D.   Research Area Head, Preclinical CNS Research, F. Hoffmann-La Roche Ltd., Basel, Switzerland

Charles Kennedy, Ph.D.   Department of Physiology and Pharmacology, University of Strathclyde, Glasgow, Scotland

James N. C. Kew, Ph.D.   Laboratory Head, Preclinical CNS Research, F. Hoffmann-La Roche Ltd., Basel, Switzerland

Wilma Kuipers   Solvay Pharma, Weesp, The Netherlands

Sebastian Lazareno, Ph.D.   MRC Collaborative Centre, London, England

Eduardo R. Lazarowski, Ph.D.   Research Associate, Department of Medicine, University of North Carolina at Chapel Hill, Chapel Hill, North Carolina

Graeme R. Martin, Ph.D.   Institute of Pharmacology, Roche Bioscience, Palo Alto, California

Michael E. Parsons   S. B. Professor of Pharmacology, Biosciences Division, University of Hertfordshire, Hatfield, Hertfordshire, England

Mark J. Robertson, Ph.D. Department of Pharmacology, Astra Charnwood, Loughborough, Leicestershire, England

Robert R. Ruffolo, Jr., Ph.D. Vice President and Director, Department of Pharmacological Sciences, SmithKline Beecham Pharmaceuticals, King of Prussia, Pennsylvania

A. Donny Strosberg, Ph.D. Professor, Molecular Immunopharmacology Laboratory, Institut Cochin de Génétique Moléculaire (ICGM), Paris, France

Eleonora M. van der Wenden, Ph.D. Division of Medicinal Chemistry, Leiden/Amsterdam Center for Drug Research, Leiden, The Netherlands

William C. Watt Department of Pharmacology, University of Washington, Seattle, Washington

Introduction

James Black
King's College, London, England

Pharmacologists are addicted to receptors, to their subdivision and classification, to their explanatory properties, and to their use as targets and templates for designing new drugs. However, pharmacologists have no monopoly on receptors. Physiologists have an established position on chemoreceptors, baroreceptors, telereceptors, and so on. Virologists, immunologists, botanists, and, indeed, any other biologists who wish to convey a recognition concept also invoke appropriate receptors. Arguably, "receptor" needs a prefix or adjective to be informative. Traditionally, pharmacologists have used the receptor concept to support the ideas of affinity and efficacy. Affinity is the general biological concept of chemical recognition but the idea of efficacy is unique to pharmacology. Efficacy is a property of agonists. "Agonist" is an umbrella term for a hormone, or neurotransmitter, or autacoid, or mediator, or growth factor, or chemotactic agent, or whatever has the ability to induce an effect in a responsive cell by activating its own, specific, or conjugate, receptor. All of these agents can be regarded as chemical messengers. At the beginning of the century Bayliss and Starling coined the word "hormone" for a chemical messenger in physiology. It happens that secretin is a circulating hormone. It was also the first to be described, so it seems unlikely that the inventors of the word meant to restrict its use to circulating chemical messengers. During the 1930s A. J. Clark in his classic text *The Mode of Action of Drugs on Cells,* used the term "hormone" to cover acetylcholine and epinephrine as well as vasopressin, insulin, thyroxine, and so on. I believe that pharmacologists should use "hormone" as the catchall adjective for the receptors that are of particular interest to them. It is these hormone receptors that are the subject of this book.

1

Receptors, now recognized as chemical entities, began life as a late-19th-century concept, proposed independently by Ehrlich and Langley. There seems to have been little enthusiasm for the concept at the beginning. Indeed, Henry Dale, the doyen of British physiologists, thought the idea was redundant. However, the concept was kept alive by a small coterie of theoretically minded pharmacologists who were interested in trying to interpret dose–response relations. Hill, a pupil of Langley, sought to interpret the dose–response curves generated by nicotine on the frog rectus abdominis preparation by applying the Law of Mass Action. To make the algebra work, he had to turn the receptor concept into a mathematical operator. That was in 1906. Over the next 50 years, quantitative pharmacology was quietly developed on the basis of the receptor-as-mathematical-operator concept. Although the theoretical work of Clark, Gaddum, Schild, Ariens, Furchgott, and others was not ignored by mainstream pharmacologists and textbook writers, neither was it incorporated into their work until after 1950. Ahlquist complained that he failed to get his 1948 paper on adrenotropic receptors published in any of the main pharmacology journals and only succeeded in getting it published in the *American Journal of Physiology* because the editor, W. F. Hamilton, was a friend and colleague. Yet Ahlquist's paper was the first time that "receptor" was used as an explanatory model, was the start of the enthusiasm among pharmacologists for classification of receptor subtypes, and, as it happens, was the foundation of my own work on beta-blockers.

Receptors have turned out to be good targets for drug hunters. After all, the highly selective interaction between a hormone and its receptor gives the hormone drug-like properties. Indeed, one of the greatest discoveries of analytical pharmacology has been that the selective activity of drugs is usually determined by their interaction with molecular sites—receptors, enzymes, pumps, and so on—where natural ligands interact. Medicinal chemists discovered that they were able to turn natural hormones into drugs by making analogs and derivatives of the natural ligand. The process was particularly successful when the complementary hormones were small, nonpeptidic, molecules. Then, pharmacologists used bioassays to detect changes in the efficacy of these modified hormones. A favorite strategy was to use the exposure of a partial loss of efficacy to identify the molecular substrate for agonist activity. The iterative process of synthesis-assay-redirected synthesis was slow but effective. An important advance came when the bioassays used for drug screening were coupled to the neglected theoretical models of drug–receptor interactions, models that had been developed 30 years earlier. The models were applicable because the inherent and inscrutable complexity of intact tissues was avoided by using null methods, i.e., comparing concentrations that produced equal responses, and by making the test ligand interact with the natural one, on the assumption that only a single class of proteins was being activated. The use of these models enhanced intellectual property by allowing much more explicit definition of a drug's action. Thus, systematic medicinal chemistry coupled to model-interpreted bioassay provided the platform for receptor-based new drug discovery for over 30 years. However, that was yesterday's technology. First came radioligand-binding techniques and then came the spectacular advances in molecular biology. Together, they have revolutionized receptor-based drug discovery.

Comparative bioassay had been the rate-limiting step in the iterative screening loop. So, I remember that the arrival of ligand-binding technology was greeted with a lot of enthusiasm. However, the technique was found to have some limitations. Radioligand-binding works best when the labeled molecule has a high affinity and high selectivity for the particular receptor. Many hormones, particularly nonpeptide hormones, we now know, do not have a high affinity (or even high efficacy) for their conjugate receptors. Perhaps low affinities derive from high dissociation rate constants and high turnover rates, reminding us of Paton's Rate Theory, which has been neglected rather than discredited. Low efficacy in hormones allows for powerful potentiating interactions with other hormones reaching the responding cells at the same time, interactions that amplify the informativeness of the excitatory process. Whatever the reason, the ideal of these assays, namely measurement of the displacement of the "hot" hormone by "cold" ligands, was often not achieved. The assays worked best when high-affinity antagonists had already been discovered and could be labeled. The other disadvantage compared to intact-tissue bioassay is the lack of a physiologically relevant measurement of efficacy, particularly of partial agonists. Tweaking the incubation conditions to disclose receptor distribution, the index of ternary complex formation, can estimate efficacy potential but without necessarily predicting the expression of efficacy in intact tissues. Indeed, in our hands, ligand-binding is most informative when it is combined with classic bioassays.

The impact of molecular biology on receptor-based drug discovery has been much more profound than radioligand-binding. In less than 25 years, the hormone receptor has gone from idea to image, from nebulous concept to chemical structure. From the point of view of screening ligands, recombinant DNA technology can be used to express receptors onto all kinds of cells harvested ubiquitously from bacteria to mammals. These expression systems are the basis of most of the robot-driven, high-throughput screens, which are such a feature of pharmaceutical research today. However, these are still ligand-binding assays and, as far as efficacy detection is concerned, they have the same shortcomings as conventional radioligand-binding assays. Currently, there is a lot of effort and some success in devising screens that will detect the existence of potentially efficacious conformational changes in receptors. Nevertheless, the same problem arises as in conventional radioligand-binding assays. The question is, to what extent have the artificial conditions of the expression system disturbed the relationship between receptor activation and effect? As far as I can see at this time, intact-tissue bioassay will always be needed to calibrate artificially constructed assays.

The new robotic assays have huge appetites for ligands. Once existing libraries of compounds have been tested, these appetites can only be satisfied by high-yielding combinatorial chemistry. In combinatorial chemistry there is an inverse relationship between constraints and yields. So, the voracity of the new high-throughput screens puts a premium on combinatorial synthetic programs, which are designed to be unconstrained and unsystematic. In this respect, the new synthetic programs are quite unlike the classic systematic approach, which was based on hormone structure. The design of the screening program will, of course, ensure that only highly selective ligands will be picked out. However, there are no systematic or modeling constraints to direct ligands to active sites. The chemical nature of active sites, a 3D array of amino acid side chains and backbone, is not inherently different from any other part of the protein. There would seem to me to be a fair chance that foreign ligands will find all sorts of accessible, sticky, parts of proteins to combine to so as to alter their function. Selectivity seen in a screening pool of 50 proteins may not necessarily hold for 50,000 proteins. We will have to watch the development of the ligands of the new era with great interest.

However, random synthesis might turn out to succeed where molecular modeling has still to win its spurs. Molecular modeling has given me much second-hand pleasure and excitement during 40 years of working with medicinal chemists. I started with ball and stick models that did not "talk" to me very much. However, it was the arrival of CPK models that switched me on. Even today, I like to have a CPK model of a molecule of current interest on my desk. I use it like a Japanese jade feeling piece, twisting and turning it in my hands to help to concentrate my imagination. Although I can twist the model, it is essentially a static structure, a model of possibilities. However, once the number crunchers were big enough, fast enough, and accessible enough, chemists with a mathematical bent were able to move from static structures to molecular dynamics. Identification of low-energy conformations and calculation of properties, such as electrostatic fields, has introduced a whole new dimension to molecular modeling. Today, modeling profoundly stimulates the imagination of medicinal chemists. Deeply satisfying insights are discovered in structure–activity analysis. So does this mean that molecular modeling can be prescriptive for new drug activity? The jury is still out. In my experience, modeling can be very helpful about making extrapolations from structure–activity analysis. However, totally new structures designed to embody desirable activity are still remote. The molecular modeler is working in a rarefied space with the molecule influenced only by its own internal forces. The real molecule, made to the modeler's design, will have to interact with its intended target when it is being buffeted in the molecular density of the biophase. A whole new set of constraints on molecular conformations are likely to obfuscate the theoretical calculations. So, we have a long way still to go but the traveling is very exciting.

I have spent most of my life involved, in some way or another, with receptor-based drug design. During this time, I have been continuously amazed and excited by the inventiveness of pharmaceutical research. The beauty of it all is that sick people have been a significant beneficiary of the discovery of new drugs. Some of these success stories are recounted in this book. Taken together, the contributions to this treatise are an eloquent testament to the fecundity of receptor-based drug design.

The Characterization and Classification of Receptors

Patrick P. A. Humphrey
University of Cambridge and Glaxo Institute of Applied Pharmacology,
Cambridge, England

I. INTRODUCTION

At the beginning of this century, Ehrlich and Langley put forward the idea that intercellular chemical signaling was mediated by receptive sites on cell surfaces (1). It was not until nearly the end of the century that the first amino acid structure of a receptor (the nicotinic receptor for the neurotransmitter, acetylcholine) became known (2). In this period leading up to the revolutionary effect of molecular biology on pharmacology, pharmacologists learned much about the variety of receptor types, their recognitory characteristics, and their degree of selectivity for different chemicals, both endogenous (hormones, neurotransmitters, etc.) and synthetic (drugs), but they were only able to guess at their substance and structure. Amusingly, De Jongh metaphorically compared this process to a man exchanging letters with a beautiful lady whom he does not know; with time he would learn much about her, despite never having seen her (3). Today, in reference to De Jongh's metaphor, the "picture" more frequently comes first; amino acid sequences (deduced from the nucleotide sequence of cloned cDNA or genes) now abound for receptors, not only those that are familiar but also those for which there are no good ligands or indeed the endogenous ligand is unknown, so-called "orphan receptors." (4,5). With this explosion of new information about receptor structure and even their distribution it becomes essential to reconsider the way we approach the characterization and classification of receptors for neurotransmitters, hormones, and autacoids. Only two receptor superfamilies will be exemplified here, the hepta-helical G-protein-linked receptors and the multimeric ligand-gated ion channel receptors, which in mediating the wide variety of effects of numerous neurotransmitters have been the most studied by pharmacologists to date. The two other major structural classes of receptor are the enzyme-linked receptors (e.g., for insulin and growth factors) and the nuclear receptors (e.g., for steroids).

Even today, the precise definition of a receptor is problematic. Certainly Stephenson's long-established definition of "that spatial arrangement of atoms to which a substance endogenous to the organism attaches itself as an essential step in modifying cellular function" no longer seems appropriate (6). Nevertheless it can be argued that it is still essential to classify receptors according to their natural ligands, because of the critical importance of such information in understanding the functional role of a given receptor or receptors in the whole body. In the light of modern knowledge, a receptor must now be regarded as a macromolecule, which may or may not be a single molecular entity, with multiple sites of interaction (7–9). The individual proteins can be defined in terms of their corresponding genes, but there may be local molecular differences according to the host cell, in terms of variable patterns of glycosylation and different transcriptional splice variants (10–14). A modern definition of a receptor has been suggested as "the entire protein molecule or cluster of molecules which can selectively recognize and be collectively activated by an endogenous ligand (agonist) to mediate a cellular event" (15). Synthetic drugs may recognize different (amino acid) sites on the same receptor, and even if only a very localized region on the macromolecule is required for agonist binding, adjacent proteins (specific G-proteins or ion channel components) will be essential for transducing the effect produced by agonist binding (16–21).

II. CRITERIA FOR RECEPTOR CHARACTERIZATION

Receptors for hormones and neurotransmitters can be characterized by the effects of synthetic ligands or drugs, with selective actions at one or more receptors. Such ligands display either agonist or antagonist actions, which can be quantified to define the characteristics of receptor function, often providing powerful discriminatory data between receptors of interest. By this approach good evidence for different types of receptor for a single hormone was provided, long before the molecular structures of receptors were known. Thus, Ahlquist suggested that norepinephrine acted at two types of receptor, which he called α and β, on the basis of two distinct rank orders of agonist potencies for norepinephrine, epinephrine, and some synthetic analogs in various tissue preparations (22). This proposal found wider acceptance with the discovery of selective α- or β-adrenoceptor blocking drugs (antagonists), which further distinguished between the two types of receptor for norepinephrine (23–26). Agonists were also shown to differentiate between putative β_1- and β_2- adrenoceptor subtypes and hence there was no surprise when their molecular structures were determined and found to be distinct but related, sharing about 50% homology (27,28). In many other areas of pharmacology too, receptor subtypes have been identified with the aid of selective drug tools, well in advance of identifying the nature and amino acid composition of the receptors. Examples are numerous and include receptors for histamine, 5-hydroxytryptamine (5-HT), prostaglandin, and adenosine (29–34). Clearly, functional characteristics are predictive of structural differences and therefore scientifically valuable but also importantly provide invaluable information toward drug discovery. There is no better illustration of this than the work of the Nobel Laureate, Sir James Black, whose focus on functional characterization of β-adrenoceptors and H_2 histamine receptors led to the development of important medicines for the treatment of cardiovascular diseases and gastric ulceration, respectively (24,26,30).

However, with the ready availability of structural information today, the approach to receptor characterization and classification requires revision. This exercise has been systematically pioneered over more than a decade with regard to receptors for 5-HT, and some interesting considerations have come to light (32,35–39). Many of the known receptor types for 5-HT had been identified functionally prior to their gene cloning (e.g., 5-$HT_{1B/D}$, 5-HT_2, 5-HT_3, 5-HT_4), but others had not. To simplify a potentially confused literature regarding the true identities of the growing number of receptors for 5-HT, an attempt was made to establish a framework of classification that would rationalize existing data and yet provide a basis for incorporation of newly identified recombinant 5-HT receptor types for which there might be little or no functional data (37). This led to the concept that information on receptor function (operation) and receptor structure were equally important and also that transductional information should be included (later seen as providing additional operational characteristics) in an integrated scheme (see Fig. 1).

III. RECEPTOR FUNCTION

Unquestionably data on receptor function are important, and although often indicative of receptor structure, the corollary is not always so, such that in some cases differences in receptor structure seem to have little functional consequence whereas in others a single amino acid change can make a major difference to the functional characteristics (16,18,39,41,42).

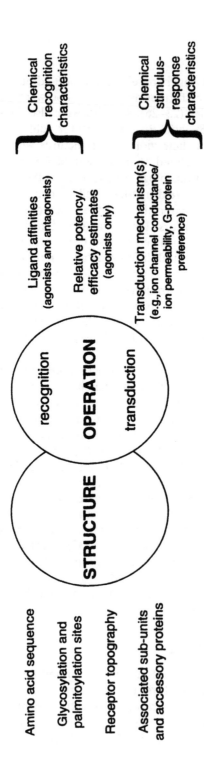

Figure 1 An integrated approach toward receptor characterization. It is proposed that both structural and operational data are essential for receptor characterization in an integrated pharmacological scheme. The importance of transductional data is increasingly being recognized, but rather than treat this as a third set of data for classification purposes, it is recommended that such information be considered as a component of receptor operation (see text).

However, it must be recognized that it is not qualitative functional information that is critical to *receptor characterization,* rather it is the rigorously derived quantitative data from functional studies (e.g., dissociation equilibrium constants for antagonists) which provide the essential fingerprint relating to the *chemical* identity of a given receptor, defined by the chemicals that functionally interact with it. Equivalent data for both agonists and antagonists can be obtained from radioligand-binding studies but these cannot be referred to as "functional studies" as only binding interactions are involved. The term *operational* has therefore been introduced to collectively describe data from both functional and radioligand-binding studies, which together provide estimates of drug affinity, which reflect the *recognitory* characteristics of a given receptor (37,43). However, it is important to emphasize that attempts to measure affinity constants (more usually their reciprocal as dissociation equilibrium constants) are fraught with difficulties at both the technical and the theoretical level, requiring rigor and understanding of the potential pitfalls in the determination of meaningful estimates of any such parameter of drug-receptor interaction. Detailed considerations of these problems will not be considered here as they have been well reviewed in the literature and continue to occupy the minds of pharmacologists interested in drug-receptor theory (44–48). It is worth noting that differences in functional characteristics may also arise with different receptor locations due to different host cell factors, differences in levels of receptor expression, stimulus-response coupling efficiencies, and so forth. This may be particularly so when comparing recombinant receptors in a host cell line with native receptors in a tissue and needs to be systematically addressed experimentally to ascertain the full extent of the implications for quantitative pharmacology (49–52).

IV. RECEPTOR TRANSDUCTION

It is still a matter of debate as to whether transductional data can be useful for receptor characterization and hence receptor classification purposes. Undoubtedly, knowledge about *transductional* characteristics has been invaluable in the subclassification of 5-HT receptors. It remains to be seen whether this is coincidental or whether the concept is generally applicable. In this regard, it should be appreciated that transduction in the context of receptor characterization is intended to refer to specific changes in the receptor protein(s) itself or changes in the adjacent proteins that are essential to transduce the effects of agonists. Transductional data should not involve the categorization of downstream second-message cascades per se although such information might be used judiciously to infer events upstream. Even if tightly defined, the use of transduction data to characterize receptors is controversial but it has been invaluable nevertheless in the classification of 5-HT receptors (32). Thus 5-HT$_{1C}$ (now called 5-HT$_{2C}$) and 5-HT$_2$ (now called 5-HT$_{2A}$) receptors were predicted to belong to the same receptor group on the basis of shared transduction mechanisms. This was later fully confirmed when both receptor genes were cloned and the respective proteins were shown to share a very high degree of homology (see Table 1) (32). In contrast, although both the 5-HT$_3$ and the 5-HT$_4$ receptor can be blocked by tropisetron, it was obvious on the basis of transduction that the two receptors were quite distinct even before both genes had been cloned (13,32,37). Thus, at the very least, consideration of the transduction mechanism involved will delineate between a ligand-gated cation channel and a G-protein-linked receptor (i.e., the 5-HT$_3$ and 5-HT$_4$, respectively). A knowledge of receptor transduction mechanisms is therefore undoubtedly useful, but how much value should be attached to such data for receptor characterization purposes as such remains to be determined.

However, there is a growing belief that the unique intracellular face of each receptor protein will dictate specific stoichiometric interactions with adjacent proteins, which will be characteristic of the receptor involved. It remains to be seen to what degree the intracellular receptor structure and topography dictate G-protein preference (53–57). There is certainly good evidence using mRNA antisense techniques to indicate functional preferences for all three G-protein subunits for different receptors in the same cell (see Fig. 2) (58). The general applicability of these findings will dictate the value of transductional information for G-protein receptor classification. For ion channel receptors much of the transductional data will be determined from electrophysiological studies. This will provide information on the consequences of receptor activation regarding ion fluxes and associated charge carried, resulting from conformational changes and molecular rearrangements of the multimeric channel proteins. Analogously it might be argued that for activation of heptahelical receptors, the essential G-protein involvement allows one to consider the molecular combination as an ephemeral tetrameric receptor (i.e., receptor protein plus three complexed G-protein molecules, α, β, and γ).

Regardless, transductional data will be generated if only to understand how receptors mediate the cellular effects of the agonists that activate them. Much of this work will be carried out on recombinant receptors transfected into stable cell lines (54,59–62). This could theoretically provide data of little relevance to endogenous receptors in whole tissues because of differences in receptor densities, transfection into unnatural cell environments, and other consequences of heterologous transfection (see above) (56,57). However, it is evident that it will provide important information

Table 1 Percentage Amino Acid Identity in Transmembrane Domains of Rat 5-Hydroxytryptamine Receptors

	5-HT$_{1A}$	5-HT$_{1B}$	5-HT$_{1D}$	5-ht$_{1e}$	5-ht$_{1f}$	5-HT$_{2A}$	5-HT$_{2B}$	5-HT$_{2C}$	5-HT$_4$	5-ht$_{5a}$	5-ht$_{5b}$	5-ht$_6$	5-HT$_7$
5-HT$_{1A}$	**100**												
5-HT$_{1B}$	**54**	**100**											
5-HT$_{1D}$	**58**	**73**	**100**										
5-ht$_{1e}$	**53**	**63**	**64**	**100**									
5-ht$_{1f}$	**54**	**59**	**63**	**68**	**100**								
5-HT$_{2A}$	39	40	38	42	42	**100**							
5-HT$_{2B}$	43	40	40	45	38	**68**	**100**						
5-HT$_{2C}$	39	40	43	44	42	**80**	**71**	**100**					
5-HT$_4$	46	43	41	42	40	43	40	43	100				
5-ht$_{5a}$	49	48	48	48	49	38	37	40	36	**100**			
5-ht$_{5b}$	48	49	49	49	45	37	39	41	38	**84**	**100**		
5-ht$_6$	40	38	40	37	36	43	41	42	35	39	37	**100**	
5-HT$_7$	50	52	49	53	49	40	37	39	46	44	44	39	100

Note the bold numbering to highlight the high homology within the larger receptor groups with subtypes (5-HT$_1$, 53–73%; 5-HT$_2$, 68–80%) and the range of 36–53% homology between the different groups or receptor types. Note too the use of lower-case letters to denote recombinant receptor types whose full operational profile has yet to be determined fully in whole tissues (79).
Source: Data modified from Ref. 13.

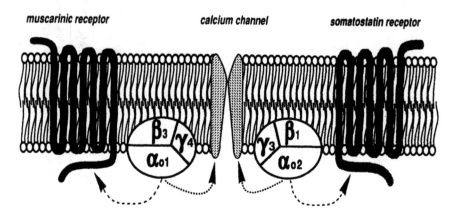

muscarinic receptor   *calcium channel*   *somatostatin receptor*

Figure 2 Selective transduction mechanisms for muscarinic and somatostatin receptors. Schematic representation of the G-proteins that couple and mediate receptor transduction for endogenous muscarinic and somatostatin receptors in pituitary GH_3 cells. Electrophysiological studies using anti-sense indicate a unique trimeric G-protein requirement for the function of each inhibitory neurotransmitter receptor. (58)

about how receptors function and are activated and could ultimately lead to an understanding of efficacy at the molecular level (63). It should be acknowledged that appropriate tranductional data is technically much more difficult to obtain for endogenous receptors in tissues, which may limit its value for classification purposes. Thus, as defined above, transductional data does not specifically include the cataloguing of second-messenger involvement but rather the definition of trimeric G-protein involvement in the cellular response, which must restrict the availability of such data from whole tissue studies. Nevertheless, the circumspect use of data on second-messenger activation (i.e., activation of adenylate cyclase or protein kinase C) may give a general indication of the type of G-protein involved (i.e., G_s and Gq, respectively), albeit far from definitive. In the case of ligand-gated ion channel receptors, transductional data may be more easily obtained directly from electrophysiological studies on freshly dispersed cells as well as tissue slices.

V. RECEPTOR STRUCTURE

The value of structural information is obvious. However, it should be acknowledged that the receptor protein composition is invariably deduced from the cloned cDNA sequence and that local tissue variations may pertain such as differential glycosylation, transcriptional splice variations, or differential polymerisation patterns for heteromeric receptors (10,11,64,65). The latter particularly relates to multiunit ion channel receptors but might also conceptually include heptahelical receptor-trimeric G-protein couplings.

Nevertheless, at the level of the gene a definitive descriptor can be provided that unequivocally defines the receptor identity. Such information may provide useful clues to receptor function (e.g., whether it is likely to be G-protein linked or a ligand-gated ion channel) but little about ligand recognition, unless other closely related receptors are already well known. Structural information is valuable not only for receptor characterization purposes but also in their classification (see Fig. 1 and Table 1). Structural homologies can be quantified to provide information about interspecies differences as well as differences between receptor families (defined by the endogenous transmitter substance) and within families between receptor types (or groups) and subtypes. Unfortunately, there seem to be no consistent quantitative demarcations between these divisions, but species variants of receptors (the products of orthologous genes) commonly display 85–95% homology, whereas interreceptor family homologies are usually in the 15–35% region (40,64,66,67). Similar limitations apply to the differentiation between receptor types and subtypes. In the case of 5-HT receptors, there is between 53 and 80% identity between subtypes within a given receptor group (see Table 1). However, somatostatin receptors have been functionally classified into two groups, $SRIF_1$ and $SRIF_2$ (68), yet the receptor amino acid identity within a group (46–61%) is not much greater than the identities of subtypes between the two groups (44–50%) (69). However, the structures of the hydrophobic transmembrane regions of G-protein-coupled receptors are highly conserved (64,66), such that, for example, there is 72–96% homology within either the $SRIF_1$ or $SRIF_2$ receptor groups and somewhat less (59–64%) between the groups in the second and third transmembrane regions of the five different recombinant somatostatin receptor subtypes (70). In an integrated scheme other criteria are also considered in deciding on a pragmatic and practical scheme of subclassification into types and subtypes.

VI. AN INTEGRATED APPROACH TO RECEPTOR CLASSIFICATION

It remains to be shown whether or not all neurotransmitter receptor families can be classified using a similar integrated approach to that used for 5-HT receptors (see above; Fig. 1). It would seem important to systematically analyze all the structural and operational data for all receptor families to determine the value of such an approach in pharmacologically classifying receptors generally. A cursory consideration of a number of receptor families in the G-protein-coupled receptor superclass suggests that it is generally applicable (see Fig. 3). In some cases, however, the approach has been ignored despite its obvious merit. Thus, for example, both dopamine and muscarinic receptors have been well characterized structurally as well as functionally but no attempt has been made to subclassify them, with the receptors in both families simply being numbered from one to five according to the chronology of receptor gene cloning. However, it is arguably more useful and illuminating to pharmacologists to subclassify according to the principles discussed here. Thus D_1 and D_5 dopamine receptors clearly fall into a completely different group to D_2, D_3, and D_4 receptors on structure, operational, and transductional grounds, with the former group mediating activation of adenylyl cyclase and the latter, inhibition. Similarly, muscarinic M_2 and M_4 receptors can be grouped together on the basis of structure (see Fig. 3) and transduction with preferential coupling to G-protein a subunits of the G_i/G_o type. In contrast, M_1, M_3, and M_5 receptors are much more closely related structurally and all commonly mediate their respective responses through G-proteins of the $G_{q/11}$ type (see Fig. 3) (71,72).

An integrated approach to receptor classification can also be used to subclassify large receptor families other than that for 5-HT, such as the norepinephrine and prostaglandin receptor families (33,73). Thus for the adrenoceptors, there is good agreement between the structural and operational data and the transductional characteristics also appear to be broadly diagnostic for the intrafamily receptor groups (see Fig. 4). The division into α_1, α_2, and β-adrenoceptor classes is further supported by the findings that in whole tissues α_1-adrenoceptors mediate increases in intracellular calcium while α_2-adrenoceptors inhibit adenylyl cyclase and β-adrenoceptors activate adenylyl cyclase (73). However, although α_2-adrenoceptor activation is consistently associated with inhibition of adenylyl cyclase, it does not always appear to represent the signal transduction mechanism responsible for the tissue response mediated. Despite the distinct transductional characteristics for each of the three groups there are no obvious differences within each group of adrenoceptor (73). This might either suggest that transductional data do not provide a sensitive enough indicator to differentiate between closely related receptor subtypes or, alternatively, the problems of collecting the appropriate data in whole tissues, alluded to above, may mask any transductional differences.

In the case of prostanoid receptors, there is again good reason for their subclassification on both structural and operational grounds and each receptor group also has different transductional characteristics (33,74). However, in the EP_3 receptor group multiple second-messenger systems appear to be activated but each appears to be differentially mediated by distinct splice

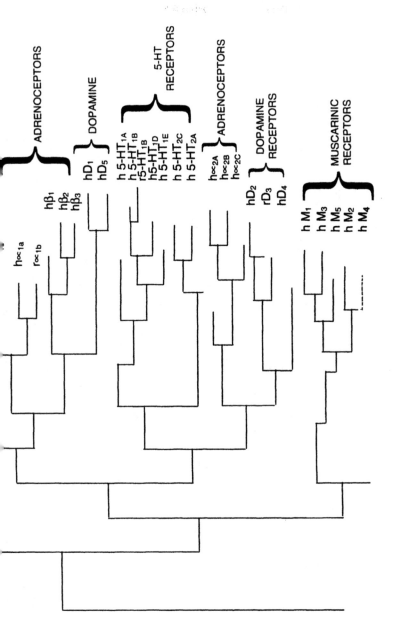

Figure 3 A dendogram analysis of the structural homologies of some common G-protein-coupled receptor families, showing the clustering into groups and receptor subtypes according to amino acid sequence similarities. The data were obtained from the GCRD$_b$ database established by Frank Kolakowski and can be viewed on the internet at the URL, http://receptor.mgh.harvard.edu/GCRDB.html.

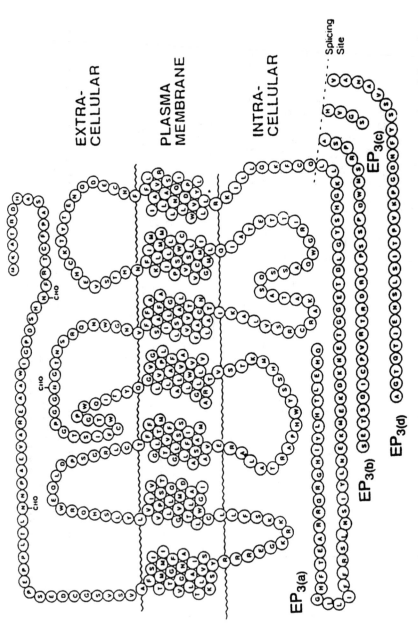

Figure 4 Prostaglandin EP$_3$ receptor splice variants. A schematic two dimensional representation of the amino sequence of the bovine prostaglandin EP$_3$ receptor to show the four alternative splice variants, EP$_{3(a)}$, EP$_{3(b)}$, EP$_{3(c)}$, and EP$_{3(d)}$, which only differ in the C-terminal intracellular tail. Each receptor variant appears to activate different transduction mechanisms such that they can be differentiated functionally (Refs. 33 and 74).

variants, of which at least four in any given species have been described (see Fig. 4) (74). It is apparent that the variations, which all occur in the C-terminal tail, lead to differential G-protein coupling (33). Such splice variants for both G-protein-linked receptors (e.g., 5-HT$_4$, somatostatin sst$_2$, and EP$_3$) and ion channel receptors (e.g., 5-HT$_3$) cannot usually be distinguished on the basis of agonist or antagonist potencies but can be distinguished on the basis of transductional information (12,13,33,75). This provides further justification for the pharmacological importance of studying transductional mechanisms.

VII. CONCLUDING REMARKS

There appears to be an incipient view among molecular biologists and pharmacologists interested in receptor classification that receptors should be characterized and classified using an integrated approach based on both receptor structure and operation. As structural information is definitive, it will always be preeminent in terms of receptor identification and archival reference. If only to avoid ambiguity, the primary archival data will be the structure of the receptor gene or genes (for multimeric receptors) and the inferred amino acid sequence of the corresponding protein(s). The recognitory characteristics of the receptor derived from operational data will also be essential and diagnostic for a given receptor. Regardless, the identification of selective ligands, be they agonists or antagonists, will be a *sine qua non* for drug design. Whether transductional data should be considered as important or essential data for characterization purposes remains to be determined. The current consensus favors integrating transductional with recognitory data, which collectively relates to receptor operation.

Receptor transduction mechanisms are increasingly being studied for recombinant receptors transfected into host cell lines. The emerging picture appears much more complex than was first thought. First, although the intracellular domain of each distinct receptor type is unique, it remains to be demonstrated whether there is sufficient repertoire of trimeric G-protein permutations to allow distinct transduction pathways for each. It would seem better to refer to G-protein preference(s) rather than selectivity, for as with ligand-receptor interactions, selectivity is not absolute and stoichiometric considerations are paramount (76,77). Second, the fundamentals of agonist efficacy remain to be understood and the concept of a two-state conformational model for receptor activation now seems less certain. Indeed, there is now evidence to suggest that there may be multiple conformational states, differentially dictated by different agonists, as well as antagonists and inverse agonists (51,57,78). These different states might theoretically command different G-protein preferences and hence initiate different transduction pathways (57,61). Given the controversy already alluded to concerning the use of transductional data to characterize receptors it would seem prudent to include such data under operational characteristics as further descriptors of receptor operation. This in no way devalues the importance of studying transduction mechanisms. Indeed, they will be different for different types of receptor and their detailed study with proper application of the principles of sound quantitative pharmacology will lead us closer to understanding the molecular properties responsible for agonist receptor activation, which pharmacologists abstractly refer to as "efficacy."

REFERENCES

1. Himmelweit F. The Collected Papers of Paul Ehrlich. Ed. and compiled by Himmelweit F, Vol III, Chemotherapy. London: Pergamon Press, 1960.
2. Claudio T. Recombinant DNA technology in the study of ion channels. Trends Pharmacol Sci 1986; 7:308–312.
3. De Jongh DK. A molecular approach to general pharmacology. In: Ariens EJ, ed. Molecular Pharmacology: The Mode of Action of Biologically Active Compounds. Vol I. New York: Academic Press, 1964:16
4. Mills A, Duggan MJ. Orphan seven transmembrane domain receptors: reversing pharmacology. Trends Pharmacol Sci 1993; 14:394–396.
5. Meunier J-C, Mollereau C, Toll L, Suaudeau C, Moisand C, Alvinerie P. Butour J-L, Guillemot J-C. Ferrara P, Monsarrat B, Mazarguil H, Vassart G, Parmentier M, Costentin J. Isolation and structure of the endogenous agonist of opioid receptor-like ORL_1 receptor. Nature 1995; 377:532–534.
6. Stephenson RP. Interaction of agonists and antagonists with their receptors. INSERM 1975; 50:15–28.
7. Role LW. Diversity in primary structure and function of neuronal nicotinic acetylcholine receptor channels. Curr Opin Neurobiol 1992; 2:254–262.
8. Monyer H, Sprengel R, Schoepfer R, Herb A, Higuchi M, Lomeli H, Burnashev N, Sakmann B, Seeburg PH. Heteromeric NMDA receptors: molecular and functional distinction of subtypes. Science 1992; 256:1217–1221.
9. Schwartz TW. Locating ligand-binding sites in 7TM receptors by protein engineering. Curr Opin Biotechnol 1994; 5:434–444.
10. Giros B, Sokoloff P, Martres M-P, Riou J-F, Emorine LJ, Schwartz JC. Alternative splicing directs the expression of two D_2 dopamine receptor isoforms. Nature 1989; 342:923–926.
11. Epelbaum J, Dournaud P, Fodor M, Viollet C. The neurobiology of somatostatin. Crit Rev Neurobiol 1994; 8:25–44.
12. Downie DL, Hope AG, Lambert JJ, Peters JA, Blackburn TP, Jones BJ. Pharmacological characterization of the apparent splice variants of the murine 5-HT_3 R-A subunit expressed in xenopus laevis oocytes. Neuropharmacology 1994; 33:473–482.
13. Gerald C, Adham N, Kao H-T, Olsen MA, Laz TM, Schechter LE, Bard JA, Vaysse PJ-J, Hartig PR, Branchek TA, Weinshank RL. The 5-HT_4 receptor:

molecular cloning and pharmacological characterization of two splice variants. EMBO J 1995; 14:2806–2815.

14. Simon J, Kidd EJ, Smith FM, Chessell IP, Murrell-Lagnado R, Humphrey PPA, Barnard EA (1997). Localization and functional expression of splice variants of the P2X$_2$ receptor. Mol Pharmacol, 52:237–248.

15. Humphrey PPA. The characterization and classification of neurotransmitter receptors. Trist DG, Humphrey PPA, Lett P, Shankley NP, Eds. NY Acad Sci 1997; 812:1–13.

16. Regoli D, Boudon A, Fauchere J-C. Receptors and antagonists for substance P and related peptides. Pharmacol Rev 1994; 46:551–599.

17. Birnbaumer L, Abramowitz J, Yatani A, Okabe K, Mattera R, Graf R, Sanford J, Codina J, Brown AM. Roles of G proteins in coupling of receptors to ionic channels and other effector systems. Crit Rev Biochem Mol Biol 1990; 25:225–244.

18. Oksenberg GD, Marsters SA, O'Dowd BF, Jin H, Havlik S, Peroutka SJ, Ashkenazi A. A single amino-acid difference confers major pharmacological variation between human and rodent 5-HT$_{1B}$ receptors. Nature 1992; 360: 161–163.

19. Gether U, Johansen TE, Snider RM, Lowe JA, Nakanishi S, Schwartz TW. Different binding epitopes on the NK$_1$ receptor for substance P and a nonpeptide antagonist. Nature 1993; 362:345–347.

20. Gingrich JA, Caron MG. Recent advances in the molecular biology of dopamine receptors. Annu Rev Neurosci 1993; 16:299–321.

21. Sargent PB. The diversity of neuronal nicotinic receptors. Annu Rev Neurosci 1993; 16:403–443.

22. Ahlquist RP. A study of the adrenergic receptors. Am J Physiol 1948; 153:586–600.

23. Nickerson M. The pharmacology of adrenergic blockade. Pharmacol Rev 1949; 1:27–101.

24. Black JW, Stephenson JS. Pharmacology of a new adrenergic β-receptor blocking compound. Lancet 1962; 2:311–314.

25. Doxey JC, Smith CFC, Walker JM. Selectivity of blocking agents for pre- and postsynaptic α-adrenoceptors. Br J Pharmacol 1977; 60:91.

26. Black JW. Development of ideas about noradrenaline receptors. In De Schaepdryver AF, ed., The Heyman Institute of Pharmacology 1890–1990. Ghent, Belgium: Heymans Foundation, 1990:109–115.

27. Lands AM, Arnold A, McAuliff JP, Luduena FP, Brown TG Jr. Differentiation of receptor systems activated by sympathomimetic amines. Nature 1967; 214:597–598.

28. Frielle T, Daniel KW, Caron MG, Lefkowitz RJ. The structural basis of β-adrenergic receptor subtype specificity: studies with chimeric β$_1$/β$_2$-adrenergic receptor. Proc Natl Acad Sci USA 1988; 85:9494–9498.

29. Ash ASF, Schild HO. Receptors mediating some actions of histamine. Br J Pharmacol 1966; 27:427–439.

30. Black JW, Duncan WAM, Durant CJ, Ganellin CR, Parsons EM. Definition and antagonism of histamine H$_2$-receptors. Nature 1972; 236:385–390.

31. Arrang JM, Garbarg M, Lancelot JC, Lecomte JM, Pollard H, Robba M, Schunack W, Schwartz JC. Highly potent and selective ligands for histamine H3-receptors. Nature 1987; 327:117–123.

32. Hoyer D, Clarke DE, Fozard JR, Hartig PR, Martin GR, Mylecharane EJ, Saxena PR, Humphrey PPA. VII. International Union of Pharmacology classification of receptors for 5-hydroxytryptamine (serotonin). Pharmacol Rev 1994a; 46:157–203.

33. Coleman RA, Smith WL, Narumiya S. VIII. International Union of Pharmacology classification of prostanoid receptors: properties, distribution, and structure of the receptors and their subtypes. Pharmacol Rev 1994; 46:205–229.

34. Fredholm BB, Abbracchio MP, Burnstock G, Daly JW, Harden TK, Jacobson KA, Leff P, Williams M. VI. International Union of Pharmacology nomenclature and classification of purinoceptors. Pharmacol Rev 1994; 46:143–156.

35. Humphrey PPA. Peripheral 5-hydroxytryptamine receptors and their classification. Neuropharmacology 1984; 23:1503–1510.

36. Bradley PB, Engel G, Feniuk W, Fozard JR, Humphrey PPA, Middlemiss DN, Mylecharane EJ, Richardson BP, Saxena PR. Proposals for the classification and nomenclature of functional receptors for 5-hydroxytryptamine. Neuropharmacology 1986; 25:563–576.

37. Humphrey PPA, Hartig P, Hoyer D. A proposed new nomenclature for 5-HT receptors. Trends Pharmacol Sci 1993; 14:233–236.

38. Martin GR, Humphrey PPA. Receptors for 5-hydroxytryptamine: current perspectives on classification and nomenclature. Neuropharmacology 1994; 33:261–273.

39. Hartig PR, Hoyer D, Humphrey PPA, Martin GR. Alignment of receptor nomenclature with the human genome: classification of 5-HT_{1B} and 5-HT_{1D} receptor subtypes. Trends Pharmacol Sci 1996; 17:103–105.

40. Hartig PR, Branchek TA, Weinshank RL. A subfamily of 5-HT_{1D} receptor genes. Trends Pharmacol Sci 1992; 13:152–159.

41. Sachais BS, Snider RM, Lowe JA III, Krause JE. Molecular basis for the species selectivity of the substance P antagonist CP-96,345. J Biol Chem 1993; 268:2319–2323.

42. Adham N, Tamm JA, Salon JA, Vaysse PJ-J, Weinshank RL, Branchek TA. A single point mutation increases the affinity of serotonin $5\text{-HT}_{1D\alpha}$, $5\text{-HT}_{1D\beta}$, 5-HT_{1E} and 5-HT_{1F} receptors for β-adrenergic antagonists. Neuropharmacology 1994; 33:387–391.

43. Humphrey PPA, Spedding M, Vanhoutte PM. Receptor classification and nomenclature: the revolution and the resolution. Trends Pharmacol Sci 1994; 15:203–204.

44. Furchgott RF. The classification of adrenoceptors (adrenergic receptors). An evaluation from the standpoint of receptor theory. In: Blaschko H, Muscholl E, eds. Catecholamines. Handbook Exp Pharmac NS. Berlin: Springer Verlag, 1972:33:283–335.

45. Humphrey PPA. Pharmacological characterisation of cardiovascular 5-hydroxytryptamine receptors. In: Bevan JA, et al, eds. Vascular Neuroeffector Mechanisms, IVth International Symposium. New York: Raven Press, Chapter 17, 1983:237–242.

46. Kenakin TP. The classification of drugs and drug receptors in isolated tissues. Pharmacol Rev 1984; 36:165–222.

47. Kenakin TP. Challenges for receptor theory as a tool for drug and drug receptor classification. Trends Pharmacol Sci 1989; 10:18–22.

48. Kenakin TP, Bond RA, Bonner TI. Definition of pharmacological receptors. Pharmacol Rev 1992; 44:351–362.

49. Schwarz RD, Davis RE, Jaen JC, Spencer CJ, Tecle H, Thomas AJ. Characterization of muscarinic agonists in recombinant cell lines. Life Sci 1993; 52(5–6):465–472.

50. Richards MH, van Giersbergen PL. Differences in agonist potency ratios at human M_1 muscarinic receptors expressed in A9L and CHO cells. Life Sci 1995; 57(4):397–402.

51. Kenakin T. The classification of seven transmembrane receptors in recombinant expression systems. Pharmacol Rev 1996; 48:413–463.

52. Bockaert J, Brand C, Journot L. Do recombinant receptor assays provide affinity and potency estimates? Trist DG, Humphrey PPA, Lett P, Shankley NP, eds. NY Acad Sci 1997; 812:55–70.

53. Jackson T. Structure and function of G protein coupled receptors. Pharmacol Ther 1991; 50:425–442.

54. Milligan G. Mechanisms of multifunctional signaling by G protein-linked receptors. Trends Pharmacol Sci 1993; 14:239–244.

55. Watson S, Arkinstall S. The G-Protein Linked Receptor. Factsbook. Orlando, FL: Academic Press, 1994.

56. Kenakin T. Agonist-receptor efficacy. I. Mechanisms of efficacy and receptor promiscuity. Trends Pharmacol Sci 1995; 16:188–192.

57. Kenakin T. Agonist-receptor efficacy. II. Agonist trafficking of receptor signals. Trends Pharmacol Sci 1995; 16:232–238.

58. Kleuss C, Scherubl H, Hescheler J, Schultz G, Wittig B. Selectivity in signal transduction determined by gamma subunits of heterotrimeric G proteins. Science 1993; 259:832–834.

59. Vallar L, Muca C, Magni M, Albert P, Bunzow J, Meldolesi J, Civelli O. Differential coupling of dopaminergic D_2 receptors expressed in different cell types. J Biol Chem 1990; 265:10320–10326.

60. Birnbaumer L, Birnbaumer M. Signal transduction by G proteins. J Rec Sig Trans Res 1994; 15:213–252.

61. Spengler D, Waeber C, Pantaloni C, Holsboer F, Bockaert J, Seeburg PH, Journot L. Differential signal transduction by five splice variants of the PACAP receptor. Nature 1993; 365:170–175.

62. Castro SW, Buell G, Feniuk W, Humphrey PPA. Differences in the operational characteristics of the human recombinant somatostatin receptor types, sst_1 and sst_2, in mouse fibroblast (Ltk⁻) cells. Br J Pharmacol 1996; 117:639–646.

63. Stephenson RP. A modification of receptor theory. Br J Pharmacol 1956; 11:379–393.

64. Probst WC, Snyder LA, Schuster DI, Brosius J, Sealfon SC. Sequence alignment of the G-protein coupled receptor superfamily. DNA Cell Biol 1992; 11:1–20.

65. Lewis C, Neidhart S, Holy C, North RA, Buell G, Surprenant A. Coexpression of P2X₂ and P2X₃ receptor subunits can account for ATP-gated currents in sensory neurons. Nature 1995; 377:432–435.
66. Dohlman H, Caron M, Lefkowitz RJ. A family of receptors coupled to guanine nucleotide regulating proteins. Biochemistry 1987; 26:2657–2664.
67. Hall JM, Caulfield MP, Watson SP, Guard S. Receptor subtypes or species homologues: relevance to drug discovery. Trends Pharmacol Sci 1993; 14:376–383.
68. Hoyer D, Bell GI, Berelowitz M, Epelbaum J, Feniuk W, Humphrey PPA, O'Carroll A-M, Patel YC, Schonbrunn A, Taylor JE, Reisine T. Classification and nomenclature of somatostatin receptors. Trends Pharmacol Sci 1995; 16:86–88.
69. Reisine T, Bell GI. Molecular biology of somatostatin receptors. Endocrine Rev 1995; 16:427–442.
70. Hoyer D, Lübbert H, Bruns C. Molecular pharmacology of somatostatin receptors. Naunyn-Schmiedeberg's Arch Pharmacol 1994b; 350:441–453.
71. Caulfield MP. Muscarinic receptors—characterization, coupling and function. Pharmacol Ther 1993; 58(3):319–379.
72. Wess J. Molecular basis of muscarinic acetylcholine receptor function. Trends Pharmacol Sci 1993; 14:308–313.
73. Bylund DB, Eikenberg DC, Hieble JP, Langer SZ, Lefkowitz RJ, Minneman KP, Molinoff PB, Ruffolo RR, Trendelenburg U. IV. International Union of Pharmacology nomenclature of adrenoceptors. Pharmacol Rev 1994; 46:121–136.
74. Narumiya S. Prostanoid receptors and signal transduction. Prog Brain Res 1996; 113:231–241.
75. Vanetti M, Vogt G, Hollt V. The two isoforms of the mouse somatostatin receptor (mSSTR2A) and mSSTR2B) differ in coupling efficiency to adenylate cyclase and in agonist-induced receptor desensitization. FEBS Lett 1993; 331:260–266.
76. Offermanns S, Wieland T, Homann D, Sandmann J, Bombien E, Spicher K, Schultz G, Jakobs KH. Transfected muscarinic acetylcholine receptors selectively couple to G₁-type G proteins and G_{q/11}. Mol Pharmacol 1994; 45:890–898.
77. Gilman AG. G proteins and regulation of adenylyl cyclase. Biosci Rep 1995; 15:65–97.
78. Schwartz TW, Gether U, Schambye HT, Hjorth SA. Molecular mechanism of action of non-peptide ligands for peptide receptors. Curr Pharm Design 1995; 1:325–342.
79. Vanhoutte PM, Humphrey PPA, Spedding M. X. International Union of Pharmacology Recommendations for nomenclature of new receptor subtypes. Pharmacol Rev 1996; 48:1–2.

2

Functional Methods for Quantifying Agonists and Antagonists

Iain G. Dougall
Astra Charnwood, Loughborough, Leicestershire, England

I. INTRODUCTION

Pharmacologists study the effects of drugs on living systems (cells, isolated tissues, or whole animals). The responses observed, are in most cases, the result of the drug interacting with receptors, which have the capacity to convert chemical information into biological information. In all such studies, irrespective of the endpoint measured, the experimenter aims to quantify effect as a function of agonist concentration; that is, the prime objective of all pharmacological experiments is (or should be) to generate agonist concentration-effect (E/[A]) curve data. This chapter concerns itself with the analysis of data of this type, illustrating how it can be used to quantify the parameters that describe the interactions of agonists and antagonists with receptors. Such information is useful in the process of receptor classification, the interpretation of structure-activity relationships, and the rational design of new medicines (1).

II. DRUG-RECEPTOR THEORIES

Many theories (2–9) have been proposed to describe the interaction of drugs with receptors. At present, the most widely accepted of these is occupancy theory (see Ref. 10 for review), in which a response is thought to emanate from a receptor only when it is occupied by an appropriate agonist molecule; that is, response is some function of the equilibrium concentration of agonist-occupied receptors. Despite all the assumptions

(11) that underlie occupancy theory, it has been remarkably successful in describing a vast amount of pharmacological data. All of the following discussion presumes this theory.

Although occupancy theory is the preeminent description of drug-receptor interaction, it is worth emphasizing that its development (2,3,4,6,12) occurred to a large extent along nonmechanistic lines. This was the result of relative ignorance about the events surrounding agonist-induced activation of receptors and the necessarily indirect nature of the measurements made. As receptor mechanisms are gradually unraveled and thus better understood, the appropriateness of occupancy theory as a description of drug-receptor interactions may be seriously challenged. Indeed, the recent findings that under certain conditions, unoccupied receptors (both ion-channel linked and G-protein linked) exist in two states, resting and activated (see Chapter 16), may have important consequences with regard to the conceptual definition and quantification of the actions of agonists and antagonists (13).

III. ASSAY SYSTEMS

Quantitative analysis of drug-receptor interactions necessitates the generation of reliable E/[A] curve data. This in turn requires accurate knowledge of drug concentration in the receptor compartment (biophase); that is, generation of reliable E/[A] curve data is dependent on precise knowledge of the independent experimental variable, agonist concentration. The level of confidence that the experimenter can ascribe to this quantity diminishes as the system becomes more complex. Thus cell-based assay systems are generally accepted to be the most reliable in this respect and, conversely, the majority of in vivo model systems are inherently very unreliable. Between these two extremes lies the isolated tissue, which represents the system of maximal complexity that can be used with any degree of confidence for quantitative pharmacological analysis. Indeed, the development of traditional receptor theory (2–4,6) was largely dependent on the use of various isolated smooth muscle preparations. Typically, in such experiments, the effects measured, muscle contraction or relaxation, are distantly removed from the initial drug-receptor interaction. However, as cell systems become increasingly utilized as bioassays, the endpoints measured (for example, changes in intracellular calcium concentration) have become more proximal to the initial binding event. Despite these changes in the nature of the responses measured, generation of reliable E/[A] curve data remains the principal goal of the pharmacologist.

IV. AGONIST CONCENTRATION-EFFECT CURVES

The agonist concentration-effect (or dose-response) curve has become one of the hallmarks of modern pharmacology. Such curves are typically sigmoidal when plotted in semilogarithmic form ($E/\log_{10}[A]$) and are described by four parameters: (1) a lower asymptote (β), which represents the basal state of the system, (2) an upper asymptote (α), which represents the maximum effect that the agonist produces in the system, (3) a location ($[A]_{50}$), which represents the concentration of agonist that produces an effect equal to 50% of α-β, and (4) a slope parameter (m), which is a measure of the gradient of the curve at the $[A]_{50}$ level. Today, it is common practice to employ computers to estimate these parameters by fitting experimental $E/[A]$ curve data to the following form of the Hill equation (a saturable function that adequately describes curves of varying gradients):

$$E = \beta + \frac{\alpha[A]^m}{[A]^m + [A]_{50}^m} \tag{1}$$

An example of this pragmatic fitting of data is illustrated in Figure 1. In practice, in the majority of cases $\beta = 0$; that is, the basal effect level is ascribed a value of zero, and therefore most $E/[A]$ curve data can be

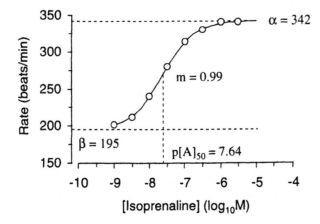

Figure 1 Agonist concentration-effect curve data. An isoprenaline $E/[A]$ curve obtained in the guinea pig isolated, spontaneously beating right atrium preparation. The data was fitted to Eq. (1) yielding the Hill function parameters shown. Note the curve is a rectangular hyperbola ($m = 1$); $p[A]_{50}$ ($-\log_{10}[A]_{50}$). (Unpublished data.)

adequately described by a three-parameter Hill equation. It is analysis of how these three curve parameters (α, $[A]_{50}$, and m) are affected by experimental manipulation that allows drug-receptor interactions to be quantitatively described in terms of affinity (binding) and efficacy (response-eliciting capacity).

V. EXPERIMENTAL DESIGN

Two different kinds of experimental design are commonly used in pharmacological experiments: a single-curve design or a multiple (usually two)-curve design. The latter design is usually considered favorable, as it allows comparisons of E/[A] curves to be made that are uncomplicated by intertissue differences. Such an experimental design is often possible in isolated tissue systems, thereby allowing each piece of tissue to provide an agonist (or antagonist) affinity and efficacy estimate. However, in many other assay systems (cell based and isolated tissues) a multiple-curve design is not feasible. In this situation, a single piece of tissue (or cuvette of cells, etc.) provides only part of the information required to analyze an agonist (or antagonist). Estimation of the parameters of interest therefore necessitates combining data obtained in different pieces of tissue. This makes for a less straightforward statistical treatment of the data than that used in the multiple-curve design (see Ref. 14 for details). For this reason, where possible, the experimenter should adopt a multiple-curve design.

VI. ANALYSIS OF AGONISTS

A. Concepts of Affinity and Efficacy

The first step in agonist action is the formation of a reversible agonist-receptor (AR) complex, a process that is generally assumed to be governed by the Law of Mass Action. Accordingly, the equilibrium concentration of agonist occupied receptors is a rectangular hyperbolic (a special case of the Hill function where the midpoint slope parameter = 1) function of the agonist concentration. This curve is defined by a maximal value of $[R_0]$, the total functional receptor concentration, and a midpoint value of K_A, the agonist dissociation constant. The K_A is a purely drug-dependent parameter and it determines how well the agonist binds; that is, it is a measure of the affinity (the reciprocal of the dissociation constant) of the agonist for its receptors. Agonist occupancy is subsequently converted into functional effect by the biochemical/biophysical machinery of the cell/tissue and this is what is measured experimentally in the form of an E/[A] curve. The efficiency of this transduction process can vary between agonists and across tissues; that is, it is both drug- and tissue-dependent. Agonist efficacy is a measure of the efficiency of the transduction process. Full agonists have high efficacies and can therefore elicit the maximum effect (E_m) that the receptor-transducer system can generate. Partial agonists have low efficacies and generate maximum effects significantly less than E_m.

The following sections describe how estimates of these two quantities, affinity and efficacy, can be obtained from E/[A] curve data.

B. The Operational Model of Agonism

As outlined above, agonist occupancy is assumed to be a rectangular hyperbolic function of agonist concentration, and the shape of the agonist concentration-effect relationship can be defined experimentally in the form of an E/[A] curve. Knowing the shape of the functions at the beginning and end of the process of agonist action allows the shape of the intervening transducer function to be mathematically deduced. This function needs to have a smooth saturable shape, to account for the phenomenon of receptor reserve (15) and a flexible gradient. These are characteristics of the Hill function [see Eq. (1)]. It can also be proved (16,17) that if, as is commonly found, experimental E/[A] curves for full agonists follow the shape of the Hill equation (hence the use of this function for pragmatic-fitting of E/[A] curve data), then the transducer function must also be of the Hill form. Thus the mapping of the hyperbolic agonist occupancy ([AR]/[A]) function, through a Hill equation defining the E/[AR] relationship, allows an explicit description of agonist action to be made. The result of this procedure was the operational model of agonism (12,18), which describes agonist action in terms of four parameters: (1) E_m, the maximum response that the receptor-transducer mechanism can generate, (2) n, the slope index of the transducer relation, which is a measure of the sensitivity of the system, (3) K_A, the agonist equilibrium dissociation constant, and (4) τ, the transducer ratio, which is equivalent to agonist efficacy. The mathematical formulation of this model is shown below and in Figure 2.

$$E = \frac{E_m \tau^n [A]^n}{([A] + K_A)^n + \tau^n [A]^n} \qquad (2)$$

This model can be used to estimate agonist affinities and efficacies by direct fitting of raw experimental E/[A] curve data. The three commonly used pharmacological methods of analyzing data using the operational model are described below. However, to understand these analyses, it is necessary first to describe the relationship between operational model parameters and E/[A] curve parameters.

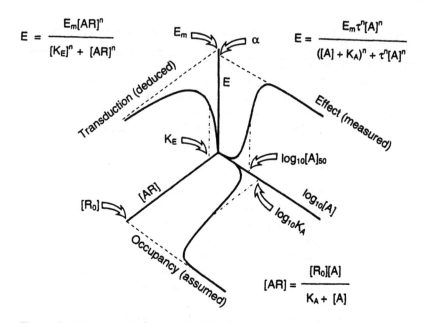

$$E = \frac{E_m[AR]^n}{[K_E]^n + [AR]^n}$$

$$E = \frac{E_m \tau^n [A]^n}{([A] + K_A)^n + \tau^n [A]^n}$$

$$[AR] = \frac{[R_0][A]}{K_A + [A]}$$

Figure 2   The operational model of agonism. The model is based on the traditional pharmacological idea that agonists (A) produce their effects (E) by initially binding to receptors (R) to form agonist-receptor complexes (AR). This occupancy function (shown on the base of the three-dimensional diagram) is assumed to follow the Law of Mass Action and is therefore described by a rectangular hyperbola, defined by a maximum value of $[R_0]$, the total functional concentration of receptors, and a midpoint value of K_A, the concentration of A, required for 50% occupancy. Agonist binding is subsequently converted into functional effect and measured in the form of an E/[A] curve (right-hand face of the diagram), defined by a maximal effect value of α and a midpoint value of $[A]_{50}$, the concentration of A that elicits an effect value of 50% of α. The shape of the deduced transducer (E/[AR]) function (left-hand face of the diagram) follows the Hill equation and is defined by a maximum effect value of E_m and a midpoint of K_E, the concentration of AR needed to produce half-maximal effect. The ratio $[R_0]/K_E$ determines the efficiency of transduction and is therefore a measure of agonist efficacy (τ). The shape of the transducer function is fixed for a particular receptor-transducer system in terms of E_m and n but K_E can vary; that is, a series of agonists can generate AR complexes with different response-eliciting capacities. Variations in $[R_0]$ also affect τ, such that a given agonist will be more efficacious in a tissue with a higher $[R_0]$ value. In the particular example shown, $\tau = 10$ and therefore α approaches E_m and $[A]_{50}$ is to the left of K_A on the $\log_{10}[A]$ axis [see Eq. 3 and 4].

C. The Relationship Between Operational Model Parameters and E/[A] Curve Parameters

Experimental E/[A] curves are characterized by three parameters, α, $[A]_{50}$, and m. Using Eq. (2), these experimental parameters can be written in terms of operational model parameters. This is illustrated below for the case where $n = 1$; that is, when E/[A] curves are founded to be hyperbolic (a relatively common experimental finding).

$$\alpha = \frac{E_m \tau}{1 + \tau} \qquad (\alpha = E \text{ as } [A] \to \infty) \qquad (3)$$

$$[A]_{50} = \frac{K_A}{1 + \tau} \qquad ([A]_{50} = [A] \text{ when } E = 0.5\alpha) \qquad (4)$$

These simplified equations will be used as a basis for the following discussion (the more general equations describing nonhyperbolic E/[A] curves [18] reveal subtly different behavior). It can be seen from Eq. (3) that high values of τ (>10) result in the agonist generating a maximum response ($\alpha > 0.91E_m$) experimentally indistinguishable from E_m; that is, full agonism is exhibited (see Fig. 2). Low τ values (<10) are associated with partial agonism; for example, an agonist with a τ value of 1 will produce a maximum response that is 50% of E_m. Antagonists, which produce no effect, have τ values of 0.

Equation (4) highlights the often ignored fact that agonist potency ($[A]_{50}$) is a two-parameter quantity, being dependent on both affinity and efficacy. It also illustrates that as the efficacy of an agonist decreases (τ tends to 0), then K_A is the limit that $[A]_{50}$ approaches. In other words, the lower the efficacy of an agonist, the better the $[A]_{50}$ of the E/[A] curve serves as an estimate of agonist affinity.

Finally, the midpoint slope of hyperbolic E/[A] curves is constant ($m = 1$) and is independent of τ.

D. Methods of Estimating Agonist Affinity and Efficacy

The following discussion assumes that the agonist(s) being analyzed are interacting with only a single receptor population in the system under study. It is therefore necessary to experimentally verify that this condition applies by using appropriate antagonists. Any other potentially interfering receptor interactions must also be blocked.

1. Conditions

Affinity (K_A) and efficacy (τ) can only be estimated by direct operational model-fitting methods under the following conditions. First, the agonist in question must behave as a partial agonist or must be made to do so by appropriate experimental interventions. This is because τ can only be estimated for an agonist when the asymptote (α) of its E/[A] curve is measurably less than E_m [see Eq. (3)]. Second, it must be possible to determine the shape of the transducer relation for the system in question; that is, E_m and n must be estimable; and third, the curves must be monophasic.

2. The Comparative Method

This method, which was developed by Barlow and colleagues (19), relies on the comparison of the effects of a partial agonist with those of a standard full agonist. The first condition outlined above is therefore naturally fulfilled. The second condition is satisfied by analysis of the full agonist E/[A] curve, the shape of which is determined by the shape of the transducer relation (16,17). Accordingly, E_m and n can be estimated by fitting the full agonist E/[A] curve data to a Hill equation of the form:

$$E = \frac{E_m[A]^n}{[A]^n + [A]_{50}^n} \tag{5}$$

Then assuming these values, the operational model [Eq. (2)] can be fitted to the partial agonist E/[A] curve data to estimate K_A and τ. In practice, both steps of the analysis are carried out simultaneously. An example of such an analysis is shown in Figure 3.

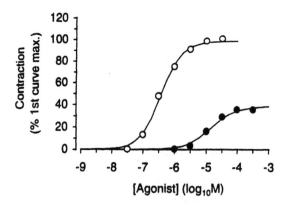

Figure 3 The comparative method: operational model fitting. Analysis of the muscarinic receptor partial agonist McN-A-343 (●) in comparison with the reference full agonist carbachol (○) in the rat isolated trachea preparation. The data shown are from a single tissue and the lines drawn through them are the results of operational model fitting using Eq. (2) (McN-A-343) and (5) (carbachol). The model parameter estimates obtained were as follows: $E_m = 97.78$; $n = 1.34$; $p[A]_{50}$ (Carbachol) = 6.43; $\tau = 0.72$ ($\log_{10}\tau = -0.14$); pK_A ($-\log_{10}K_A$) = 4.81. (Unpublished data.)

3. The Receptor Inactivation Method

The receptor inactivation method, as developed by Furchgott (6), relies on the use of irreversible competitive antagonists, agents that bind covalently to receptors thereby preventing their subsequent activation by an agonist. The resultant reduction in the functional receptor concentration ($[R_0]$) of the system reduces the efficiency of the transduction process (τ decreases) until the full agonist behaves as a partial agonist. In effect, the control curve (that obtained in the absence of irreversible antagonist treatment) serves the same purpose as the standard full agonist in the comparative method; that is, E_m and n can be estimated from it. However, the control curve and the other curves generated at increasing concentrations or exposure times to the irreversible antagonist belong to a "family" of curves that differ only by $[R_0]$ and, therefore, τ. So, in practice, all the curves are fitted simultaneously to Eq. (2) allowing only τ to vary between them. In this way all the curves contribute curve shape information to the estimation of E_m and n. This analysis yields a single estimate of K_A and a τ value corresponding to each curve, the one of interest being that of the control curve. An example of such an analysis is shown in Figure 4.

Although the inactivation method is usually used to analyze full agonists, it can also be applied to the analysis of partial agonists. Practically, the

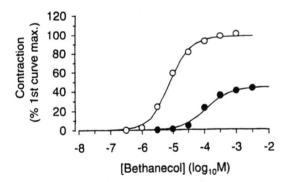

Figure 4 The inactivation method: operational model fitting. Analysis of the muscarinic receptor agonist bethanecol in the rat isolated trachea using the irreversible antagonist phenoxybenzamine (Pbz). Bethanecol E/[A] curves were obtained before (O) and following 30-min exposure to 1 μM Pbz ($\bullet$). The data shown are from a single tissue and the lines drawn through them are the results of operational model fitting [Eq. (2)]. The model parameter estimates obtained were as follows: $E_m = 99.59$; $n = 1.38$; $\tau_1 = 16.60$ ($\log_{10}\tau_1 = 1.22$); $\tau_2 = 0.85$ ($\log_{10}\tau_2 = -0.07$); $pK_A = 3.90$. (Unpublished data.)

reliance of this method on the availability of suitable alkylating agents precludes its use in many receptor systems.

4. The Interaction Method

The interaction method developed by Stephenson (4) represents another means of analyzing partial agonists. It involves studying how a full agonist E/[A] curve is affected by the presence of a partial agonist. This method has the advantage of allowing confirmation, in a single experiment, that the effects of the full and partial agonist are mediated by the same receptor population. The comparative method (see above) makes this assumption but does not test it. As for the comparative method, the necessity for the agonist being studied to be partial is naturally fulfilled in the interaction method. Similarly, the standard full agonist again serves the purpose of defining the shape of the transducer function in terms of E_m and n. Direct model fitting of E/[A] curve data of this form, however, does require extension of Eq. (2). The modified equation (20), which describes the interaction between a full agonist (A) and a partial agonist (B), is:

$$E = \frac{E_m([A]K_B + \tau_B[B][A]_{50})^n}{[A]_{50}^n(K_B + [B])^n + ([A]K_B + \tau_B[B][A]_{50})^n} \tag{6}$$

in which E_m and n are as defined previously, $[A]_{50}$ is the midpoint location of the full agonist E/[A] curve obtained in the absence of the partial agonist, and K_B and τ_B are the dissociation constant and efficacy of the partial agonist, respectively.

The analysis involves simultaneously fitting the E/[A] curve data for the full agonist obtained in the absence ([B] = 0) and presence of the partial agonist to Eq. (6). This allows the parameters of interest, namely K_B and τ_B, to be estimated and also provides estimates of E_m, n, and $[A]_{50}$. Figure 5 illustrates the fitting procedure when a single concentration of partial agonist is used. The interaction method has the added advantage that in situations where multiple concentrations of partial agonist have been used, it is possible to test if the interaction between the full agonist and partial agonist is consistent with simple competition. That is, the model provides an estimate of a parameter (m) that is analogous to the slope parameter of a Schild plot (see below). This fitting procedure involves raising the concentration of partial agonist to the power m, each time it appears in Eq. (6) (i.e., $[B]^m$ replaces $[B]$) and then simultaneously fitting the control curves and the curves in the presence of the different concentrations of partial agonist to this modified equation (20). An example of an analysis of this type is shown in Figure 6.

E. Null Methods

The operational model contains an assumption about the algebraic form of the relationship between agonist-receptor occupancy and functional effect,

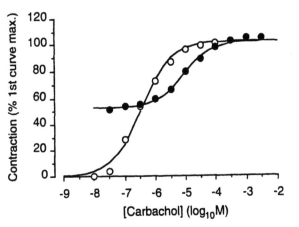

Figure 5   The interaction method: operational model fitting. Analysis of the interaction between the partial agonist pilocarpine and the full agonist carbachol at M_3-muscarinic receptors in the guinea pig isolated trachea. Carbachol E/[A] curves were obtained in the absence (○) and presence of 30 μM pilocarpine (●). The data shown are from a single tissue and the lines drawn through them are the results of operational model fitting [Eq. (6)]. The model parameter estimates obtained were as follows: E_m = 101.84; n = 0.90; p[A]$_{50}$ (control) = 6.47; τ_B = 1.17 ($\log_{10}\tau_B$ = 0.07); pK_B = 5.49. (From Ref. 20.)

which allows an explicit equation to be written down for the E/[A] relation (see above). In theory, this assumption about postreceptor events may restrict the use of the operational model (21,22). Such assumptions are deliberately avoided in the conventional null approach to agonist analysis (4,6,19). These null methods are based on the idea of pharmacological stimulus (S), which was originally defined by Stephenson (4) to be the product of efficacy (e) and fractional occupancy (y). This leads to a series of null equations that relate concentrations of drugs that produce equal responses under different experimental conditions. This procedure is illustrated below for the receptor inactivation method [similar methods have been described for the comparative (19) and interaction (4) methods].

1.   The Receptor Inactivation Method

This null method relies on comparing equal effects of the agonist before and after treatment with a receptor-inactivating agent. Such effects are considered to be due to the generation of equal stimuli, which in turn result from equal fractional receptor occupancies under the two conditions. These relationships are shown algebraically below:

$$\frac{[AR]}{[R_0]} = \frac{[A]}{K_A + [A]} = \frac{q[A']}{K_A + [A']} \tag{7}$$

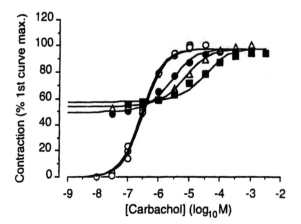

Figure 6 The interaction method: testing for competition. Analysis of the interaction between the partial agonist pilocarpine and the full agonist carbachol at M_3-muscarinic receptors in the guinea pig isolated trachea. Carbachol E/[A] curves were obtained in the absence (O) and presence of 10 μM ($\bullet$), 30 μM ($\triangle$), or 100 μM pilocarpine ($\blacksquare$). The data shown are from a single experiment in which three tissues were used. In each case the first curve was a carbachol control and the second curve was in the presence of pilocarpine. The lines drawn through the data are the results of simultaneously fitting all the data to the operational model [modified version of Eq. (6)]. The model parameter estimates obtained were as follows: $E_m = 97.59$; $n = 1.29$; $p[A]_{50}$ (control$_1$) = 6.50; $p[A]_{50}$ (control$_2$) = 6.54; $p[A]_{50}$ (control$_3$) = 6.48; $\tau_{B1} = 1.23$ ($\log_{10}\tau_{B1} = 0.09$); $\tau_{B2} = 1.26$ ($\log_{10}\tau_{B2} = 0.10$); $\tau_{B3} = 1.35$ ($\log_{10}\tau_{B3} = 0.13$); $pK_B = 6.05$; m (slope) = 1.07. The estimated slope parameter was not significantly different from unity, indicating that the interaction is consistent with simple competition. (From Ref. 20.)

where [AR]/[R$_0$] represents fractional receptor occupancy, [A] and [A'] are equieffective concentrations of agonist before and following receptor inactivation, K_A is the agonist dissociation constant, and q is the fraction of receptors remaining after receptor inactivation. Rearrangement of the above equations allows the following linear relationship between [A] and [A'] to be derived (6):

$$\frac{1}{[A]} = \left(\frac{1}{q}\right)\frac{1}{[A']} + \frac{1-q}{qK_A} \tag{8}$$

Thus a plot of 1/[A] versus 1/[A'] yields estimates of the slope (1/q) and the intercept [(1 − q)/qK_A] from which K_A can be estimated as slope-1/intercept. After K_A has been estimated, the efficacy of the agonist can be estimated from the [A]$_{50}$ of the control E/[A] curve.

Although this null method avoids making an assumption about the nature of the transducer relationship, the raw data has to be manipulated before it can be analyzed. This involves interpolation of equieffective points on the pre- and postinactivated E/[A] curves, extrapolation to estimate the corresponding values of [A] and [A'], and finally linear regression analysis. Data handling using the null method is therefore considerably more cumbersome than with the operational model, where the raw E/[A] curve data is used (see Figs. 3–6).

2. Nested Hyperbolic Method

The indirectness of data handling and the statistical weakness of double reciprocal plots (undue weight may be given to values generated from the smallest observed responses) inherent in the conventional null method have been addressed in later developments of this analysis. Thron (23) proposed methods of statistically weighting the double reciprocal regression, and Parker and Waud (24) proposed a hyperbolic function relating [A] and [A'] that obviated the need to perform such regressions. More recently, development of the nested hyperbolic method (25) has allowed direct fitting of experimental data. This procedure involves fitting the control data to a Hill equation of the form shown previously [Eq. (5)] while simultaneously fitting the postinactivation E/[A] curve to the following equation:

$$E = \cfrac{E_m}{\left(\cfrac{[A]_{50}}{qK_A[A']}(K_A + [A'](1 - q))\right)^n + 1} \qquad (9)$$

where E_m, n, $[A]_{50}$, $[A']$, q, and K_A are as defined previously (see above). The inclusion of a slope factor (n) in this method allows it to fit nonhyperbolic E/[A] curve data (Furchgott's original method is valid only when E/[A] curves are hyperbolic). Figure 7 illustrates this direct-fitting procedure using the same data set as was employed in Section VI.D.3 (Fig. 4), thus allowing a comparison of this null method of analysis with operational model fitting.

3. Operational Model Versus Null Methods

Very few studies have been published where the results of operational model fitting have been directly compared with those obtained by null methods. Where this has been done, the results obtained are essentially identical (18,20,26–30) (small differences may, in some cases, be the result of applying hyperbolic methods to nonhyperbolic E/[A] curve data) indicating that both procedures provide reliable estimates of agonist affinity and efficacy.

Operational model fitting has the advantage over conventional null methods in terms of ease of data handling (see above). This direct treatment of

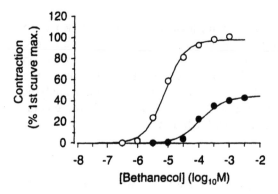

Figure 7 The inactivation method: nested hyperbolic null method. The data set shown is the same as that in Figure 4. The lines drawn through the data are the results of fitting the control curve (○) to Eq. (5) and the postinactivation curve (●) to Eq. (9). The model parameter estimates obtained were as follows: E_m = 97.18; n = 1.32; $p[A]_{50}$ (control) = 5.11; q = 0.048; pK_A = 3.87. Note that the estimates obtained are essentially identical to those obtained using the operational model (see Fig. 4); q, the fraction of receptors remaining after inactivation, is equivalent to τ_2/τ_1 ($[R_0]_2/[R_0]_1$) from the operational model (0.051).

the data means that they can be displayed along with the model-fitted lines on the scale on which the responses are recorded (in contrast to plots of equieffective concentrations for the null methods). This facilitates assessment of goodness-of-fit (see Figs. 3–6). Subsequent developments of the null methodology have largely addressed these issues (21,24,25,31,32), but in general such advances have been achieved by making similar assumptions to those underlying the operational model (the shape of the transducer function has been explicitly defined). The methodological convergence of these means of analysis has now reached the point where the choice between operational model and "null" method has become a matter of personal preference.

VII. ANALYSIS OF ANTAGONISTS

Several different classes of antagonists with distinct mechanisms of action (irreversible competitive, reversible competitive, allosteric, functional, and chemical) have been identified. The following sections concentrate on the analysis of reversible competitive antagonists, the most important of these classes.

As with agonists (see above), the first step in the action of these drugs is the formation of a reversible, short-lasting, drug-receptor complex governed by the Law of Mass Action. However, in this case no effect is generated ($\tau = 0$) and therefore the interaction can be described by a single parameter, the antagonist equilibrium dissociation constant (K_B).

The following sections outline how this binding parameter is measured experimentally.

A. Competitive Antagonism

In reversible competitive antagonism, the binding of the agonist and antagonist is mutually exclusive. The presence of the antagonist therefore decreases the probability that an agonist-receptor interaction will occur. To achieve the same degree of agonist occupancy and therefore the same effect (see above), in the presence of the antagonist as in its absence, the agonist concentration must be increased. The factor (r) by which it must be increased depends on both the concentration of antagonist ([B]) used and how well it binds (K_B). This relationship, which was first described by Schild (33), is shown below:

$$r - 1 = [B]^n/K_B \tag{10}$$

where $r = [A]_{50}/[A]_{50}^c$ (location parameter of the E/[A] curve in the presence of the antagonist/location parameter of the E/[A] curve in the absence of the antagonist) and n represents the order of reaction between the antagonist and the receptors (for a first order reaction $n = 1$; that is, one molecule of antagonist binds to one receptor molecule).

1. Schild Analysis

Experimentally, a K_B is estimated by studying the interaction of an agonist and antagonist over a wide range of antagonist concentrations (the wider, the better). This is necessary because drugs that are not reversible competitive antagonists may appear to be so within a narrow range of concentrations. If the antagonist is truly competitive, it should produce parallel rightward displacement [i.e., no change in midpoint slope (m) occurs] of the E/log[A] curves with no change in the maximal response (α) (see Fig. 8a–e). This is intuitively obvious, since the antagonist is merely decreasing the probability that an agonist-receptor interaction will occur, an effect that can always be overcome by increasing the agonist concentration. The analysis involves fitting experimentally derived values of r at different concentrations of antagonist to the following form of Eq. (10) (34) (See Fig. 8f):

$$\log_{10}(r - 1) = n\log_{10}[B] - \log_{10}K_B \tag{11}$$

Adherence of the data to this equation is judged by the finding of a linear plot with unit slope. Under these conditions the intercept on the X-axis

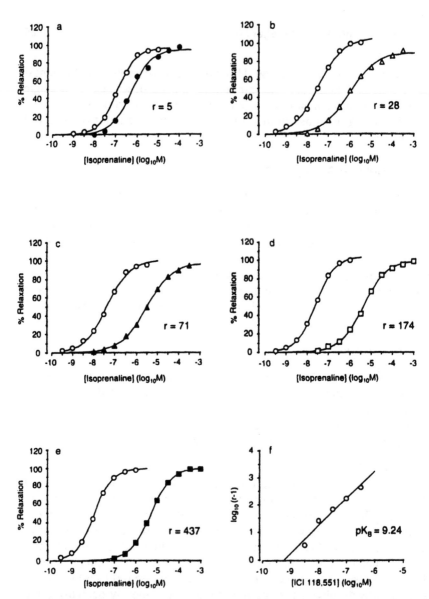

Figure 8 Schild analysis: homogenous receptor system. Antagonism of the β_2-adrenoceptor-mediated relaxant effects of isoprenaline by the competitive antagonist ICI 118, 551, in the guinea pig isolated trachea. The data shown is from a single experiment. In each tissue (a–e), the first isoprenaline E/[A] curve performed was a control (O); the second, was in the presence of ICI 118,551 at concentrations of 3 nM (●), 10 nM (△), 30 nM (▲), 100 nM (□), and 300 nM (■). Note the concentration-dependent parallel rightward displacement of the control curves. (f) illustrates these displacements (r values) in Schild plot form. The plot has a slope of unity and the intercept on the X-axis yields an estimate of 9.24 for the pK_B ($-\log_{10}K_B$).

($\log_{10}[B]$) gives an estimate of K_B. If the slope (n) is not statistically different from the theoretically expected value of 1, then the data should be refitted to Eq. (11) with the slope constrained to unity. When n is significantly different from 1, the intercept gives an estimate of pA_2 ($-\log_{10}K_B/n$). The pA_2 is an empirical estimate of antagonist affinity and equates to the negative logarithm of the concentration of antagonist that produces a twofold rightward shift ($r = 2$) of the control E/[A] curve. Nonlinearity and slopes other than unity can result from many causes. For example, a slope of greater than 1 may indicate incomplete equilibration with the antagonist or removal of the antagonist from the biophase. A slope that is significantly less than 1 may indicate removal of the agonist by a saturable uptake process, or it may result from the interaction of the agonist with more than one receptor. In the latter case the Schild plot may be nonlinear with a clear inflexion. All of these potential complicating factors have been described in detail previously by Kenakin (35) as have experimental manipulations and mathematical models that allow reliable antagonist affinity information to be extracted from such "nonclassic" data (see next section).

It should be emphasized that the finding that experimental data adheres to the Schild equation over a wide range of concentrations does not prove that the agonist and antagonist act at the same site. All that may be concluded is that the data is consistent with the hypothesis of mutually exclusive binding. Nevertheless, the estimation of antagonist affinities by Schild analysis currently forms the basis of classification schemes for the major hormone receptors, and its application will undoubtedly contribute significantly to future classifications.

2. Schild Analysis in heterogeneous Receptor Systems

As outlined above, a relatively common experimental finding in Schild analyses is a nonlinear plot with an inflexion, the result of an agonist activating two distinct receptor types mediating the same response, for which the antagonist under study has differential affinities. Lemoine and Kaumann (36) extended Eq. (11) to produce a model that mathematically describes such data for the specific case where an agonist has high efficacy at both receptor types (Q and R). The equation shown below is identical in form to their model, but expresses agonist activity in terms of location parameters rather than in terms of fractional stimuli:

$$\log_{10}(r - 1) = \log_{10}[B] - \log_{10}\left(\frac{(aK_{BQ} + K_{BR})[B] + K_{BQ}K_{BR}(1 + a)}{[B](1 + a) + aK_{BR} + K_{BQ}}\right) \quad (12)$$

where r is defined above, [B] is the concentration of antagonist, K_{BQ} and K_{BR} are the equilibrium dissociation constants of the antagonist for receptor

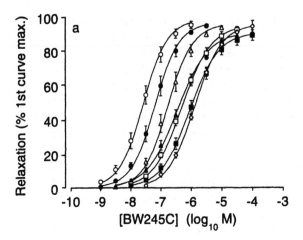

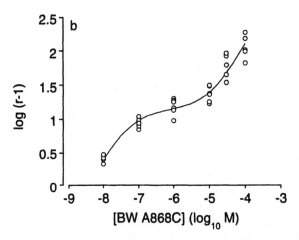

Figure 9 Schild analysis: heterogeneous receptor system. Antagonism of the relaxant effects of the prostaglandin DP-receptor agonist BW245C by the antagonist BW A868C in the rabbit isolated saphenous vein preparation. (a) shows the mean second curve data (5–7 replicates) with vertical lines indicating SE. The concentrations of BW A868C used were zero (O), 0.01 μM (●), 0.1 μM ($\triangle$), 1 μM ($\blacktriangle$), 10 μM ($\square$), 30 μM (■), and 100 μM ($\diamond$). The antagonist caused rightward displacements of the curves but these were clearly not consistent with simple competitive behavior. (b) illustrates the data in Schild plot form. The line drawn through them is the result of fitting to a two-receptor model [Eq. (12)], which assumes that the observed relaxations are mediated by interaction with two distinct receptor types. This analysis yielded estimates of the affinity of the antagonist for the two receptor types of $pK_{BQ} = 8.50 \pm 0.07$; $pK_{BR} = 4.89 \pm 0.08$; and an estimate of the relative locations

types Q and R, respectively, and $a = [A]_{50R}/[A]_{50Q}$, the ratio of the midpoint locations of the two "individual" E/[A] curves that contribute to the experimentally measured E/[A] curve. Equation (12) can be fitted directly to experimentally determined values of $\log_{10}(r - 1)$ as a function of $\log_{10}[B]$, allowing estimates of K_{BQ}, K_{BR}, and a to be made. The position of the inflexion of the plot depends on the values of $[A]_{50R}$ and $[A]_{50Q}$. For an antagonist that has higher affinity for receptor type Q than for receptor type R, the inflexion will be below the abscissa of the Schild plot when $a < 1$, above it when $a > 1$, and will lie on the abscissa when $a = 1$. Thus for a given antagonist the position of the inflexion point would be expected to vary when a series of agonists with different values of $[A]_{50R}$ and $[A]_{50Q}$ are studied. Only in the case where $a > 1$ is the inflexion likely to be experimentally observable. An example of this type of analysis is shown in Figure 9 and illustrates how mathematical modeling can be used to extract antagonist affinity information, from data that deviates from simple competition.

B. Clark Analysis

Schild analysis should only be used as a method of K_B estimation when a concentration ratio (r) can be estimated from a single experiment, that is, when a multiple curve design is employed (see Section V). This allows an intratissue comparison of the E/[A] curves obtained in the absence and presence of the antagonist to be made. In contrast, when a single-curve design is used, calculation of r values necessitates referring the location of each displaced curve to the same control curve location. Schild analysis of such data is statistically flawed because it overweights the control data. This problem can be overcome by using a modified version of Eq. (11), which analyzes location ($[A]_{50}$) values rather than concentration-ratios:

$$\log_{10}[A]_{50} = \log_{10}[A]_{50}^c + \log_{10}(1 + [B]^n/K_B) \tag{13}$$

This so-called Clark analysis (37–39) involves fitting E/[A] curve location parameters obtained in the absence ($[A]_{50}^c$) and presence ($[A]_{50}$) of antago-

---

of the two individual agonist E/[A] curves that make up the overall relaxant curve of $a = 13.7 \pm 1.5$. The antagonist, which has a higher affinity for receptor type Q, shifts the curve to the right until it reaches the concentration range in which receptor type R contributes significantly to the observed response. The E/[A] curve then appears to hit a barrier (corresponding to the inflexion in the Schild plot), which is overcome only when the antagonist concentration exceeds that needed to block the second receptor. (From Ref. 42, reprinted by permission of the publisher. Copyright 1996 by Elsevier Science Inc.)

nist (B) directly to Eq. (13) (38). This allows estimates of K_B, n (equivalent to the Schild slope parameter), and $[A]_{50}^c$ to be made. As for the Schild analysis, if the slope parameter is not significantly different from unity, then the data should be refitted to Eq. (13) with n constrained to 1. Data of this type should be displayed graphically as a Clark plot (37,39), the purpose of which is merely to allow a visual assessment of goodness-of-fit to be made (compare with the Schild plot, which is used to estimate K_B and n). An example of a Clark analysis is shown in Figure 10.

Clark Analysis in Heterogeneous Receptor Systems

In a similar manner to that outlined above for the Schild analysis, Eq. (13) can be extended (40) to account for the interaction of a full agonist with two receptor types for which an antagonist has different affinities. This model, shown below, should be used to analyze data obtained from experiments in which a single-curve design has been used:

$$\log_{10}[A]_{50} = \log_{10}[A]_{50}^c + \log_{10} \left(1 + \frac{[B][(aK_{BR} + K_{BQ} + [B] (1 + a)]}{[B](aK_{BQ} + K_{BR}) + K_{BQ} K_{BR} (1 + a)} \right) \quad (14)$$

where $[A]_{50}$, $[A]_{50}^c$, K_{BQ}, K_{BR}, and a are as defined previously. Fitting experimentally derived values of $[A]_{50}$ as a function of $[B]$ to Eq. (14) allows estimates of K_{BQ}, K_{BR}, a, and $[A]_{50}^c$ to be made.

VIII. SUMMARY

The development of occupancy theory has allowed the formulation of a series of mathematical models that describe the interaction of agonists and antagonists with their receptors, in terms of affinity and efficacy. These models provide a framework for the analysis and interpretation of E/[A] curve data and have proved to be useful tools in quantitative pharmacology. Unfortunately, despite the proven utility of this approach and the widespread availability of powerful computer-based curve-fitting programs [BMDP (41), Microsoft Excel, etc.], which greatly facilitate analysis, the application of mathematical modeling remains the exception rather than the rule in pharmacological studies.

IX. FUTURE DIRECTIONS

The application of occupancy theory to pharmacological studies has undoubtedly contributed to receptor-based drug design despite the fact that such analyses are based on concepts that were established 40 years or more ago. Since then, a great deal has been learned about receptor mechanisms

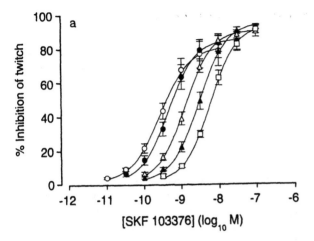

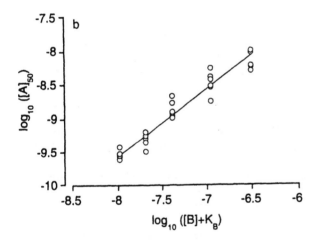

Figure 10 Clark analysis: homogeneous receptor system. Antagonism of the effects of the DA_2-receptor agonist SKF 103376 (see Ref. 43) by (−) sulpiride in the rabbit isolated field-stimulated ear artery preparation. (a) shows the mean E/[A] curve data (5–6 replicates) with vertical lines indicating SE. The concentrations of (−) sulpiride used were zero (O), 10 nM (●), 30 nM (△), 100 nM (▲), and 300 nM (□). (b) illustrates the data in Clark plot form. The adherence of the data with the unit slope line drawn through them indicates consistency with simple competition. The pK_B value obtained by fitting the $[A]_{50}$ values to Eq. (13) (with n constrained to 1) was 7.98 ± 0.07 (SE 27 d.f.). (Unpublished data.)

(for example, two-state behavior) and the bioassay systems that are routinely used by pharmacologists. To date, such knowledge has had limited impact in altering the way that pharmacologists analyze data. Nevertheless, there seems little doubt that future drug design will be significantly influenced by such information.

REFERENCES

1. Leff P, Dougall IG. Is pharmacological analysis of agonist action useful in medicinal chemistry? In: Angeli P, Gulini U, Quaglia W, eds. Trends in Receptor Research. Amsterdam: Elsevier Science Publishers BV, 1992:43–60.
2. Clark AJ. General pharmacology. In: Heffter's Handbuch der experimentallen Pharmakologie Erganzungswerk. Vol 4. Berlin: Springer-Verlag, 1937:1–223.
3. Ariëns EJ, Affinity and intrinsic activity in the theory of competitive inhibition. Arch Int Pharmacodyn Ther 1954; 99:32–49.
4. Stephenson RP. A modification of receptor theory. Br J Pharmacol 1956; 11:379–393.
5. Mackay D. A flux-carrier hypothesis of drug-action. Nature 1963; 197:1171–1173.
6. Furchgott RF. The use of β-haloalkylamines in the differentiation of receptors and in the determination of dissociation constants of receptor-agonist complexes. In: Harper NJ, Simmonds AB, eds. Adv Drug Res. Vol 3. New York: Academic Press, 1966:21–55.
7. Paton WDM. A theory of drug action based on the rate of drug-receptor combination. Proc R Soc Lond B 1961; 154:21–69.
8. Colquhoun D. The relationship between classical and cooperative models for drug action. In: Rang HP, ed. Drug Receptors. Baltimore: University Park Press, 1973:149–182.
9. Gosselin RE. Drug-receptor inactivation: a new kinetic model. In: Van Rossum JM, ed. Kinetics of Drug Action. Berlin: Springer-Verlag, 1977:323–356.
10. Ruffolo RR. Important concepts of receptor theory. J. Autonom Pharmacol 1982; 2:277–295.
11. Mackay D. A critical survey of receptor theories of drug action. In: Van Rossum JM, ed. Kinetics of Drug Action. Berlin: Springer-Verlag, 1977:255–322.
12. Black JW, Leff P. Operational models of pharmacological agonism. Proc R Soc Lond B 1983; 220:141–162.
13. Leff P. The two-state model of receptor activation. Trends Pharmacol Sci 1995; 16:89–97.
14. Leff P, Prentice DJ, Giles H, Martin GR, Wood J. Estimation of agonist affinity and efficacy by direct operational model-fitting. J Pharmacol Methods 1990; 23:225–237.
15. Nickerson M. Receptor occupancy and tissue response. Nature 1956; 178:697–698.

16. Waud DR. On the measurement of the affinity of partial agonists for receptors. J Pharmacol Exp Ther 1969; 170:117–122.

17. Leff P. Can operational models of agonism provide a framework for classifying hormone receptors? In: Black JW, Jenkinson DH, Gerskowitch VP, eds. Perspectives on Receptor Classification. New York: Alan R Liss, 1987:157–167.

18. Black JW, Leff P, Shankley NP, Wood J. An operational model of pharmacological agonism: the effect of E/[A] curve shape on agonist dissociation constant estimation. Br J Pharmacol 1985; 84:561–571.

19. Barlow RB, Scott NC, Stephenson RP. The affinity and efficacy of onium salts on frog rectus abdominis. Br J Pharmacol 1967; 31:188–196.

20. Leff P, Dougall IG, Harper D. Estimation of partial agonist affinity by interaction with a full agonist: a direct operational model-fitting approach. Br J Pharmacol 1993; 110:239–244.

21. MacKay D. Concentration-response curves and receptor classification: null method or operational model? Trends Pharmacol Sci 1988; 9:202–205.

22. Leff P. Analysis of agonist action using the operational model. Trends Pharmacol Sci 1988; 9:395–398.

23. Thron CD. Graphical and weighted regression analyses for the determination of agonist dissociation constants. J Pharmacol Exp Ther 1970; 175:541–553.

24. Parker RB, Waud DR. Pharmacological estimation of drug-receptor dissociation constants. Statistical evaluation. I. Agonists. J Pharmacol Exp Ther 1971; 177:1–12.

25. James MK, Morgan PH, Leighton HF. A new method for estimation of agonist dissociation constants (K_A); directly fitting the postinactivation concentration-response curve to a nested hyperbolic equation. J Pharmacol Exp Ther 1989; 249:61–69.

26. Kramer TH, Davis P, Hruby VJ, Burks TF, Porreca F. In vitro potency, affinity and agonist efficacy of highly selective delta opioid receptor ligands. J Pharmacol Exp Ther 1993; 266:577–584.

27. Wiener LH, Thalody GP. Differential attenuation of the responses to adenosine and methoxamine in isolated rabbit aorta. J Pharmacol Exp Ther 1993; 267:828–837.

28. Sallés J, Giraldo J, Badia A. Analysis of agonism at functional prejunctional α_2-adrenoceptors of rat vas deferens using operational and null approaches. Eur J Pharmacol 1994; 258:229–238.

29. Palea S, Corsi M, Rimland JM, Trist DG. Discrimination by benextramine between the NPY-Y_1 receptor subtypes present in rabbit isolated vas deferens and saphenous vein. Br J Pharmacol 1995; 115:3–10.

30. Zernig G, Issaevitch T, Woods JH. Calculation of agonist efficacy, apparent affinity and receptor population changes after administration of insurmountable antagonists: comparison of different analytical approaches. J Pharmacol Toxicol Methods 1996; 35:223–237.

31. Zaborowsky BR, McMahan WC, Griffin WA, Norris FH, Ruffolo RR. Computerized graphic methods for determining dissociation constants of agonists, partial agonists and competitive antagonists in isolated smooth muscle preparations. J Pharmacol Methods 1980; 4:165–178.

32. McPherson GA, Molenaar P, Raper C, Malta E. Analysis of dose-response curves and calculation of agonist dissociation constants using a weighted non-linear curve fitting program. J Pharmacol Methods 1983; 10:231–241.

33. Schild HO. pA_x and competitive drug antagonism. Br J Pharmacol 1949; 4:277–280.

34. Arunlakshana O, Schild HO. Some quantitative uses of drug antagonists. Br J Pharmacol 1959; 14:48–58.

35. Kenakin TP. Pharmacological Analysis of Drug-Receptor Interaction, 2nd ed. New York: Raven Press, 1993.

36. Lemoine H, Kaumann AJ. A model for the interaction of competitive antagonists with two receptor subtypes characterized by a Schild plot with apparent slope unity. Naunyn-Schmiedeberg's Arch Pharmacol 1983; 322:111–120.

37. Stone M, Angus JA. Developments of computer-based estimation of pA_2 values and associated analysis. J Pharmacol Exp Ther 1978; 207:705–718.

38. Trist DG, Leff P. Quantification of H_2-agonism by clonidine and dimaprit in an adenylate cyclase assay. Agents Actions 1985; 16:222–226.

39. Lew MJ, Angus JA. Analysis of competitive agonist-antagonist interactions by nonlinear regression. Trends Pharmacol Sci 1995; 16:328–337.

40. Dougall IG, Harper D, Lydford SJ, Leff P. Unpublished.

41. Dixon WJ, ed. BMDP Statistical software. Berkeley: University of California Press, 1981.

42. Lydford SJ, McKechnie KCW, Leff P. Interaction of BW A868C, a prostanoid DP-receptor antagonist, with two receptor subtypes in the rabbit isolated saphenous vein. Prostaglandins 1996; 52:125–139.

43. DeMarinis RM, Hieble JP. Dopamine receptor agonists: chemical and biological studies of the aminoethylindolones. Drugs Fut 1989; 14:780–797.

3

Quantification of Receptor Interactions Using Binding Methods

Sebastian Lazareno
MRC Collaborative Centre, London, England

I. INTRODUCTION

Since a drug acting at a receptor must bind to the receptor before exerting an effect, it may seem obvious that measuring directly the binding of radio-labeled drugs, alone and in combination with other drugs, would provide valuable pharmacological information. As recently as 15 years ago, however, radioligand-binding experiments, and the conclusions drawn from them, were regarded with deep suspicion by a large section of classic pharmacologists. There were good reasons for this, since binding assays often produced complex and unexpected results that were often interpreted, perhaps all too easily, as indicating the existence of receptor subtypes. Today, many of these complexities are understood, either as artifacts or as a true phenomena that contribute to our understanding of receptor transduction and heterogeneity, and radioligand-binding assays are now central to many drug discovery projects.

Radioligand-binding studies are used in many different ways, from high-throughput screening for active compounds to detailed mechanistic studies of receptor function. There are many excellent books and reviews covering every aspect of radioligand-binding (1–10). In this chapter I will consider briefly some theoretical and practical aspects of radioligand-binding assays, and some current models of drug action at G-protein-coupled receptors and the information that binding studies may provide about the parameters of these models.

II. PRINCIPLES AND CONCEPTS UNDERLYING THE ANALYSIS OF BINDING DATA

In a radioligand-binding assay, receptors, or tissue containing receptors, are incubated with a radiolabeled compound for some time, perhaps in the presence of unlabeled ligands or other modulators, and then the label bound to the tissue is separated from the unbound ligand, allowing the bound label to be quantified. The source of receptors may be whole tissues or tissue fragments (e.g., brain sections), whole cells, cell fragments (e.g., synaptosomes), membranes from fragmented cells, or receptors in solution. The method for separating bound from free radioligand will depend on the receptor preparation and the kinetic characteristics of the radioligand (see below). The different receptor preparations and methods of separation will have their own technical problems and limitations, and assay systems should be optimized so that the assumptions underlying the model used to analyze the data are reasonably accurate. In some cases only qualitative data are required, i.e., the presence of binding or inhibition of binding, but more usually the data will be interpreted in the context of an empirical or mechanistic model describing in mathematical terms the relationship between the concentrations of labeled and/or unlabeled ligand and observed binding.

There are two components of radioligand-binding: a "specific" component of binding to the receptor(s) of interest, and a "nonspecific" component to other sites. The nonspecific component of radioligand-binding is measured in the presence of a concentration of unlabeled defining ligand (which ideally binds to the receptors of interest but is chemically different from the radioligand) sufficient to prevent binding of the radioligand to the binding sites under study, and only to those sites. Nonspecific binding is usually, but not always, a linear function of radioligand concentration, reflecting binding to nonsaturable components (e.g., lipid) or to high-capacity–low-affinity binding sites. Specific binding is measured as the difference between "total" binding observed in the absence of the defining agent and nonspecific binding.

Binding studies are often used in an attempt to quantify the parameters of drug-receptor interaction according to a model of drug action. The simplest model is the Law of Mass Action, in which ligands bind reversibly to a single binding site. With one type ligand and one type of receptor:

$$L + R \underset{k_{off}}{\overset{k_{on}}{\rightleftharpoons}} L.R \tag{1}$$

According to this model, a ligand, L, binds to a receptor, R, with an association rate constant k_{on}, to form a receptor-ligand complex, L.R, which is stable for a time but eventually dissociates to free, unchanged receptor and ligand, with a dissociation rate constant k_{off}. Values of k_{on} and k_{off} can be measured empirically (see below). The units of k_{off} are time^{-1}, and k_{on} has units of M^{-1}.time^{-1}.

Binding reaches equilibrium when the rate of formation of bound receptor is equal to the rate of dissociation of bound receptor:

$$[L]*[R]*k_{on} = [L.R]*k_{off} \tag{2}$$

This equation is rearranged to give two equilibrium constants:

$$\frac{[L]*[R]}{[L.R]} = \frac{k_{off}}{k_{on}} = K_d \tag{3}$$

where K_d is the equilibrium dissociation constant with units of M, and

$$\frac{[L.R]}{[L]*[R]} = \frac{k_{on}}{k_{off}} = K_a \tag{4}$$

where K_a is the equilibrium association constant, or affinity constant, with units of M^{-1}.

The dissociation constant, K_d, is a rather special concentration of ligand, the concentration that results in binding to 50% of the receptors. This is easy to see once we make the distinction between free and total receptor concentrations. The ligand will usually be present at a much higher concentration than the receptors, so that only a very small proportion of the total ligand will be bound to receptors and we can assume that the free ligand concentration, [L], is equal to the total ligand concentration (this assumption may be untrue for high-affinity radioligands, see below). The situation is different with receptors, however. The total receptor concentration, [Rt], exists in two species, free receptor [R] and ligand-bound receptor [L.R], and any changes in the concentration of bound receptor will alter the concentration of free receptor.

$$[Rt] = [R] + [L.R] \tag{5}$$

When 50% of the receptors are bound with ligand, [R] = [L.R], and the left-hand side of Eq. (3) becomes [L], so the K_d concentration of a ligand results in 50% receptor occupancy.

The association, or affinity, constant, K_a, is the reciprocal of the K_d. Although the meaning of the K_a is less intuitively clear than that of the K_d, the K_a has the advantage that, as a measure of the tightness of the binding between ligand and receptor, its numerical value increases as the tightness of binding, or potency, increases. In practice, the affinity of a ligand for a receptor is often expressed as a positive logarithm, which can be regarded as either the log of the affinity constant or the negative log of the equilibrium dissociation constant.

The measurement of the K_d of a radioligand can therefore be approached in two quite different ways: by measuring the concentration of radioligand giving 50% receptor occupancy, and by measuring the rates of association and dissociation of the radioligand.

Equation (5) can be combined with Eq. (3) or (4) to give an equation describing the binding of any concentration of L:

$$[L.R] = [Rt]* \frac{[L]}{[L] + K_d} = \frac{[Rt]}{1 + \frac{K_d}{[L]}}$$

and   (6)

$$[L.R] = [Rt]* \frac{[L]* K_a}{1 + [L]* K_a}$$

in terms of K_d and K_a, respectively.

The fraction of the total receptor population bound with radioligand is obtained by dividing Eq. (6) by Rt:

$$\text{Fractional occupancy} = \frac{[L]}{[L] + K_d}$$

and   (7)

$$\text{Fractional occupancy} = \frac{[L]*K_a}{1 + [L]*K_a}$$

in terms of K_d and K_a, respectively.

The binding of unlabeled ligands is measured indirectly, by the reduction in radioligand-binding in the presence of the unlabeled ligand. If the unlabeled ligand interacts competitively with the radioligand, i.e., both ligands bind reversibly to the same site, then fractional inhibition of radioligand-binding by antagonist has the same form as Eq. (7):

$$\text{Fractional inhibition} = \frac{[I]}{[I] + IC_{50}}$$ (8)

where [I] is the concentration of antagonist, and IC_{50} is the antagonist concentration causing 50% inhibition. Just as the inhibitor competes with the radioligand, so the radioligand competes with the inhibitor, and the IC_{50} will usually be larger than the dissociation constant of the inhibitor. This value is calculated from the IC_{50} using the equation of Cheng and Prusoff (11):

$$K_i = \frac{IC_{50}}{1 + \frac{[L]}{K_d}}$$ (9)

where K_i is the dissociation constant of the inhibitor, and [L] and K_d are the concentration and dissociation constant, respectively, of the radioligand.

These equations are the starting point for the analysis of equilibrium binding data. They define a specific shape for the binding curve. At a ligand concentration of $K_d/10$ the fractional receptor occupancy will be 0.09, and at a ligand concentration 10 times higher than K_d the occupancy will be 0.91, so a Law of Mass Action occupancy curve goes between about 10% and 90% over a hundredfold range of ligand concentrations. Similarly, a Law of Mass Action inhibition curve spans about 10% and 90% inhibition between antagonist concentrations of $IC_{50}/10$ and $10*IC_{50}$. Equation (8) is a particular case of a more general Hill function

$$\text{Fractional inhibition} = \frac{[I]^b}{[I]^b + IC_{50}^b} \tag{10}$$

where b is the Hill slope factor. Law of Mass Action curves have Hill slopes of 1. Curves that cover 10–90% occupancy over a smaller range of ligand concentrations are said to have "steep" slopes, >1, and curves that extend over more than a hundredfold range of ligand concentrations are said to be "shallow," with slopes <1. Steep or shallow slopes indicate that the mechanism of binding is more complex than the Law of Mass Action, and possible explanations for such complexity are discussed below.

III. ASSAY SYSTEMS—PROBLEMS AND PITFALLS

Binding assays are relatively easy to perform, but a number of factors should be considered in the design of a successful binding assay.

A. Equilibration

An equilibrium binding study should, of course, reach equilibrium. In order to ensure this, binding is often measured at a single radioligand concentration over a range of times in order to select a suitable incubation time, but it is still quite easy for nonequilibrium conditions to occur.

Possible scenarios are:

1. Perhaps there is a fast and a slow component of binding, but the range of times chosen for the assay only allows clear detection of the fast component. This situation may arise if the receptors exist in more than one compartment with differential access to the ligand, such as membranes and vesicles, or if there are diffusion barriers in tissue fragments. Other possible causes of slow binding components are receptor isomerization and receptor heterogeneity.

2. The observed rate of radioligand-binding depends on the radioligand concentration relative to the K_d and on the dissociation rate constant:

$$k_{obs} = k_{off}*(1 + [L]/K_d) \tag{11}$$

Equation (11) is quite useful. It allows an estimate of the observed association rate constant, K_{obs}, from a knowledge of the radioligand concentration, [L], its dissociation rate constant, k_{koff}, and its K_d. The binding half-life, or time for half the free receptors (relative to equilibrium binding) to become bound, is $-ln(0.5)/k_{obs}$, and five half-lives gives about 97% of true equilibrium binding, which is usually considered to be adequate. The following generalizations can be made from Eq. (11):

1. If $[L] \ll K_d$ then $k_{obs} \approx k_{off}$.
2. If $[L] = K_d$ then $k_{obs} = k_{off}* 2$.
3. If $[L] \gg K_d$ then $k_{obs} \approx k_{off}* [L]/K_d$.

If the radioligand concentration used to determine the appropriate incubation time was equal to or greater than the K_d, then lower concentrations of radioligand, as are used in saturation studies (see below) may not reach binding equilibrium. For example, the muscarinic antagonist radioligand $^3H-(+)QNB$ has a K_d of about 30 pM and a dissociation rate constant of about 0.003 min^{-1}, corresponding to a dissociation half-time of 220 min [at m1 receptors and at 25°C (12)]. This radioligand is often used at a concentration of about 200 pM for routine studies. At this concentration, equilibrium binding will be achieved in about 2.5 hr, but saturation studies using concentrations around the K_d, and inhibition studies with a competing ligand (see below), will not be in equilibrium.

3.   The presence of a competing ligand will affect the time to equilibrium (13). If the competing ligand has a slower k_{off} than the radioligand, then radioligand-binding will "overshoot" and then decrease to the equilibrium level as the antagonist binding reaches its own equilibrium. If the competing ligand has much faster kinetics than the radioligand, as is often the case, then the rate of radioligand-binding will decrease with increasing antagonist concentrations. This can be understood intuitively when it is considered that the effect of a competitive antagonist is to increase the apparent K_d of the radioligand [Eq. (9) with terms swapped between radioligand and competitor]: from Eq. (11) it can be seen that the observed rate of radioligand-binding, k_{obs}, will decrease with increasing antagonist concentrations, with k_{off} as the limiting value. So, in order for an inhibition assay to be completely in equilibrium, the incubation time should be at least five *dissociation* half-lives. In the case of ^3H-QNB, with a dissociation half-life of 220 min, inhibition assays will never be in equilibrium. It is necessary and useful to measure the rate of association, but it is the dissociation rate that should guide the choice of incubation time.

B. Binding Stability

It is desirable for equilibrium binding to be stable, but this is not always possible, especially for receptors in solution. Binding becomes more stable at lower temperatures, but also slower, so time course measurements may need to be taken over a range of temperatures to find the optimal one. Tables of protease inhibitors and other protecting agents are given in Ref. 5.

Radioligand-binding to whole cells will often not be in equilibrium, though it may be at a steady state, because receptors are constantly being synthesized, degraded, internalized, and recycled. Exposure to agonists will often alter these processes, leading to a rapid loss of cell surface receptors, but the particular effects of exposure to agonist depend on both the receptor and its cellular environment (7,10,14). Cell surface receptors are thought to be stable at 4°C.

C. Radioligand Concentration and Depletion

In inhibition assays the radioligand concentration should be as low as possible, ideally well below the K_d, for two reasons. First, inspection of Figure 3 shows that the best specific:nonspecific binding ratio is found at low radioligand concentrations, and second, at these concentrations the IC_{50} of the inhibitor provides a good estimate of its K_d, while at higher radioligand concentrations a correction factor must be applied to the IC_{50}, which requires an accurate knowledge of the K_d of the radioligand [see Eq. (9)]. In practice, however, this is often not possible. The choice of radioligand concentration will often be a compromise based on a number of factors: the specific activity of the radioligand, the availability of radioligand, the availability of receptors, the K_d of the radioligand, the practical limit of the incubation volume, and the specific:nonspecific binding ratio in the receptor preparation. The objective is an assay with an acceptable incubation volume (e.g., 0.1–1 ml), sufficient signal, i.e., radioactive dpm, for accurate detection of changes in binding, and with bound radioligand accounting for not more than about 15% of added radioligand.

This last requirement may pose a problem, especially with very high affinity tritiated radioligands. Such radioligands bind at low (i.e., pM) concentrations, which means that receptors must be present at even lower concentrations to avoid depletion of free radioligand (see below). But the concentration of bound receptors must provide an adequate signal. For example, 1 ml of bound receptors at a concentration of 1 pM and a specific activity of 80 Ci/mmol will give 178 dpm, which is not sufficient for accurate data unless very long counting times are used. If the radioligand were iodinated, its specific activity would be >2000 Ci/mmol and 4450 dpm would be recovered, which would solve the problem. But for a tritiated radioligand the only practical solution is to increase the receptor and radioligand concentrations.

If a sizable fraction of added radioligand is bound, this will reduce the concentration of free radioligand. Depletion of radioligand by more than about 15% leads to problems in the interpretation of saturation and inhibition curves. The depletion is greatest at low radioligand concentrations, so saturation curves will be "steep" and the radioligand K_d will be overestimated. The same effects will be seen with inhibition curves—steepening and overestimation of IC_{50}—because, especially at low inhibitor concentrations, radioligand displaced by the inhibitor increases the free radioligand concentration and partially rebinds.

The problem of radioligand depletion in saturation studies can partly be overcome by estimating the free concentration as the difference between added and bound radioligand, but, for filtration assays, this does not allow for dissociation from receptors or for the loss of loosely bound, "washable" nonspecific binding during the washing procedure, which may be substantial at high membrane concentrations (15), or for the loss of membranes through the filter, which will usually occur to some extent. These losses of binding will cause the degree of radioligand depletion to be underestimated, leading to an increase in K_d under some conditions (15). The problem with saturation studies can be solved completely by conducting a parallel assay that is centrifuged, rather than filtered, allowing the free radioligand concentration in the supernatant to be measured directly.

There are no empirical or computational adjustments to solve the problem of ligand depletion in inhibition studies. If there is >15% observed radioligand depletion in inhibition studies, one should be aware of the problem and treat experimental values with due caution.

D. Separating Free from Bound

The choice of a separation method will depend on the radioligand and receptor preparation. Methods used with soluble receptors are described by Hulme (16). Tissue adhering to a surface may simply be rinsed with buffer. Cell membranes and whole cells in suspension are usually collected by filtration or, less commonly, by centrifugation.

Filtration involves passing the incubation mixture over filters, usually glass fiber, using a filtration manifold or cell harvester. Glass fiber filters allow rapid filtration while trapping most of the membranes. Many radioligands will bind to the filters themselves, which should always be tested for and which may be a serious problem. The binding of positively charged radioligands to filters may be substantially reduced by presoaking the filters in a 0.1% solution of polyethylenimine (PEI), but if the problem is still serious, an alternative filter material such as cellulose should be tried. After the initial filtering of the incubation mixture, the filters are usually washed one or more times by passing a few milliliters of cold buffer or water over them to remove residual free radioligand and washable nonspecific binding: this washing step will often substantially improve the specific:nonspecific binding ratio, but will also incur some loss of specific binding through dissociation. For this reason filtration is not feasible with low-affinity, rapidly dissociating radioligands. As a rule of thumb, filtration is unsuitable for radioligands with $K_d > 10^{-8}$ M (see Refs. 17 and 18 for comparisons of separation procedures). Finally, the radioactivity trapped in the filters is measured using liquid scintillation spectrometry. For tritiated ligands it is important that the filters are either dried before immersion in scintillation fluid or immersed in scintillation fluid for sufficient time (e.g., overnight) and then shaken vigorously to allow the water in the filter to dissolve fully in the fluid, leaving a translucent filter and a homogeneous medium.

Centrifugation assays involve spinning the sample for a few minutes, usually in a benchtop microfuge generating 14,000 g (19). The supernatant is aspirated or decanted, the surface of the pellet may optionally be rinsed with cold buffer or water, and then the pellet is dispersed or dissolved and the radioactivity counted. A simple method for dispersing/dissolving the pellet is to incubate it with 10 μl of 1 M NaOH at 65°C for about 40 min, agitating the tube to ensure that the pellet has been liquefied, and then neutralizing with 10 μl 10% HCl before adding scintillation fluid. The centrifugation assay has the advantage that the binding equilibrium is undisturbed during the separation of bound from free radioligand, so it can be used with rapidly dissociating radioligands. The main disadvantage of this assay is the high level of nonspecific binding, caused partly by free radioligand trapped in the pellet and partly by the presence of washable nonspecific binding, which would have been removed in a filtration assay. Another point to bear in mind, especially if using a benchtop microfuge, is that the recovery of membranes in the pellet may be less than complete and can be strongly affected by the viscosity and ionic composition of the incubation mixture (19).

E. Receptor Criteria

If specific binding is to a receptor protein, it should be a linear function of tissue concentration until depletion of free radioligand occurs, and it should be eliminated after the tissue is boiled. If either criterion is not met, some or all of the binding is probably an artifact. There are other criteria for establishing that a binding site corresponds to a receptor, but it is difficult to lay down hard and fast rules.

Ideally, the radioligand should show mass action binding to a single site. This requirement can be tested quite stringently: saturation studies should detect saturable binding to a single site, association and dissociation assays should display monoexponential kinetics and full reversibility of specific binding, and the K_d measured in saturation assays should be the same as that calculated from the ratio k_{off}/k_{on} obtained from the kinetic assays. There are, however, many cases of radioligands, especially agonists, binding to genuine receptors but without satisfying some or all of the above criteria. For example, binding of agonist radioligands to cloned adenosine A_1 receptors and somatostatin sst_2 receptors under some conditions is complex and only partially reversible (20,21). The identification of a binding site as a receptor is made simpler if the radioligand has mass action binding, but this cannot be an absolute requirement.

Another criterion is that the binding site exists in tissue known to contain functionally defined receptors, and does not exist in tissue without the functional receptor. Again, there are exceptions. One example is that untransfected CHO cells show $5HT_{1B}$ functional responses but barely detectable specific radioligand-binding (22). Another example concerns the snake venom α-bungarotoxin, a potent antagonist and affinity label at nicotinic receptors in muscle. Neuronal nicotinic receptors, in contrast, are relatively insensitive to α-bungarotoxin, yet α-bungarotoxin does bind with high affinity in the brain and with a nicotinic pharmacology, though with a different regional distribution to the binding of ^{3}H-ACh and ^{3}H-nicotine (23). To confuse matters further, there were reports that thymopoetin, a polypeptide thymic hormone, bound with high affinity to the neuronal α-bungarotoxin site and might be the endogenous ligand (24). The problems are now resolved. The neuronal α-bungarotoxin site, once an enigma, is now known to be an α7 subunit containing nicotinic receptor, and the inhibitory material supplied as "thymopoetin" was actually α-cobratoxin, another snake venom (25,26).

The most important criterion is that used in establishing the identity of any pharmacologically defined receptor—the affinities of a range of chemically diverse agonists and antagonists for the site should be consistent with the functionally derived affinities characteristic of the receptor. This criterion is often not met with agonists (see above and Section V), but usually antagonists do show similar affinities in binding and functional studies, and the use of stereoselective antagonists, when available, adds stringency to the test. A binding site that fails to meet this criterion, especially with antagonists, is probably not the pharmacological receptor, but the converse need not be true: there are many examples of binding sites that initially appeared to have the "correct" pharmacology but which, on further study, were found not to be receptors (8,27,28).

IV. STANDARD ASSAYS AND THE PARAMETERS THEY MEASURE

There are four general designs of assay for the characterization of labeled and unlabeled ligands: dissociation, association, saturation, and inhibition. The assays are interpreted initially by fitting the data to a mechanistic model, the Law of Mass Action or, in the case of inhibition curves, to an empirical model, a logistic function. Before computers were generally available, it was necessary to transform the data with a function that yields a straight line when applied to Law of Mass Action data, and then use linear regression analysis with the transformed data. The most commonly used transformations will be presented for historical reasons, and because the linear plots are useful intuitively and for detecting deviations from the model. Wherever possible, however, parameter estimates should be obtained by fitting the untransformed data directly to the model using nonlinear regression analysis. The appropriate equations will be presented, although most curve-fitting programs have the equations already set up.

A. Dissociation Analysis

1. Objective

The objective of dissociation analysis is to measure the rate, pattern, and extent of radioligand dissociation.

2. Parameter to Be Estimated

The dissociation rate constant k_{off} is to be estimated.

3. Procedure

Receptors are incubated with radioligand, usually, though not necessarily, until equilibrium is reached. Then any further binding of radioligand is prevented, either by dilution of the sample (at least 100-fold) or by adding a high concentration of unlabeled competing ligand, which will instantly occupy receptors as they become free and thus prevent radioligand reassociation. Binding is measured at various times thereafter. The assay contains, in addition, samples that have not dissociated and that provide a measure of binding at time zero, and a measure of nonspecific binding. If the prelabeled sample is diluted, then measures of nonspecific binding at various times should be made, at least in initial studies.

If a filtration manifold is used, the assay is straightforward: the above procedure is carried out in a beaker and aliquots are removed and filtered at various times. With a cell harvester, however, all the samples are filtered at the same time, so some thought must be given to timing. Ideally, especially with slowly dissociating ligands, the assay should be planned so that all the samples are exposed to radioligand for the same time before dissociation is initiated, which means that the assay will be started at different times for the different time points. If it has been shown that "equilibrium" binding remains stable over the total time used for the longest dissociation time point, as may be the case with rapidly dissociating radioligands, then the assay can be initiated with all the tubes at the same time and only the initiation of dissociation need be timed individually.

4. Linearizing Transformation

Subtract nonspecific binding from all data points. Plot $\ln(B_t/B_0)$ against time (see Fig. 1) where B_0 is the specific binding before dissociation and B_t is the binding remaining at time t. The negative slope of the straight line is the dissociation rate constant k_{off}.

5. Equations for Nonlinear Regression

Nonspecific binding can be handled in two ways. It can be subtracted from the other data points, as above, or it can be included in the analysis as an observation at a very large time.

With nonspecific binding subtracted, the equation is

$$B_t = B_0*\exp(-k_{off}*t)$$

where the symbols have the meaning given above.

If the data appear to reach an asymptotic value above zero, this level can be built into the equation:

$$B_t = (B_0 - asymptote)*\exp(-k_{off}*t) + asymptote$$

With nonspecific binding included as binding observed at a very long time point, the equation is

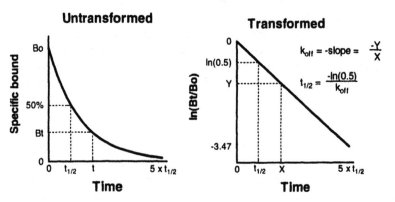

Figure 1 Dissociation assay—linearizing transform.

$$B_t = (B_0 - nsb)*exp(-k_{off}*t) + nsb$$

where B_0 and B_t refer to observed binding at times 0 and t, respectively.

6. Interpretation

These equations are empirical models to describe the data. If the data fit the simple model well without the need for an asymptote parameter, then the results are consistent with the mechanistic Law of Mass Action. The time for half the bound and displaceable radioligand to dissociate, $t_{1/2}$, is given by $-\ln(0.5)/k_{off}$.

B. Association Analysis

1. Objective

The objective of association analysis is to measure the rate and pattern of radioligand association.

2. Parameters to Be Estimated

These include the observed association rate constant k_{obs} and, by calculation, the bimolecular association rate constant k_{on}.

3. Procedure

Receptors are incubated with a certain concentration [L] of radioligand and binding is measured at various times. The assay contains, in addition, measures of nonspecific binding at some or all of the time points.

The design of the assay is straightforward, but some thought needs to be given to the concentration of radioligand to be used. On the one hand, it is important to know the time course of binding of the radioligand concentration used in other studies such as inhibition assays (see below), a concentration that is often around the K_d level or below. On the other hand, for an accurate estimate of k_{on} it is desirable to use a much higher radioligand concentration [see Eq. (11) and below] if nonspecific binding is not too large. Best practice is to use at least two radioligand concentrations.

4. Linearizing Transformation

Subtract nonspecific binding from all data points. Estmate by eye the equilibrium level of specific binding, B_{eq}. Plot $\ln(B_{eq}/(B_{eq} - B_t))$ against time (see Fig. 2), where B_t is the specific binding at time t. The slope of the straight line is the observed association rate constant k_{obs}.

5. Equations for Nonlinear Regression

Nonspecific binding can be subtracted from the other data points, as above, or, if it is the same at all measured times, it can be included in the analysis as a parameter. Only a few programs, however (e.g., SigmaPlot), allow the

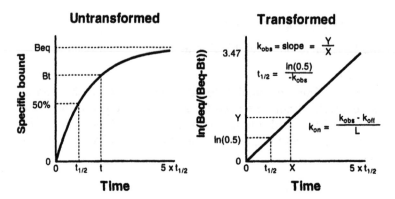

Figure 2 Association assay—linearizing transform.

user to define a model that can distinguish between total and nonspecific binding.

With nonspecific binding subtracted, the equation is

$$B_t = B_{eq}*(1 - \exp(-k_{obs}*t))$$

where the symbols have the meaning given above.

With nonspecific binding included as a time-independent parameter, the equation is

$$B_t = (B_{eq} - nsb)*(1 - \exp(-k_{obs}*t)) + nsb$$

where B_t and B_{eq} refer to observed binding at time t and total equilibrium binding.

6. Interpretation

These equations are empirical models to describe the data. If the data fit the model well, then the results are consistent with the mechanistic law of mass action, especially if similar values of k_{on} are obtained with different radioligand concentrations.

The time for specific binding of the radioligand concentration used in the assay to reach 50% of its equilibrium level, $t_{1/2}$, is given by $-\ln(0.5)/k_{obs}$.

The association rate constant k_{on} is calculated from the values of the radioligand concentration [L], the observed association rate constant at that concentration k_{obs}, and the dissociation rate constant k_{off}:

$$k_{on} = (k_{obs} - k_{off})/[L]$$

The ratio k_{off}/k_{on} defines the equilibrium dissociation constant, which should agree with the value obtained with equilibrium saturation analysis (see below) if the binding is mass action.

The value of k_{obs} approaches k_{off} at small [L] [see Eq. (11)], so for an accurate estimation of k_{on} values of [L] > K_d should be used.

C.   Saturation Analysis

1. Objective

The objective of saturation analysis is to measure the amount of specific binding with a range of radioligand concentrations.

2. Parameters to Be Estimated

These include the affinity (K_a) or dissociation constant (K_d) of the radioligand, and the total number of binding sites (Bmax).

3. Procedure

Receptors are incubated with a range of radioligand concentrations in the absence and presence of unlabeled defining ligand to measure total and nonspecific binding. The concentration of unlabeled defining ligand should be at least 100 times its apparent dissociation constant, $K_{iapparent}$, in the presence of the highest concentration of radioligand, [L], where

$$K_{iapparent} = K_i^*(1 + [L]/K_d).$$

The assay is incubated for sufficient time for binding equilibrium to be reached (see above).

If possible, concentrations of radioligand should be equally spaced on a log scale, should include the K_d, and should span at least a hundredfold range.

4. Linearizing Transformation

The calculation of specific binding depends upon whether or not the free radioligand (added dpm minus bound dpm) differs by more than about 15% between samples measuring total and nonspecific binding.

1.   When free radioligand concentration is constant: Calculate specific binding at each radioligand concentration by subtracting nonspecific from total binding.
2.   When radioligand differs in total and nonspecific samples: Calculate free radioligand as the difference between added dpm and bound dpm. Perform a linear regression of nonspecific bound on free. From the results, calculate the nonspecific binding for the values of free radioligand found in the total samples. Calculate specific binding at each radioligand concentration by subtracting the estimated nonspecific from total binding.

Calculate free radioligand as the difference between added dpm and total bound dpm. Convert the free dpm to nM.

The Scatchard plot shows specific/free versus specific (Fig. 3). The parameters K_d and Bmax can be calculated from this plot, but it is easier and statistically more accurate to calculate them from the reverse plot of specific versus specific/free, sometimes called a Hofstee plot.

Perform a linear regression analysis of specific on specific/free. The intercept is the Bmax; the slope is $-K_d$.

5. Equations for Nonlinear Regression

There are three types of nonlinear regression program that may be used to analyze saturation data.

1. LIGAND (29) (available from Biosoft, Cambridge, England) is purpose-built to analyze these types of data and automatically handles nonspecific binding and radioligand depletion; the user need only follow the instructions for data input.

2. Programs allowing a single independent variable in the analysis should be provided with values of free radioligand and specific bound, calculated as described above, and the data fitted to the equation.

Specific = Bmax/[1 + (K_d/free)].

For a two-site fit, use the equation

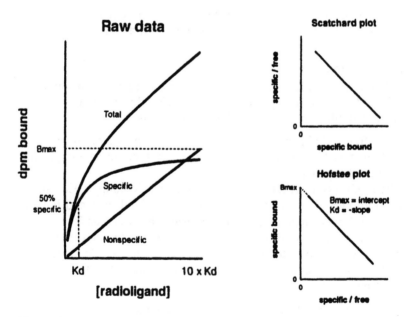

Figure 3 Saturation assay—linearizing transform.

Specific = Bmax1/(1 + K_d1/free) + Bmax2/(1 + K$_d$2/free).

3. Programs that can handle more than one independent variable and complex models, such as SigmaPlot (Jandel Scientific, Erkrath, Germany), should be provided with the added dpm and the total and nonspecific bound dpm. Calculate free dpm at each point as the difference between added and bound dpm.

Total binding is fitted to the equation

Total (free) = specific (free) + nsb (free).

Nonspecific binding is fitted to the equation

nsb (free) = k_{nsb}*free + background

and

Specific (free) = Bmax/[1 + (K_d/free)].

For a two-site fit, use the equation

Specific (free) = Bmax1/(1 + K_d1/free) + Bmax2/(1 + K_d2/free).

Alternatively, the data may be fitted to the equations of Swillens (30), which also allow for radioligand depletion.

6. Interpretation

Although it is always best to analyze experimental data using nonlinear regression, the Scatchard plot remains useful as a visual representation of the goodness of fit of the saturation data to the model. If the radioligand binds specifically according the Law of Mass Action, then the Scatchard plot will be a straight line. Curvature of the plot indicates that Hill slope is flat (i.e., <1, Scatchard plot is concave) or steep (i.e., >1, Scatchard plot is convex). A linear Scatchard plot is consistent with mass action behavior at a single site. A nonlinear Scatchard plot requires additional experiments in order to interpret the data.

All the methods described above allow for radioligand depletion, to the extent that it can be seen, but they cannot take into account any loss of binding during the filtration procedure, which may lead to overestimation of the K_d under some conditions (see above).

D. Inhibition Analysis

1. Objective

The objective of inhibition analysis is to measure the degree and pattern of inhibition of the binding of a fixed concentration of radioligand in the presence of a range of concentrations of unlabeled ligand.

2. Parameters to Be Estimated

These include the concentration of inhibitor that inhibits specific binding by 50%, IC_{50}, the slope of the inhibition curve, b, and by calculation, if possible, the dissociation constant of the inhibitor, K_i.

3. Procedure

Receptors are incubated with a single concentration of radioligand, alone and in the presence of increasing concentrations of the unlabeled ligand. Nonspecific binding is also measured. The assay is incubated with sufficient time for binding equilibrium to be reached (see above).

If possible, concentrations of inhibitor spanning at least a hundredfold range should be used.

4. Linearizing Transformation

Calculate the fractional inhibition of specific binding at each inhibitor concentration by subtracting binding in the presence of inhibitor from total binding in the absence of inhibitor, and dividing by specific binding in the absence of inhibitor.

For each inhibitor concentration, calculate

Log [fractional inhibition/(1 − fractional inhibition)].

The Hill plot shows log [fractional inhibition/(1 − fractional inhibition] versus log [inhibitor] (Fig. 4).

Using fractional inhibition values in the range 0.2–0.8, perform a linear regression analysis of values of log [fractional inhibition/(1 − fractional

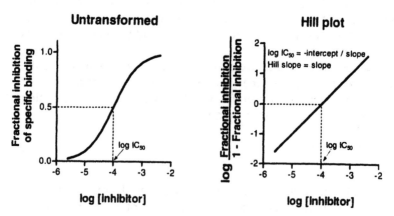

Figure 4 Inhibition assay—linearizing transform.

inhibition] on log [inhibitor]. The slope is the Hill slope. The log IC_{50} is $-$intercept/slope.

If the Hill slope is 1, then the equilibrium dissociation constant of the inhibitor, K_i, is calculated from the IC_{50} using the Cheng-Prusoff equation [Eq. (9)] with values of the radioligand concentration, [L], and the K_d of radioligand:

$$K_i = IC_{50}/(1 + [L]/K_d)$$

5. Equations for Nonlinear Regression

Two types of program may be used to analyze inhibition data.

1. LIGAND (29) is purpose-built to analyze these types of data according to a model containing one or more binding sites. It automatically handles nonspecific binding and ligand depletion, and the user need only follow the instructions for data input.

2. General nonlinear regression programs can be used to fit the data either to a logistic function or to a one- or two-site model. Nonspecific binding can either be subtracted before the analysis or be included as a parameter in the model (treating observed nonspecific binding as that obtained in the presence of a very high antagonist concentration). For simplicity it will be assumed that nonspecific binding is included.

a. Empirical Model. Fit the observed data to the logistic equation

$$\text{Observed} = \frac{(\text{maximum} - \text{nsb})}{1 + (10^{(\log([I]) - \log(IC_{50}))})^b} + \text{nsb}$$

where [I] is the inhibitor and b is the slope factor, equivalent to the Hill slope. The fitting procedure is more robust if the log IC_{50} is estimated, rather than the IC_{50}.

b. Mechanistic Model. For a one-site model, fit the observed data to the equation

$$\text{Observed} = \frac{(\text{maximum} - \text{nsb})}{1 + 10^{(\log([I]) - \log(IC_{50}))}} + \text{nsb}$$

which is the logistic function with a slope factor of 1.

For a two-site model, fit the observed data to the equation

$$\text{Observed} = (\text{maximum} - \text{nsb})*$$
$$\left(\frac{\text{fraction1}}{1 + 10^{(\log([I]) - \log(IC_{50_1}))}} + \frac{1 - \text{fraction1}}{1 + 10^{(\log([I]) - \log(IC_{50_2}))}} \right) + \text{nsb}$$

where IC_{501} and IC_{502} are the IC_{50} values of the two components and fraction 1 is the fraction of control specific binding comprising the first component.

IC_{50} values may be converted to estimates of K_i using the Cheng-Prusoff equation [Eq. (9)] with values of [L] and K_d. Alternatively, the antagonist K_i can be estimated directly by fitting the data to the equation:

$$\text{Observed} = \frac{(\text{maximum} - \text{nsb})}{1 + 10^{(\log([I]) - \log(K_i))} * \dfrac{K_d}{K_d + [L]}} + \text{nsb}$$

6. Interpretation

Inhibition curves with Hill slopes of 1 are consistent with competitive action at a single site, and use of the Cheng-Prusoff correction to estimate K_i is valid. Inhibition curves with slopes < 1 are not consistent with simple binding at a single site. One possible explanation for such curves is that the inhibitor recognizes two independent populations of labeled sites with different affinity. Note that the fact that a flat inhibition curve can be fitted to a two-site model does not, by itself, prove that this model is correct—many other models would also fit the data.

If the radioligand is known to bind to two sites, and the fractions of the two sites occupied by the fixed radioligand concentration are consistent with the fractions detected by the two-site analysis of the inhibition data, then use of the Cheng-Prusoff correction with the two K_d and IC_{50} values to estimate the two K_i values is also valid.

If the radioligand appears to bind to a single site and a two-site fit has been obtained, then the Cheng-Prusoff correction may be valid. If logistic fits have slopes $\neq$ 1 then the Cheng-Prusoff correction may be used to account for the competitive effect of the radioligand concentration, but the derived K_i value from a logistic fit should be considered as an estimate of the IC_{50} that would probably have been obtained with a vanishingly low radioligand concentration, rather than an estimate of the dissociation constant of the inhibitor.

V. THE INFLUENCE OF PHARMACOLOGICAL EFFICACY ON BINDING CHARACTERISTICS AT G-PROTEIN-COUPLED RECEPTORS

The difference between agonists and antagonists is that normally only the binding of agonists causes a conformational change in the receptor that results in a functional response (though under certain conditions some antagonists may also cause "inverse" functional effects, see below). It is therefore not surprising that the binding of agonists is often found to be more complex than that of antagonists. In particular, agonists are often found to bind at much lower concentrations that are required to cause functional effects. If agonist binding is studied by competition with an antagonist radioligand, then the complete concentration dependency of agonist binding can be observed. If agonist binding is measured directly with an agonist radioligand, then only the highest affinity states of the binding may be observed, either because of dissociation of radioligand from low-affinity sites in a filtration assay, or because of the unfavorable specific:nonspecific binding ratio of higher radioligand concentrations in a centrifugation assay. All affinity states may be detected by some peptide agonist radioligands (31).

The Law of Mass Action, the simplest model to account for antagonist binding, has been extended to the ternary complex model (32), and more recently to the extended ternary complex model (33), to provide models of agonist binding and receptor function.

According to the ternary complex model, the receptor can bind simultaneously an agonist and a second, membrane-bound ligand, a G-protein, to form a ternary complex.

$$
\begin{array}{ccccl}
& A & & A & \quad A = \text{agonist} \\
& + & & + & \quad R = \text{receptor} \\[4pt]
G \;+\; R & \overset{K_g}{\rightleftharpoons} & & R.G. & \quad G = \text{G-protein} \\
& {\scriptstyle K_a}\!\!\updownarrow & & {\scriptstyle \alpha K_a}\!\!\updownarrow & \quad K_a = \text{agonist/receptor affinity} \\[4pt]
G \;+\; A.R & \overset{\alpha K_g}{\rightleftharpoons} & & A.R.G. & \quad K_g = \text{G-protein/receptor affinity} \\
& & & & \quad \alpha = \text{allosteric change in } K_a \text{ and } K_g
\end{array}
$$

The G-protein is activated when bound to the receptor, and binding of an agonist to the receptor causes G-protein activation by increasing the affinity of the receptor for the G-protein. A corollary of the positive cooperativity between agonist and G-protein is that the agonist has a higher affinity for the G-protein-coupled receptor than for the free receptor. The efficacy of the agonist is determined by the degree to which it increases the affinity of the receptor for the G-protein (34). The G-protein contains a binding site for guanine nucleotides, and it is assumed that binding of a guanine nucleotide destablizes the ternary complex, so that in the presence of guanine nucleotide the high-affinity ternary complex is transient, and only binding of agonist to the uncoupled, low-affinity state of the receptor is detected.

At some receptors, such as the M_2 muscarinic receptor found in heart, agonists inhibit antagonist radioligand-binding with shallow slopes, and inclusion of a high concentration of a guanine nucleotide such as GTP in the assay causes the agonist concentration/effect curve to shift to the right and become steeper (35). These observations are broadly compatible with the ternary complex model. The shape of agonist inhibition curves derived from the ternary complex model in the absence of guanine nucleotide depends on the stoichiometry of receptor to G-protein (36). If there are many more receptors than G-proteins, only a small proportion of the receptors can participate in the ternary complex and agonists will bind mainly to the free receptor. If there are many more G-proteins than receptors, all receptors can form the high-affinity ternary complex, and the agonist will bind with high affinity and a slope of 1. Flat inhibition curves are only predicted by the model if receptors and G-proteins occur in roughly similar numbers. In this case, G-proteins bound in the ternary complex are a sizable fraction of the total G-protein population and result in a reduction in the number of free G-proteins and less ternary complex formation than would have occurred in the absence of free G-protein depletion.

Flat agonist inhibition curves have been analyzed by fitting the data to the ternary complex model, to estimate the agonist affinities at the free and coupled receptor [actually, the high-affinity component from such a fit is a composite value (37)], but more commonly such curves are fitted to a two-site model, yielding estimates of two affinity states and the fraction of total binding comprising each state (36). The values derived from a two-site fit may be similar to the values obtained by fitting the same data to the ternary complex model (36,37). A two-site fit of agonist inhibition curves in the absence of guanine nucleotide may be useful, allowing estimates of the low-affinity value of the agonist, which may approximate its "true" affinity, and of the ratio of high- and low-affinity values, which may reflect the "efficacy" of the agonist. Often, however, the results of two-site fits are presented with little further interpretation, and in such cases the two-site fit is probably inappropriate. There may be many causes of complex inhibition curves, and unless there is good reason to interpret flat curves according to the ternary complex model, the resuts of a two-site fit may be misleading, both because the model may be inappropriate (36), and because the shape of the curve reflects the fraction of high-affinity binding as well as the ratio of the affinity states: it is difficult to interpret differences in the fraction of high-affinity binding between ligands or assays, and it is difficult to visualize the shape of the curve from the numerical results of a two-site fit. The results of a logistic fit, i.e., IC_{50} and slope, are much easier to visualize.

The effects of guanine nucleotides to reduce the potency of agonist inhibition provide a clearer indication that agonist binding is coupling to G-proteins, and also a robust measure of this effect. The "GTP shift" is the ratio of agonist IC_{50} values in the presence and absence of guanine nucleotide, ignoring changes in slope, and in many receptor systems the size of the GTP shift is a predictor of agonist efficacy (35). We have found that the size of the GTP shift with a particular agonist may vary from day to day, but the relative shift, compared to the GTP shift of a standard agonist measured in the same assay, is very reproducible (unpublished observations using muscarinic M_2 receptors). It should be noted that the ability of an agonist to exhibit a GTP shift may be related to its chemical structure as well as its efficacy (38,39), and that not all G-protein-coupled receptor systems show GTP shifts.

If antagonists do not alter the affinity of the receptor for the G-protein, then guanine nucleotides should not affect antagonist affinity, and usually that is the case. Under conditions favoring receptor-G-protein coupling, however, typically low ionic strength with magnesium and without sodium, guanine nucleotides can sometimes increase the binding of antagonist radioligands (40). Such observations suggest that, under these conditions, (1) a proportion of the receptors are precoupled to G-proteins, (2) precoupled receptors are destabilized by guanine nucleotide, and (3) the antagonist radioligand has a lower affinity for the coupled than for the free receptor. Such an antagonist might be expected to show "inverse efficacy" in an appropriate functional assay.

Inverse efficacy is the property of an antagonist to reduce basal levels of a response in the absence of agonist. Inverse agonism has been detected using membranes from animal cells and tissues (41,42), cells and whole animals overexpressing the wild-type receptor (43,44), and in preparations expressing constitutively active mutant receptors (see Ref. 45). These observations can be explained by the ternary complex model if it is assumed that inverse agonists reduce the affinity of the receptor for the G-protein, and that this can only be detected if a significant fraction of coupled receptors exist in the absence of an agonist (46): enhanced levels of basal receptor coupling are achieved either by increasing the affinity of the receptor for the G-protein through the choice of assay conditions or mutation of the receptor, or by increasing the concentration (i.e., expression) of receptors or G-proteins (47).

Although the ternary complex model can explain the binding data and most of the functional data (see below) with inverse agonists, it cannot account for the binding behavior of agonists at constitutively active receptors. It is found that agonists bind with higher affinity to constitutively active receptors than to the wild-type, even in the absence of G-proteins, and that the size of this difference depends on the efficacy of the agonist. To account for these observations, Samama et al. (33) developed the extended ternary complex model. [Leeb-Lundberg and Mathis (31) proposed a related model to account for kinetic data with ^3H-bradykinin.] One importance of this model for the interpretation of binding studies is its implication that agonist (and inverse agonist) affinity cannot be measured directly (48).

The extended ternary complex model assumes that the receptor shuttles between a "ground" and "active" state, the proportions of which are determined by an isomerization constant. Only the active state can form a high-affinity and effective ternary complex.

$$
\begin{array}{ccccl}
A & A & A & & \\
+ & + & + & & R = \text{ground state of receptor} \\
\quad K_{iso} & & & & \\
R \rightleftharpoons & R^* + G \rightleftharpoons & R^*.G. & & R^* = \text{active state of receptor} \\
K_a \updownarrow & \updownarrow \beta K_a & \updownarrow & & K_{iso} = \text{isomerization constant} \\
A.R. \rightleftharpoons & A.R^* + G \rightleftharpoons & A.R^*.G. & & K_a = \text{agonist affinity} \\
\quad \beta K_{iso} & & & &
\end{array}
$$

β = allosteric change in K_{iso} and K_a

An agonist has higher affinity for the active than for the ground state, so the effect of binding an agonist is to increase the value of the isomerization constant, shifting the equilibrium of the receptor from being mainly in the ground state to being mainly in the active state. The greater the increase in isomerization constant, the greater the efficacy of the ligand (inverse agonists have higher affinity for the ground state and reduce the isomerization constant). This model implies that the apparent affinity of a ligand for the receptor, in the absence of G-proteins, depends on the both the true affinity and the efficacy of the ligand, and also on the value of the isomerization constant. Samama et al. (33) could account for the efficacy-related increase in agonist binding affinity for a constitutively active mutant β_2 adrenoceptor, as well as the functional changes, by assuming that the isomerization constant determining the fraction of receptors in the ground and active states is larger in the mutant receptor. This model, unlike the ternary complex model, is also consistent with functional inverse effects that do not involve G-proteins (49). Although the model is presented in terms of only one active state, it is possible that a receptor can adopt a number of different active states, which might differ in their affinity for different classes of G-proteins, and that different agonists could stabilize different subsets of active states (45). In support of this possibility, a constitutively active mutant receptor may show different changes in affinity with different chemical classes of agonist (50).

The ternary complex model is a simplified form of the extended ternary complex model, but in many cases experimental data may be conceptualized, and can only be analyzed, with the simpler model. Although the extended model assumes that the free receptor can exist in a low- and a high-affinity state, these states are freely interconverting, so the equilibrium model generates mass action binding curves in the absence of G-protein interactions, as does the simpler model. In the extended model the apparent binding affinity of an agonist is actually a function of the receptor isomerization constant and the agonist efficacy, as well as the true affinity, but these are all receptor-specific parameters and should be constant for a particular ligand-receptor pair, giving the same apparent affinity value as the simpler model. The models do not differ in their capacity to accommodate the binding of many classes of G-protein to a single receptor, or the possiblity of different active receptor states, with differing affinity for various G-proteins, being stabilized (or induced) by different ligands (45).

At first sight it might seem that the simple and extended ternary complex models are fundamentally different, the former being an "induced fit" model, in which the effect of agonist binding is to alter the shape (i.e., affinity) of the receptor, and the latter being a "selection" model, in which the effect of agonist binding is to alter the equilibrium of the receptor between two preexisting states by virtue of a higher affinity of an agonist for the active receptor state. In fact, in terms of equilibrium binding these two types of model are equivalent (46). One way in which the simple and extended models differ is that efficacy is explained in the simpler model by an allosteric effect of the agonist on receptor/G-protein affinity, whereas the extended model accounts for efficacy by an allosteric effect on the isomerization constant between ground and active states of the receptor, which in turn leads to changes in R/G affinity. In many, perhaps most, cases, experimental manipulations modify the number of receptors or G-proteins or the ligands to which the receptor is exposed at its agonist-binding and G-protein-binding domains, rather than modifying the receptor itself, and interpretation of such studies may not require the extended model. The extended model may be useful in interpreting the results of studies in which the receptor itself has been experimentally manipulated, perhaps by mutation, chemical modification, allosteric modulation, or changes in environment.

Another difference between the simple and extended models is that the extended model contains many more receptor states than the simple model, and the extended model, as presented here, is only a small subset of the full model (45). If the full model is integrated with models considering the interactions between G-protein subunits and guanine nucleotide binding to the various species of G_α subunit (51), then the number of receptor states becomes bewildering. These models provide a rich framework for interpreting the results of receptor mutagenesis studies. The analysis of changes of agonist affinity with a number of different constitutively active receptors and chemical classes of agonist should provide information on the mechanisms of agonism of the various chemical classes, and may provide a screening tool for detecting receptor-state-selective agonists.

REFERENCES

1. Yamamura HI, Enna SJ, Kuhar MH. Neurotransmitter Receptor Binding. New York: Raven Press, 1978.
2. Nahorski SR. What are the criteria in ligand-binding studies for use in receptor classification? In: Black JW, Jenkinson DH, Gerskowitch VP, eds. Perspectives on Receptor Classification. New York: Alan R Liss, 1987:51–59.
3. Yamamura HI, Enna SJ, Kuhar MH. Methods in Neurotransmitter Receptor Analysis. New York: Raven Press, 1978.
4. Hulme EC. Receptor-Ligand Interactions: A Practical Approach. Oxford: Oxford University Press, 1992.
5. Hulme EC, Birdsall NJM. Strategy and tactics in receptor-binding studies. In: Hulme EC, ed. Receptor-Ligand Interactions: A Practical Approach. Oxford: Oxford University Press, 1992:63–176.
6. Bylund DB, Toews ML. Radioligand-binding methods: practical guide and tips. Am J Physiol 1993; 265:L421–L429.
7. Lauffenburger DA, Linderman JJ. Receptors: Models for Binding, Trafficking, and Signalling. New York: Oxford University Press, 1993.
8. Keen M. The problems and pitfalls of radioligand-binding. In: Kendall DA, Hill, SJ, eds. Methods in Molecular Biology, Vol 41: Signal Transduction Protocols. Totawa, NJ: Humana Press, 1995:1–16.
9. Motulsky H. Analysing radioligand-binding data. Available on the Internet: http://www.graphpad.com, 1995.
10. Limbird LE. Cell Surface Receptors: A Short Course on Theory and Methods, 2nd ed. Boston: Kluwer Academic Publishers, 1996.
11. Cheng Y, Prusoff WH. Relationship between the inhibition constant (K_i) and the concentration of an inhibitor which causes 50 per cent inhibition (I_{50}) of an enzymatic reaction. Biochem Pharmacol 1973; 22:3099–3108.
12. Waelbroeck M, Tastenoy M, Camus J, Christophe J. Binding kinetics of quinuclidinyl benzilate and methyl-quinuclidinyl benzilate enantiomers at neuronal (M1), cardiac (M2), and pancreatic (M3) muscarinic receptors. Mol Pharmacol 1991; 40:413–420.
13. Motulsky HJ, Mahan LC. The kinetics of competitive radioligand-binding predicted by the law of mass action. Mol Pharmacol 1984; 25:1–9.
14. Koenig JA, Edwardson JM. Intracellular trafficking of the muscarinic acetylcholine receptor: importance of subtype and cell type. Mol Pharmacol 1996; 49:351–359.
15. Lazareno S, Nahorski SR. Errors in K_d estimates related to tissue concentration in ligand-binding assays using filtration. Br J Pharmacol 1982; 77:571P (abstract).

16. Hulme EC. Gel-filtration assays for solubilised receptors. In: Hulme EC, ed. Receptor-Ligand Interactions: A Practical Approach. Oxford: Oxford University Press, 1992:255–263.

17. Bennett JP Jr. Methods in binding studies. In: Yamamura HI, Enna SJ, Kuhar MH, eds. Neurotransmitter Receptor Binding. New York: Raven Press, 1978:57–90.

18. Wang j-x, Yamamura HI, Wang W, Roeske WR. The use of the filtration technique in in vitro radioligand-binding assays for membrane bound and solubilized receptors. In: Hulme EC, ed. Receptor-Ligand Interactions: A Practical Approach. Oxford: Oxford University Press, 1992.

19. Hulme EC. Centrifugation binding assays. In: Hulme EC, ed. Receptor-Ligand Interactions: A Practical Approach. Oxford: Oxford University Press, 1992: 235–246.

20. Cohen FR, Lazareno S, Birdsall NJM. The effects of saponin on the binding and functional properties of the human adenosine A1 receptor. Br J Pharmacol 1996; 117:1521–1529.

21. Koenig JA, Edwardson JM, Humphrey PPA. Somatostatin receptors in neuro2A neuroblastoma cells: operational characteristics. Br J Pharmacol 1997; 120:45–51.

22. Giles H, Landsdell SJ, Bolofo ML, Wilson HL, Martin GR. Characterization of a 5-HT(1B) receptor on CHO cells: functional responses in the absence of radioligand-binding. Br J Pharmacol 1996; 117:1119–1126.

23. Clarke PBS. Recent progress in identifying nicotinic receptors in mammalian brain. Trends Pharmacol Sci 1997; 8:32–35.

24. Quik M, Babu U, Audhya T, Goldstein G. Evidence for thymopoietin and thymopoietin/alpha-bungarotoxin/nicotinic receptors within the brain [retracted by Venkatasubramanian K, Audhya T, Goldstein G, Revah F, Mulle C, Pinset C, Changeux JP, Quik M, Babu U. In: Proc Natl Acad Sci USA 1993; 90(21):10409]. Proc Natl Acad Sci USA 1991; 88:2603–2607.

25. McGehee DS, Role LW. Physiological diversity of nicotinic acetylcholine receptors expressed by vertebrate neurons. Annu Rev Physiol 1995; 57:521–546.

26. Quik M, Cook RG, Revah F, Changeux JP, Patrick J. Presence of alpha-cobratoxin and phospholipase A2 activity in thymopoietin preparations. Mol Pharmacol 1993; 44:678–679.

27. Cuatrecasas P, Hollenberg MD. Binding of insulin and other hormones to non-receptor materials: saturability, specificity and apparent "negative cooperativity." Biochem Biophys Res Commun 1975; 62:31–41.

28. Snyder SH. Overview of neurotransmitter receptor binding. In: Yamamura HI, Enna SJ, Kuhar MH, eds. Neurotransmitter Receptor Binding. New York: Raven Press, 1978:1–11.

29. Munson PJ, Rodbard D. LIGAND: a versatile computerized approach for characterization of ligand-binding systems. Anal Biochem 1980; 107:220–239.

30. Swillens S. Interpretation of binding curves obtained with high receptor concentrations: practical aid for computer analysis. Mol Pharmacol 1995; 47:1197–1203.

31. Leeb-Lundberg LM, Mathis SA. Guanine nucleotide regulation of B2 kinin receptors. Time-dependent formation of a guanine nucleotide-sensitive receptor state from which [^{3}H]bradykinin dissociates slowly. J Biol Chem 1990; 265:9621–9627.

32. De Lean A, Stadel JM, Lefkowitz RJ. A ternary complex model explains the agonist-specific binding properties of the adenylate cyclase-coupled beta-adrenergic receptor. J Biol Chem 1980; 255:7108–7117.

33. Samama P, Cotecchia S, Costa T, Lefkowitz RJ. A mutation-induced activated state of the beta 2-adrenergic receptor. Extending the ternary complex model. J Biol Chem 1993; 268:4625–4636.

34. Tota MR, Schimerlik MI. Partial agonist effects on the interaction between the atrial muscarinic receptor and the inhibitory guanine nucleotide-binding protein in a reconstituted system. Mol Pharmacol 1990; 37:996–1004.

35. Ehlert FJ. The relationship between muscarinic receptor occupancy and adenylate cyclase inhibition in the rabbit myocardium. Mol Pharmacol 1985; 28:410–421.

36. Lee TW, Sole MJ, Wells JW. Assessment of a ternary model for the binding of agonists to neurohumoral receptors. Biochemistry 1986; 25:7009–7020.

37. Page KM, Curtis CA, Jones PG, Hulme EC. The functional role of the binding site aspartate in muscarinic acetylcholine receptors, probed by site-directed mutagenesis. Eur J Pharmacol 1995; 289:429–437.

38. Sibley DR, Creese I. Interactions of ergot alkaloids with anterior pituitary D-2 dopamine receptors. Mol Pharmacol 1983; 23:585–593.

39. van Rampelbergh J, Gourlet P, de Neef P, Robberecht P, Waelbroeck M. Properties of the pituitary adenylate cyclase-activating polypeptide I and II receptors, vasoactive intestinal peptide1, and chimeric amino-terminal pituitary adenylate cyclase-activating polypeptide/vasoactive intestinal peptide1 receptors: evidence for multiple receptor states. Mol Pharmacol 1996; 50:1596–1604.

40. Schutz W, Freissmuth M. Reverse intrinsic activity of antagonists on G protein-coupled receptors. Trends Pharmacol Sci 1992; 13:376–380.

41. Leeb Lundberg LM, Mathis SA, Herzig MC. Antagonists of bradykinin that stabilize a G-protein-uncoupled state of the B2 receptor act as inverse agonists in rat myometrial cells. J Biol Chem 1994; 269:25970–25973.

42. Hilf G, Jakobs KH. Agonist-independent inhibition of G protein activation by muscarinic acetylcholine receptor antagonists in cardiac membranes. Eur J Pharmacol 1992; 225:245–252.

43. Barker EL, Westphal RS, Schmidt D, Sanders Bush E. Constitutively active 5-hydroxytryptamine 2C receptors reveal novel inverse agonist activity of receptor ligands. J Biol Chem 1994; 269:11687–11690.

44. Bond RA, Leff P, Johnson TD, Milano CA, Rockman HA, McMinn TR, Apparsundaram S, Hyek MF, Kenakin TP, Allen LF, et al. Physiological effects of inverse agonists in transgenic mice with myocardial overexpression of the beta 2-adrenoceptor. Nature 1995; 374:272–276.

45. Kenakin T. The classification of seven transmembrane receptors in recombinant expression systems. Pharmacol Rev 1996; 48:413–451.

46. Kenakin T. Receptor conformational induction versus selection: all part of the same energy landscape. Trends Pharmacol Sci 1996; 17:190–191.
47. Burstein ES, Spalding TA, Brauner Osborne H, Brann MR. Constitutive activation of muscarinic receptors by the G-protein Gq. FEBS Lett 1995; 363:261–263.
48. Colquhoun D. Affinity, efficacy, and receptor classification: is the classical theory still useful? In: Black JW, Jenkinson DH, Gerskowitch VP, eds. Perspectives on Receptor Classification. New York: Alan R Liss, 1987:103–114.
49. Samama P, Pei G, Costa T, Cotecchia S, Lefkowitz RJ. Negative antagonists promote an inactive conformation of the beta 2-adrenergic receptor. Mol Pharmacol 1994; 45:390–394.
50. Perez DM, Hwa J, Gaivin R, Mathur M, Brown F, Graham RM. Constitutive activation of a single effector pathway: evidence for multiple activation states of a G protein-coupled receptor. Mol Pharmacol 1996; 49:112–122.
51. Onaran HO, Costa T, Rodbard D. Beta gamma subunits of guanine nucleotide-binding proteins and regulation of spontaneous receptor activity: thermodynamic model for the interaction between receptors and guanine nucleotide-binding protein subunits. Mol Pharmacol 1993; 43:245–256.

4

Drug Analysis Based on Signaling Responses to G-Protein-Coupled Receptors

T. Kendall Harden and José L. Boyer
University of North Carolina at Chapel Hill, Chapel Hill, North Carolina

Robert W. Dougherty
Inspire Pharmaceuticals, Inc., Durham, North Carolina

I. INTRODUCTION

The discovery of cyclic AMP by Sutherland and Rall (1) in 1956 provided the initial impetus from which the scaffolding of drug receptor pharmacology with the discipline of physiology was extended to include biochemical assessment of receptor-promoted drug action. Receptor activation subsequently has been described in increasingly molecular terms, and the often inexact approach of measuring drug-receptor interactions at the intact animal or intact tissue level now can be complemented by inherently more precise approaches associated with direct biochemical assay of receptor responses. A major chapter in the history of drug development in the first half of this century involved discovery of molecules that mimic or block the effects of biogenic amines at their cognate receptors. The initial biochemical observation of an agonist-receptor interaction by Sutherland and Rall (1) was made for the biogenic amine epinephrine, and the elucidation of the biochemical basis of the glycogenolytic effects of β-adrenoceptor activation in the liver proved a harbinger of eventual identification of similar adenylyl cyclase–activating responses to many other receptors (2). Promotion of other second-messenger signaling responses, e.g., activation of guanylyl cyclase or phospholipase C and inhibition of adenylyl cyclase, subsequently

79

was associated with neurotransmitter and hormone action, and the receptors for the biogenic amines again were prominent among the receptors initially associated with these newly discovered signaling mechanisms.

The focus of this chapter primarily will be on biochemical analyses associated with assessment of drug-receptor interactions at metabotropic receptors. Similar analyses can be applied in varying degrees to analysis of receptors that primarily signal as regulators of ion channels. Since by far the largest group of the so-called metabotropic receptors are known to signal through guanine nucleotide regulatory proteins (G proteins), we will limit our discussion and examples of biochemically assessed receptors to those that are G-protein-coupled receptors (GPCR). The radioligand-binding properties of these cell surface signaling molecules are considered elsewhere in this volume, and thus the focus here will be entirely on the basic principles associated with study of the immediate biochemical responses emanating from GPCR activation. Many excellent monographs (3–9) have been published on the actual technical aspects of these measurements, which will not be detailed here.

The approaches for analysis of drug-receptor interactions that are described are applicable to studies of receptors in intact organ, tissue, and cell preparations. These analyses often can be readily extended to cell-free preparations, and arguments can be made for significant advantages of cell-free preparations in certain instances. Our discussion largely focuses on the biochemical analysis of GPCR in homogeneous cell preparations. The advantages of studying receptors in such preparations are obvious and will not be extensively delineated here. The principles that are discussed hold for receptors natively expressed in homogeneous cell lines or strains. However, they most clearly apply with cloned receptors stably or transiently expressed in a cell line that does not endogenously express that receptor or preferably also does not express other members of that receptor family. The empty-vector transfected cell can then serve as an ideal negative control to confirm that measured responses are entirely a reflection of interaction at the cloned target receptor of interest. The application of recombinant receptors as screens for drug discovery is discussed in detail in Chapter 15. A brief consideration of the general principals of G-protein-mediated signaling is followed by discussion of more specific aspects associated with the utilization of individual second-messenger signaling responses in the analysis of drug-receptor interaction.

II. G-PROTEIN-COUPLED RECEPTORS

Agonist-occupied GPCR selectively interact with one or more members of a group of approximately 20 G proteins, which in turn selectively activate individual effector enzymes or ion channels (10,11). The large degree of amplification provided by this three-protein signaling cohort is further enhanced downstream at the level of the involved second-messenger-activated protein kinase or in the sequelae of events emanating from specific changes in intracellular ion concentrations (Fig. 1). Thus, low levels of receptor occupancy by agonists often maximally activate metabotropic receptor signaling responses.

Rhodopsin was the first GPCR for which primary amino acid sequence was delineated (12). Molecular cloning of cDNA encoding the β_2-adrenoceptor (13) and a turkey erythrocyte β-adrenoceptor (14) revealed a remarkable similarity of sequence to that of rhodopsin and provided the first suggestion of the existence of a superfamily of receptors that produce their physiological effects through activation of G-proteins. The cloned members of the GPCR group of signaling proteins now number more than 300 (see Table 1), and with inclusion of the large group of receptors involved in olfaction, will eventually number over 1000 different receptor sequences (15). Analyses of the conserved amino acids in their primary sequence suggest that GPCR share a common structural motif, which in turn likely reflects commonality in their mechanism of activation (15,16). These conserved structural features include seven hydrophobic regions that likely exist as α-helical domains that span the plasma membrane and connect three hydrophilic extracellular loops with three hydrophilic cytoplasmic loops. The greatest overall homology of GPCR resides in the seven transmembrane-spanning domains, which in the case of receptors for the biogenic amines apparently form the ligand binding pocket. This intramembrane binding pocket contrasts with the putative binding pocket for GPCR for large peptide hormones, which apparently also involves extracellular loops (15,16). Receptors for thrombin, glutamate, and extracellular Ca^{2+} are members of a third subfamily of GPCR that either by proteolytic cleavage, i.e., thrombin, or by ligand-binding in the extracellular amino terminus, i.e., glutamate and Ca^{2+}, bring specific sequence in the amino terminus as an activating ligand into the receptor binding pocket apparently formed primarily by the transmembrane spanning domains (17).

Structure of the GPCR mainly has been inferred from a combination of mutational, biochemical, and pharmacological approaches and relies heavily on electron diffraction data obtained for bacteriorhodopsin (18). Although not formally a GPCR, bacteriorhodopsin possesses a primary sequence that is homologous to mammalian rhodopsin and, like the GPCR involved in phototransduction, is activated by photoisomerization of retinal. A number of GPCR now have been modeled based on the known structure of bacteriorhodopsin. These models have been further refined by mutagenesis based on predictions from the models (15,16), and importantly, by

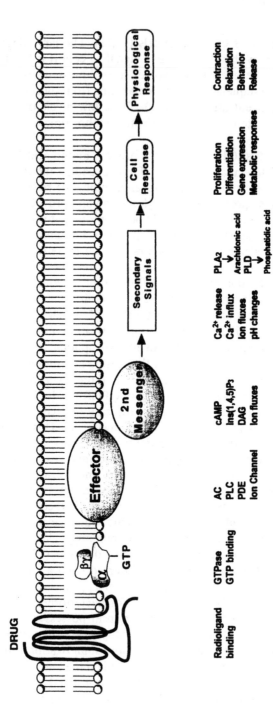

Figure 1 In this schematic representation of agonist-induced physiological responses mediated by a GPCR, direct and indirect effects of drug-receptor interactions can be evaluated at levels beginning with the binding of the drug to the receptor and extending through each step of the signaling cascade leading to the physiological response. The indicated signaling entities should be considered as examples rather than as an all-inclusive list of all the biochemical events that can occur downstream to activation of a GPCR.

Table 1 Neurotransmitters, Hormones, Growth Factors, Chemoattractants, and Other Extracellular Stimuli for Which G-Protein-Coupled Receptors Have Been Cloned

---

Biogenic Amines—5-hydroxytryptamine, acetylcholine, epinephrine, norepinephrine, dopamine, histamine, melatonin
Purines and Nucleotides—adenosine, ATP, UTP, ADP, UDP
Peptides—adrenocorticotropic hormone, melanocyte-stimulating hormone, melanocortin, neurotensin, bombesin, endothelin, cholecystokinin, gastrin, neurokinin B, substance K, substance P, neuropeptide Y, thyrotropin releasing-factor, bradykinin, angiotensin II, beta-endorphin, C5a anaphalatoxin, calcitonin, chemokines, corticotropic-releasing factor, dynorphin, endorphin, fMLP and other formylated peptides, follitropin, galanin, gastric inhibitory polypeptide, glucagon, gonadotropin-releasing hormone, growth hormone–releasing hormone, interleukin-8, leutropin, MET-enkephalin, opiod, oxytocin, parathyroid hormone, pituitary adenylyl cyclase–activating peptide, secretin, somatostatin, thrombin, thyrotropin, vasoactive intestinal peptide, vasopressin, vasotocin
Eicosanoids—prostacyclins, prostaglandins, thromboxanes
Lipids and Lipid-Based Compounds—cannabinoids, anandamide, lysophosphatidic acid, platelet activating factor, leukotriene
Excitatory Amino Acids and Ions—calcium ion, glutamate
Retinal-Based Compounds—11-*cis* retinal

This is a partial list of the natural ligands for cloned GPCR. This list is not all-inclusive and does not consider the large number of GPCR that have been cloned for which the natural agonist has not been identified.

refinements in models that have been constructed based on information provided by a projection map for mammalian rhodopsin (19). The availability of large amounts of certain purified GPCR has begun to provide avenues for probing ligand binding pockets using direct protein chemistry and biophysical approaches. Although not yet carried to completion, generation of high-diffraction crystals for atomic resolution of GPCR has been made feasible by the increasing availability of purified protein.

III. G PROTEINS

The G proteins with which GPCR interact are heterotrimeric proteins that consist of a guanine nucleotide-binding α-subunit and a tightly but noncovalently associated dimer of β- and γ-subunits (10,11,20). G-protein-mediated signaling ensues from receptor-promoted exchange of GTP for GDP and dissociation of the G-protein heterotrimer into a GTP-liganded α-subunit and free $\beta\gamma$-subunit. The lifetime of the activated state is controlled by a

GTPase activity of the α-subunit, which, at a rate that is dependent on both the identity of the α-subunit and in some cases the effector protein (21), hydrolyzes GTP and returns the activated system to the ground state. G-proteins are designated by the identity of the α-subunit in the heterotrimer, and effector protein activation largely is thought to ensue from the release and activity of GTP-liganded α-subunits. However, G-protein $\beta\gamma$-subunits also are important direct regulators of several effector proteins (22,23), and in some situations release of $\beta\gamma$-subunits from heterotrimeric G-proteins may affect the entire G-protein signaling response to G-protein activation. The three-dimensional structures of G-protein α- and $\beta\gamma$-subunits have been solved (24–27).

The G-protein α-subunit family consists of 16 different genes (splice variants are expressed from several of these genes) in mammals encoding 40–45 kDa GTP binding proteins that exhibit overall homologies of 45–90% (10,11). These can be divided into four structural families (Gs, Gi, Gq, G12) based on percent homologies and in part on function. The Gs group includes the cholera toxin substrates ($G\alpha_s$ and $G\alpha_{olf}$) and are activators of adenylyl cyclase, Ca^{2+} channels, and potentially other channels. The Gi family represents the largest group of α-subunits and includes $G\alpha_{i1}$, $G\alpha_{i2}$, $G\alpha_{i3}$, $G\alpha_0$, $G\alpha_t$, and $G\alpha_z$. With the exception of $G\alpha_z$ these are pertussis toxin substrates. The Gi family α-subunits are activators of ion channels, e.g., for K^+, and inhibitors of adenylyl cyclase. The Gq family includes the G-protein regulators of the PLC-β isoenzymes. The four α-subunits, i.e., $G\alpha_q$, $G\alpha_{11}$, $G\alpha_{14}$, and $G\alpha_{16}$, in this group are not susceptible to ADP-ribosylation by pertussis or cholera toxin. The function of the G12 group of G-protein α-subunits, which includes $G\alpha_{12}$ and $G\alpha_{13}$, has not been unambiguously defined.

Although GPCR-promoted activation of second-messenger signaling pathways primarily can be considered to occur through selective interaction with a single type of G-protein and release of its activating α-subunit, this concept and mechanism is not invariant in metabotropic receptor action. For example, Offermanns and Simon (28) recently have reported that $G\alpha_{16}$ does not exhibit the fidelity of receptor coupling that has been shown to be the case with other G-proteins including that of other members of the Gq family. That is, receptors that were previously shown to have high selectivity for Gs or Gi nonetheless markedly promoted inositol lipid hydrolysis in cells expressing $G\alpha_{16}$. The physiological significance of this phenomenon is uncertain since $G\alpha_{16}$ is primarily expressed in cells of hematopoietic origin instead of in the cell types that express many of the Gs- or Gi-linked receptors. However, advantage potentially can be taken of the promiscuity of $G\alpha_{16}$ in studies of drug-receptor interaction. For example, activity of a Gs-linked receptor like the β-adrenoceptor for catecholamines that normally is assessed by quantitation of cyclic AMP formation instead can be assessed by quantitation of downstream sequelae of activation of phospholipase C in cells overexpressing $G\alpha_{16}$. Such $G\alpha_{16}$-overexpressing cells also are of potential value in pharmacologically characterizing receptors for which identity of the associated G-protein-regulated signaling response has not been established.

Although GTP-liganded α-subunits of the Gq family can be considered the primary coupling proteins in the receptor-promoted inositol lipid signaling pathway, several receptors that couple to pertussis toxin-sensitive G-proteins, i.e., Gi and Go, also promote inositol lipid hydrolysis in certain target cells. Since α-subunits of this family of G proteins do not activate phospholipase C-β, the mechanism of this form of G-protein-mediated regulation is not unambiguously known. However, $\beta\gamma$-subunits activate certain of the phospholipase C-β isoenzymes (29–31), and receptor-promoted release of $\beta\gamma$-subunits from pertussis toxin substrate G-proteins, which are present in relatively high amounts in many target cells, likely accounts for pertussis-toxin-sensitive inositol lipid signaling.

The complexity of G-protein-mediated signaling also is increased by the existence of multiple genes encoding β- and γ-subunits. Thus, at least five G-protein β-subunits and seven γ-subunits are known to exist and most combinations of these β- and γ-subunits form functional dimers (20,22). Thus, any signaling response mediated by interaction of $\beta\gamma$-subunits with an effector protein has the possibility of being dependent on the identity of the $\beta\gamma$-subunit pair that is released from the heterotrimer upon GPCR activation.

Although differences in the capacity of various $\beta\gamma$-dimer combinations to activate effector proteins in vitro have been reported, these differences usually have been quantitative rather than qualitative (32). An important function of the $\beta\gamma$-subunit apparently resides in definition of the specificity of recognition of G-protein heterotrimers by various GPCR. For example, somatostatin and muscarinic receptors have been shown in GH3 cells to recognize G-protein heterotrimers containing $\beta_1\gamma_3$ and $\beta_3\gamma_4$ dimers, respectively (33). Such observations apparently explain why certain receptors readily activate a given G-protein signaling pathway in one target cell but not in another where the component proteins of that signaling pathway nonetheless can be shown to exist.

IV. G-PROTEIN-REGULATED EFFECTOR PROTEINS

G-protein-mediated regulation of effector proteins was recognized initially for cyclic AMP signaling and for the rhodopsin-mediated phototransduction cascade. Thus, adenylyl cyclase was shown to be activated by Gs and cyclic

GMP phosphodiesterase was demonstrated to be activated by the G-protein transducin (Gt). As intimated above, the G-protein-regulated plasma membrane effector proteins eventually were shown to include phospholipases and ion channels, and additional targets for G-proteins recently have been suggested (11,20).

The complexity introduced by multiple G-proteins selectively interacting with multiple effector-promoted signaling responses is increased considerably by the remarkably high number of isoforms that may exist for any one class of effector protein. Thus, eight adenylyl cyclase isoenzymes now are known to exist (34,35) and four isoforms of phospholipase C-β are known to be G-protein-regulated (36). These isoenzymes of very similar structure often are regulated in remarkably different ways. For example, G-protein α- and $\beta\gamma$-subunits have been shown to both activate and inhibit adenylyl cyclase in an isoenzyme-specific manner (34,35). Moreover, certain of the adenylyl cyclase isoenzymes are subject to regulation by Ca^{2+}-calmodulin and some are directly inhibited by Ca^{2+} itself. Thus, not only do receptors that couple to the Gs class of G proteins regulate cyclic AMP signaling through release of activated $G\alpha_s$, but regulation also is effected in a multistep manner by GPCR that release activated α-subunits as a consequence of coupling to the Gq class of G-proteins. Target cells expressing $\beta\gamma$-subunit-regulated forms of adenylyl cyclase potentially can exhibit cyclic AMP generating responses to any GPCR, since activation of G proteins ostensibly in all cases releases free $\beta\gamma$-subunits. A key, and to date seldom established, determinant of the complex fabric of second-messenger signaling responses resides in the relative expression of individual G proteins and effector isoenzymes in any one target cell. An even more vague and poorly understood concept centers on the distribution of signaling components within the cell. That is, regulatory protein-containing domains apparently exist, for example in calveolae, that either restrict or promote the interaction of the component proteins of a GPCR–G-protein–effector protein triumvirate (37).

V. SIGNALING RESPONSES BASED ON ACTIVATION OF ADENYLYL CYCLASE

GPCR-mediated activation of adenylyl cyclase represents a straightforward and simple means to assess drug receptor interactions. As discussed above, activation of adenylyl cyclase involves, in the simplest sense, release of GTP-liganded $G\alpha_s$ from the Gs heterotrimer and resultant increases in the rate of conversion of ATP to cyclic AMP by the enzyme. However, other messages, e.g., from $\beta\gamma$-subunit, Ca^{2+}, Ca^{2+}-calmodulin, protein kinase C, also can effect activation or inhibition of adenylyl cyclase in an isoenzyme-dependent manner (34,35). Any cell line in which a cloned Gs-coupled GPCR is expressed also will likely natively express other GPCR. Although these GPCR might not be Gs-coupled, they could nonetheless produce false positives in a drug screen. For example, these receptors might lead to promotion of adenylyl cyclase activity by a signaling message emanating from activation of phospholipase C-β. Since most target cells studied to date express multiple adenylyl cyclase isoenzymes, this concern may be amplified by a heterogeneity of enzymes subject to activation in multifarious ways by downstream mediators of G-protein activation. Perhaps the most predictably reliable cell line in which to express a Gs-coupled GPCR for a screen of ligands for that receptor would be a cell line that was genetically engineered in such a way as to express only a single adenylyl cyclase isoenzyme. This adenylyl cyclase preferably would be an isoenzyme, i.e., type II adenylyl cyclase, that is primarily regulated by GTP-liganded $G\alpha_s$.

The assumption that measurement of cyclic AMP levels will be a true reflection of drug-receptor interaction depends on several other factors. Although egress of cyclic AMP from cells has been shown to occur under various conditions in several cell types, this is a minor phenomenon in most cells (38). Nonetheless, since egress has been suggested to be a receptor-regulated event, some attention should be paid to its possible contribution to drug effects. For example, a less than maximal response to a drug could occur as a consequence of promotion of egress of cyclic AMP through activation of a second receptor (or other)-promoted mechanism in addition to the cyclic AMP–elevating effects occurring through activation of a Gs-coupled GPCR.

The fidelity to which quantitation of cyclic AMP levels reflects the rate of cyclic AMP synthesis also is a function of the extent to which cyclic nucleotide is degraded by cyclic nucleotide phosphodiesterases (2). Inhibitors of these phosphodiesterases can be included in assays to circumvent this potential problem. However, some concern exists in the relative selectivity of effect of these inhibitors. Thus, most second- and third-generation inhibitors of cyclic nucleotide phosphodiesterases have been developed on the basis of their capacity to inhibit with high affinity and selectivity individual isoenzymes of this degradative enzyme family. The molecules that are available as nonselective inhibitors of all of the cyclic nucleotide phosphodiesterases tend to be relatively low-potency methylxanthine derivatives that likely interact with many proteins in addition to the phosphodiesterases.

Cyclic AMP levels can be detected by several convenient means (3,6), and the specifics of this methodology will not be recounted here. These analyses include radioimmunoassays based on antibodies generated against a derivatized form of cyclic AMP, binding assays based on the interaction of cyclic AMP with a cyclic AMP binding protein such as the regulatory subunit of cyclic AMP-dependent protein kinase, or assessments of [³H]cyclic AMP levels after radiolabeling of intracellular ATP pools using [³H]adenine. The latter assay offers utility due to its simplicity but would not be the detection system of choice for large-through-put assays.

VI. SIGNALING RESPONSES BASED ON ACTIVATION OF PHOSPHOLIPASE C-β

A broad range of hormones, neurotransmitters, growth factors, chemoattractants, and other extracellular stimuli produce their physiological responses, at least in part, through activation of phospholipase C. This activation produces Ins(1,4,5)P$_3$ and diacylglycerol as the immediate dual second messengers and mobilized intracellular Ca^{2+} and activated protein kinase C as the second pair of messengers (39,40). These two secondary signaling species then promote a remarkably broad set of biochemical responses that include activation of other "second messenger" pathways including, for example, those concerned with activation of phospholipase A$_2$ or phospholipase D. As a consequence, interaction of molecules with receptors that couple to phospholipase C theoretically can be assessed by quantitation of many different biochemical responses. There are strong arguments that biochemical assessment of drug interaction of Gq-coupled receptors should be restricted to quantitation of the biochemical events that are most proximal to the initial drug receptor interaction. Thus, analyses of Ins(1,4,5)P$_3$ or inositol phosphate levels or diacylglycerol production may represent the clearest snapshot of drug interaction at a Gq-linked GPCR. This is certainly the case with inositol phosphate accumulation, but may be less so with measurements of diacylglycerol. That is, whereas most of the diacylglycerol produced immediately after GPCR activation predictably emanates from phospholipase C-β-catalyzed hydrolysis of polyphosphoinositides, the source of diacylglycerol becomes increasingly less clear with increased time of stimulation of the inositol lipid signaling pathway. This follows from the fact that phosphatidylcholine rather than phosphatidylinositol serves as the predominant source of diacylglycerol with extended receptor activation, and the mechanism of this formation of diacylglycerol from phosphatidylcholine involves activation of phospholipase D, and/or a phosphatidylcholine-hydrolyzing phospholipase C, and/or other mechanisms that occur as biochemical events well downstream of the initial agonist interaction at a Gq-coupled GPCR (41). Further arguments against the utilization of diacylglycerol in a biochemical assay of the activity of a GPCR apply to the assays themselves. They tend to be cumbersome, not exceptionally sensitive or reliable, and often provide signal-to-noise ratios that are not optimum.

Assays directed at the quantitation of inositol phosphate accumulation provide the most predictably useful means for assessment of Gq-linked GPCR. The immediate product of the phosphodiesteratic cleavage of PtdIns(4,5)P$_2$ and the bona fide second messenger is Ins(1,4,5)P$_3$. Thus, strong arguments can be made for its quantitation in analyses of Gq-linked GPCR. Ins(1,4,5)P$_3$ binding assays based on binding to Ins(1,4,5)P$_3$-binding proteins are available, and these exhibit considerable sensitivity, reliability, and signal-to-noise properties (40). However, a minor concern with measurement of Ins(1,4,5)P$_3$ under most conditions is that the levels of this second messenger are a function of both its formation and its rapid metabolism by Ins(1,4,5)P$_3$ kinase and Ins(1,4,5)P$_3$ phosphatases (39,40). Whereas measurement of steady-state levels of Ins(1,4,5)P$_3$ rather than its formation does not a priori cause a problem, drugs can potentially modify the measured response by effects downstream from Ins(1,4,5)P$_3$ formation, and thus, effects may be observed that are unrelated to the GPCR under measurement.

Measurement of total inositol phosphates provides an alternative method for assessment of receptor activity. This is particularly the case in assays carried out in intact cells in which the activity of inositol 1-phosphate phosphatases has been inhibited by inclusion of LiCl in the assay. Inhibition of this enzyme prevents recycling of radiolabeled inositol back into the inositol lipid pool and assures that the total inositol phosphate mass that is hydrolyzed from PtdIns(4,5)P$_2$ in response to GPCR activation is in fact quantitated (39,40). Assay of total inositol phosphate levels in LiCl-treated cells is usually accomplished with cells that have been labeled for long periods of time with [^{3}H]inositol, which in turn is incorporated into the inositol lipid pool of the plasma membrane. The disadvantage of this method is that it is not based on a simply assessed binding reaction, and therefore, has proved difficult to carry out on a large-through-put scale.

Although not as proximal to the initial event of drug-receptor interaction, measurement of Ins(1,4,5)P$_3$-promoted mobilization of intracellular Ca^{2+} also can be applied in assessment of activity at Gq-linked GPCR. The potential for misleading results in assays of intracellular Ca^{2+} is considerable and has been discussed extensively in a number of excellent reviews (41,42). Nonetheless, considerable advantages lies in the utility of fluorescent dyes for quantitation of Ca^{2+}, and therefore, physical measurement of Ca^{2+} can be applied in high-through-put analyses (see more below).

VII. THE PHOSPHOLIPASE A$_2$ SIGNALING PATHWAY

Phospholipase A$_2$ predominantly cleaves arachidonic acid from the sn-2 position of phosphatidylcholine producing lysophospholipid and free arachidonate (43). The arachidonate in turn serves as the immediate precursor substrate for the various eicosanoid-related pathways and in its own right may serve as a second messenger. Lysophosphatidylcholine and potentially other lysophospholipids that are produced from phospholipase A_2 action also may subserve signaling functions. Activation of phospholipase A_2 is not a direct consequence of G-protein activation, but, rather, occurs primarily due to membrane association of the soluble phospholipase A_2 enzyme as a consequence of increases in intracellular Ca^{2+} due to activation of phospholipase C (43,44). The process also apparently involves in part phosphorylation of phospholipase A_2 by protein kinase C, MAP kinase, and/ or other protein kinases. As a consequence of this relatively complex and multistep mechanism of activation, assay of phospholipase A_2 activity should not be considered in the primary armamentarium for quantitation of drug-receptor interactions. However, in situations where assessment of receptor-regulated phospholipase A_2 activity is of interest, assay of the enzyme at the intact cell level can be effected by measurement of radioactive arachidonate release after labeling of endogenous phospholipid pools with [^{3}H]arachidonate (7).

VIII.   THE PHOSPHOLIPASE D SIGNALING PATHWAY

Phospholipase D catalyzes the phosphodiesteratic hydrolysis of membrane phospholipids (primarily phosphatidylcholine) producing phosphatidic acid and the polar head group (41,45,46). Multiple isoforms of phospholipase D activity apparently exist that differ in subcellular distribution, in effects of ions, detergents, and fatty acids on activity, and in overall mode of regulation. Many agonists for GPCR, including peptide, monoamine, nucleotide, and lipid agents, activate phospholipase D in a broad range of cell types. Although the physiological significance of PLD activation is unclear, phosphatidic acid generation has been implicated in regulation of inflammation, secretion, cell growth, superoxide generation, and the activity of a variety of enzymes including specific isoforms of protein kinase C (47). Phosphatidic acid inhibits the activity of adenylyl cyclase through a mechanism involving Gi, activates other Gi-promoted signaling responses, and has been shown to disrupt protein:protein interactions between a GDP dissociation inhibitor and Rac2. The latter interaction may explain the enhancement by phosphatidic acid of agonist-stimulated NADPH oxidase activation in neutrophils.

    In addition to the direct action of phosphatidic acid on cellular effector systems, metabolites of phosphatidic acid likely have biological functions (47). Phosphatidic acid is rapidly broken down to diacylglycerol through the action of phosphatidate phosphohydrolase. GPCR-promoted accumulation of diacylglycerol generally occurs biphasically, and the secondary phase of accumulation occurs in large part from the breakdown produced from activation of phospholipase D. This sustained elevation of diacylglycerol serves to prolong protein kinase C activity initially promoted by activation of phospholipase C. Since phospholipase C–promoted elevation of intracellular Ca^{2+} levels is not sustained, the molecular species of PKC that is activated during long-term GPCR activation differs from that activated during the initial phases of cell stimulation when both Ca^{2+} and diacylglycerol species derived from the PLC-mediated hydrolysis of $PtdIns(4,5)P_2$ are abundant. Other metabolites of phosphatidic acid, e.g., lysophosphatidic acid, have important biological activities.

The regulation of phospholipase D activity is complex and, in spite of considerable recent progress, is not completely understood (41,45,46). Activation of phospholipase D occurs concomitantly with activation of phospholipase C, and elevation of intracellular Ca^{2+} appears to be required for maximal induction of phospholipase D activity. Activation of protein kinase C also is clearly involved (48).

Recent evidence indicates that phospholipase D activity is regulated by small monomeric GTP-binding proteins. This progress has followed from elucidation of conditions that allow observation of the regulation of phospholipase D activity in cell-free preparations and from the observation that addition of cytosolic factor(s) is necessary for GTP-promoted activation of phospholipase D. These factors include members of the ARF (49,50) and Rho (51,52) families of monomeric GTP-binding proteins, which may act synergistically in the activation of phospholipase D by GTP.

Phospholipase D activity can be quantitated in a number of ways. Cockcroft (53) provided the first evidence that phospholipase D activity could be modulated in intact cells by extracellular stimuli by demonstrating that in neutrophils, where the ATP pools were labeled with ^{32}P to a constant specific radioactivity, peptide chemoattractants stimulated a large increase in the mass of phosphatidic acid with a reciprocal decrease in its specific activity. She concluded that the bulk of the phosphatidic acid formed did not result from [^{32}P]phosphorylation of diglycerides arising from activation of inositol lipid turnover, but that rather, the phosphatidic acid was generated directly from the activation of phospholipase D. The formation of [^{32}P]phosphatidic acid in neutrophils prelabeled with [^{32}P]lysophosphatidylcholine subsequently was demonstrated directly (41,45,46). A more facile approach for quantitation of phospholipase D activity takes advantage of the unique ability of phospholipase D to catalyze a transphosphatidylation reaction between phosphatidylcholine and nucleophilic primary alcohols such as ethanol. The formation of phosphatidylethanol occurs at the expense of phosphatidic acid and forms the basis for the unambiguous detection of phospholipase D activation.

IX. BIOCHEMICAL EVALUATION OF LIGAND/RECEPTOR INTERACTION UTILIZING CELL-FREE PREPARATIONS

Ligand-promoted coupling of GPCR to G-proteins represents the initial measurable biochemical step in a GPCR signaling pathway. Two different analyses can be carried out to assess this most proximal of the biochemical sequelae of GPCR activation. Since coupling of activated receptors to G-proteins decreases GDP affinity and increases GTP affinity for the involved guanine nucleotide binding site, agonist-promoted increases in the rate of binding of radiolabeled GTP (or preferably a hydrolysis-resistant analog of GTP) are reflective of agonist-promoted GPCR/G-protein coupling (8). An alternative means of assessing coupling involves measurement of the rate of agonist-promoted hydrolysis of radiolabeled GTP by the combined effect of promotion of GDP/GTP exchange and GTPase activity of the activated G-protein (8). Both of these biochemical assays possess the advantage of proximity to the initial drug/receptor interaction and the disadvantage of high ratios of noise to signal. They are seldom considered as part of the armamentarium in routine biochemical tests of drug-receptor interaction.

Assays based on GPCR-mediated activation of effector enzyme activity in cell-free preparations have the potential advantage of more direct assessment of the effects of GPCR on second-messenger synthesis. They are also less likely to reflect the cross-talk that readily occurs between the various signaling pathways in intact cells. Whereas cell-free assay of GPCR-promoted activation or inhibition of adenylyl cyclase activity may in many instances be simpler and preferable to analogous quantitation of cyclic AMP levels in intact cells, this is not the case for cell-free assay of G-protein-regulated phospholipases, which is much more problematic in membrane preparations.

X. G-PROTEIN-MEDIATED REGULATION OF ION CHANNELS

Many ion channels are directly regulated by G proteins (22,54). This activation occurs through both G-protein α-subunits, e.g., cardiac Ca^{2+} channels, and G-protein $\beta\gamma$-subunits, e.g., cardiac K^+ channels. Due to the nature of membrane ion fluxes, assessment of ion channel activity as a direct quantitative readout of the interaction of drugs with GPCR can be cumbersome. On the other hand, electrophysical measurements provide ideal all-or-none tests of drug activity, and assessment of GPCR activity in large-through-put electrophysical assays based on activation of ion channels theoretically is both feasible and desirable.

XI. BIOCHEMICAL INDICES IN HIGH-THROUGH-PUT TESTS OF LIGAND/RECEPTOR INTERACTION: ASSESSMENT OF CA^{2+} TRANSIENTS

GPCR are important in the drug discovery process, and as is emphasized throughout this volume, a wide variety of approaches are available for studying the interaction between molecules and GPCR. Radioligand-binding assays are amenable to the high-through-put demands of pharmaceutical screening operations; these assays are discussed extensively elsewhere. An obvious shortcoming to such binding assays is that they give no insight into the functional consequences of receptor occupation. Although detailed binding curves have the potential to provide insight into the agonist versus antagonist nature of a molecule, in practice, such analyses are too cumbersome and indirect in broad screens of molecules of unknown activities. In contrast, biochemical assays of ligand/receptor interactions can yield valuable biological and pharmacological information. With the exception of certain receptors and specific responses, biochemical assays at the intact tissue or animal level should be primarily considered in the context of secondary and tertiary pharmacological analyses. On the other hand, functional assays of GPCR in homogeneous cell preparations have the versatility, sensitivity, and, to variable degrees, adaptability to be considered in primary screening and characterization of ligand/receptor interactions. However, these assays generally must live with limitations of potential low through-put, and their propensities to be labor intensive, expensive to operate, and difficult to automate.

Possible approaches for circumventing these limitations exist. For example, the measurement of Ca^{2+} transients provides a viable option for rapid functional analysis of ligand-GPCR interactions involving PLC-linked receptors. Prior to the advent of the Ca^{2+}-sensitive fluorescent dyes, Ca^{2+} transients were measured using the jellyfish bioluminescent protein aequorin. Use of the aequorin has been limited by the need for microinjection or other relatively harsh techniques to introduce the photoprotein to the cytosol. However, the recent cloning of aequorin and its constitutive expression in mammalian target cells has allowed the development of alternative assays for Ca^{2+} measurement that may offer specific advantages over fluorescent dye methods (55). Fluorescent plate readers with detection optics based on cooled charge coupled device (CCD) technologies (FLIPR and others) simultaneously allow signal detection from all wells of a 96-well plate and make high throughput screening based on a functional response a reality (55,56).

As discussed above the discovery (28) that the PLC-β-activating G-protein Gα_{16} exhibits promiscuity in coupling with GPCR that otherwise do not couple to the Gq-regulated inositol lipid signaling pathway also potentially provides a useful reagent in drug screening. That is, cells can be engineered to overexpress, for example, a Gs-linked GPCR, which then can be assessed on the basis of its capacity to activate phospholipase and mobilize Ca^{2+}. Thus, it is highly likely, although not yet unambiguously proven, that essentially all GPCR can be assayed by quantification of cytosolic Ca^{2+} concentration. Thus, it is imminently possible that high-throughput screening of molecules for most GPCR can be designed using a homogeneous cell engineered to overexpress $G\alpha_{16}$ and the GPCR of interest and utilizing the fluorescent technologies described above.

XII. NOVEL APPROACHES FOR BIOCHEMICAL ASSESSMENT OF LIGAND/GPCR INTERACTION

One novel approach for GPCR assay development has exploited the transcription-activating processes associated with these receptors (57–61). In general terms, this technology employs genetically engineered reporter constructs that are activated downstream as a result of ligand/GPCR signaling processes. Artificial promoter sequences consisting of individual or repeated copies of specific transcription response elements or the natural promoter elements of a known target gene are linked to a reporter gene capable of generating an easily quantifiable readout. For example, the cyclic AMP response element (CRE) and TPA response element (TRE) have been utilized extensively to monitor activation occurring through interactions with Gs and Gq-linked GPCR, respectively. In most cases, the reporter gene encodes the production of an enzyme or other protein, foreign to the host cell, that can be monitored by colorimetric or luminescent readouts or via immunoassay.

Lerner and co-workers have described a novel approach for detection of GPCR activity based on the redistribution of pigment within frog melanophores (62,63). Activation of the cyclic AMP or inositol lipid signaling pathway leads to a broad dispersion of pigment and the appearance of darkening; promotion of a decrease in cyclic AMP levels causes a focal aggregation of pigment and the appearance of lightening. Therefore, this response system can be utilized to detect the activation of Gs-, Gi-, and Gq-linked GPCR. This approach potentially has advantages over other technologies in that it appears to be useful for all GPCR regardless of the transmembrane signaling selectivity and, other than expression of the receptor of interest, does not require the introduction of foreign proteins (i.e., reporter constructs) into the test cell. Since there are no obvious reasons for this technology to be restricted to a conventional plate format, adaptation of the assay to a "lawn" format should be feasible given the availability of appropriate video imaging equipment for optical detection of pigment distribution.

A final technology that should be mentioned utilizes silicon-based sensor technology to detect perturbations in cellular metabolism (64). The metabolic rates of most cells are greatly enhanced as a consequence of cell activation, and the net result is an extrusion of protons to maintain intracellular pH. The Cytosensor microphysiometer (Molecular Devices, Palo Alto, CA) utilizes a light-addressable potentiometric sensor (LAPS) to measure the rate of extracellular acidification. The Cytosensor detects small alterations in the rate of proton extrusion by monitoring the pH of the immediate environment of the stimulated cell. This technology is low through-put and labor intensive in its current form. However, since the technology is not dependent on prior knowledge of the signaling characteristics of the receptor, it may represent a viable and important approach for screening ligands. This is particularly true for "orphan" GPCR for which the endogenous activating ligands have not been identified.

XIII. ANALYSIS OF LIGAND/RECEPTOR INTERACTION UTILIZING CELL SIGNALING RESPONSES

In contrast to radioligand-binding assays, assessment of drug-receptor interactions at the biochemical level not only allows study of the structure-activity relationship of agonists, but also permits assessment of the activity of molecules that antagonize activation of the signaling pathway by the cognate agonist. Therefore, biochemical analyses of drug-receptor interactions that focus on early events such as second-messenger production are potentially very useful in drug discovery programs since they allow assessment of both agonist and antagonist activities in the same assay system. Antagonism potentially can result from the interaction of a drug with the GPCR signaling system by several different mechanisms including competitive antagonism by nonactivating interaction at the agonist binding site and noncompetitive antagonism resulting from interaction with nonreceptor sites or with sites within the receptor but not within the agonist binding site. In the first case inhibition should be surmountable by increasing concentrations of agonist and in the latter case inhibition is not agonist-surmountable. Multiple types of antagonism and forms of characterization have been described in detail elsewhere, and therefore, will not be discussed further here. However, specific studies should be carried out to determine the nature of an antagonism observed with a specific drug. Model homogeneous cell systems that express nonrelated receptors (natively expressed or genetically engineered) coupled to the same signaling pathway as the GPCR of interest are particularly useful to assess the selectivity of the antagonistic effects associated with drugs of potential interest. For example, in a turkey erythrocyte model system used to study a P2Y receptor coupled to activation of phospholipase C, we have taken advantage of the realization that these cells also express a phospholipase C–activating β-adrenoceptor to evaluate the selectivity of interaction of drugs with P2 receptors (65,66).

The assessment of biochemical responses to receptor activation allows identification of partial agonists, i.e., molecules that at full receptor occupancy produce a lower maximal response than do full agonists. The inability of true partial agonists to produce a full response is not due to decreased affinity for binding to receptors since partial agonists compete (frequently with high affinities) for the full complement of receptors. Therefore, binding of partial agonists to the total receptor population results in competitive inhibition of responses produced by full agonists. A significant number of competitive antagonists of GPCR currently used as therapeutic agents were first identified as partial agonists or evolved in a structural series from an initially identified partial agonist.

The molecular mechanism that accounts for a less than maximal response of partial agonists at maximal receptor occupancy has not been established unambiguously. The simplest conceptualization is that interaction of full agonists with the receptor produces a conformational change in the GPCR that results in maximal activation of the G-protein and associated signaling cascade; competitive antagonists have been viewed to bind to the receptor at the same (or overlapping) site as the agonist without inducing a conformational change necessary for receptor activation. Partial agonists can be considered in this conceptualization to induce an alteration in the conformation of the receptor but not to the extent that results in the full efficacy resulting from receptor occupancy by full agonists. Therefore, antagonists exhibit zero efficacy, and partial agonists, in spite of the occupation of the full complement of receptors, have a lower efficacy than full agonists. Less than full efficacy of a drug in a GPCR test system does not always indicate a true partial agonist effect since drug-promoted signaling responses could be produced by effects on other components involved in the signaling cascade. The same range of concentrations of the partial agonist should produce both the partial agonist effects and competitive antagonism of the effects of a full agonist. Similarly, a partial agonist should decrease the response of the full agonist to the same level of effect produced by the partial agonist alone.

One important advantage of biochemical response-based assessment of the interaction of ligands with GPCR resides in the sensitivity that can be provided by such assay systems. For example, cloned GPCR are routinely stably expressed in heterologous cell lines at levels much higher than those natively occurring in mammalian cells. Whereas the physiological relevance of studies directed at the signaling specificity or biological regulation of heterologously overexpressed receptors is a concern, the value of such cell lines as sensitive indicators of drug interaction at GPCR is considerable. This enhanced sensitivity is most applicable for receptor agonists and follows from the high level of overexpression of receptors that can be attained with mammalian cell expression vectors.

The maximal effect produced by an agonist in a GPCR-activated signaling pathway will increase in direct proportion to receptor concentration until a level of receptor expression is attained at which complete occupancy of the available receptors also will maximally activate the G-protein-regulated signaling pathway (Fig. 2). In classic terminology, receptor reserve exists when expression of GPCR increases to a level in excess of the amount

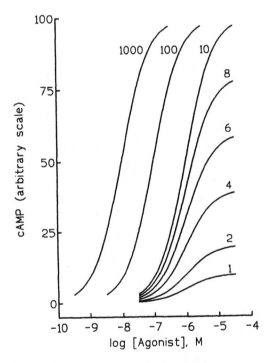

Figure 2 Increase in sensitivity of drug-receptor responses by receptor overexpression shown in hypothetical concentration response curves for a full agonist at a GPCR expressed at different densities. (Ordinate) Relative second-messenger formation. (Abscissa) Logarithm of the molar concentration of agonist. The numbers next to the curves refer to the relative expression of receptors. In this example, the maximal response of a full agonist is achieved with a relative receptor expression of 10.

necessary to maximally activate the signaling pathway. Once the number of GPCR passes this critical level of receptors, agonists fully activate the signaling response at concentrations of agonist that do not fully saturate the receptors, and the observed effect is a shift to the left of the concentration effect curve for activation. As illustrated in Figure 2, increases in receptor number result in increases in maximal effects on second-messenger production with no changes in the observed EC_{50} until a maximally activating amount of receptor is expressed. Increasing receptor levels past this point result in a progressive shift to the left of the concentration effect curve; i.e., the EC_{50} values become progressively smaller, with no further change in maximal effect occurring. The concentration effect relationships of five hypothetical molecules that range in properties from those of a classic "full" agonist to those of a classic competitive antagonist are compared in Figure 3.

The relationship illustrated in Figure 2 also emphasizes the potential value in sensitivity provided by such systems. That is, at the extreme of the highest level of receptor expression, the capacity to detect stimulatory effects of an agonist molecule is much greater than that at the extreme of low-level expression. Thus, a potentially interesting molecule that exhibited relatively low affinity for the receptor might go undetected at low levels of receptor expression, but be easily identified as an agonist in the case of high receptor expression. This relationship also can prove of value in identification of partial agonists and therefore antagonists. For example, a molecule might not exhibit measurable agonist activity at low levels of receptor, but might act as a full agonist in a GPCR signaling system that exhibits a considerable amount of receptor reserve. A "partial agonist" nature of this molecule then might be revealed in later studies carried out in this cell expression system or perhaps with the natively expressed receptor, and directed chemical syntheses could potentially evolve a molecule that binds with high affinity and no agonist activity, i.e., a competitive antagonist.

XIV. INVERSE AGONISM REVEALED BY BIOCHEMICAL TESTS OF DRUG/RECEPTOR INTERACTION

An alternative view of GPCR activation suggests that the receptor natively exists in inactive and active states and that the presence of an effective concentration of an agonist shifts the equilibrium of the receptor population toward the active state, i.e., the form coupled with the G-protein, resulting in the generation of a physiological response (67–69). This view implies that even in the absence of agonist, a fraction of receptors are in the active coupled form, resulting in activation of the G-protein and production of

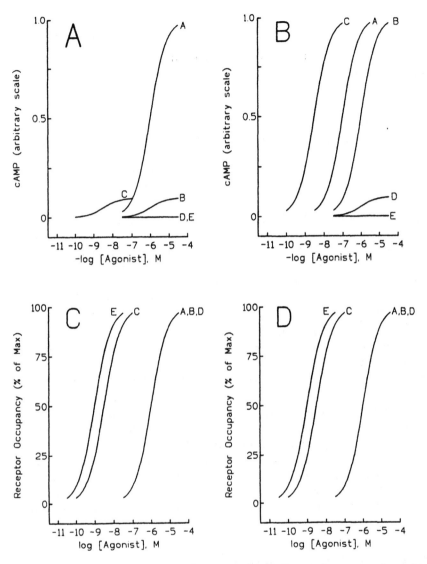

Figure 3 Effects of receptor density on maximal effect and potency of agonists. Hypothetical second-messenger responses and receptor occupancy are shown at two levels of receptor expression. A and B, the second-messenger response; C and D, the receptor occupancy curves in cells expressing no excess (A and C) and a 10-fold excess (B and D) of receptors necessary to produce a maximal response by a full agonist. Under normal expression levels, drug A is a full agonist, B and C are partial agonists with different potencies, drug D is a partial agonist with the same potency as drug B but with 10-fold less maximal effect, and drug E is a high potency competitive antagonist. Note the concentration of receptor-dependent changes in potency and relative maximal effect of the drugs compared to the receptor occupancy curves.

the second messenger. Agonist-independent activation of effector proteins has been recognized for many years. This activity has been referred to as "basal" activity without giving a formal mechanistic explanation to its occurrence and without necessarily implying that a "basal" activated state occurs even with unoccupied receptors. This basal activity is low in general and somewhat variable among different receptor subtypes and tissues of expression.

Some drugs recently have been shown to exhibit negative efficacy (67–70). That is, drugs have been identified that in a concentration-dependent manner inhibit the agonist-independent activity of GPCR. These drugs have been proposed to bind preferentially to the inactive uncoupled form of the receptor, shifting the equilibrium away from any basally active state and resulting in a decrease in agonist-independent activity. Such drugs have been called inverse agonists, or negative antagonists, and have an efficacy value below zero. The potential therapeutic application of these drugs has been widely discussed recently and will not be considered here (for a review see Refs. 68, 69).

Identification and characterization of inverse agonists requires an assay system that exhibits an easily measurable level of agonist-independent activity. There are few examples of natively expressed GPCR that exhibit sufficient basal activity to be utilized to detect and study inverse agonism. However, the availability of cloned GPCR and signaling molecules offers the possibility of generating model systems that exhibit a reasonable level of agonist-independent response driven by the overexpression of the receptor of interest. Thus, as our understanding of the therapeutic potential of inverse agonists increases, the value of biochemical analyses of the readout of GPCR activation, inhibition, and inverse agonism will grow, and assays based on the various responses emanating from G-protein activation may find new unique and crucial application in drug discovery and characterization.

XV. CONCLUDING COMMENTS

Remarkable progress in molecular understanding of receptor-regulated signaling pathways has been made in the past decade. Most of the GPCR for the commonly known extracellular signaling molecules have been cloned (Table 1), and the G-proteins and downstream components that comprise the biochemical signaling cohorts of these GPCR have also been identified. Although three-dimensional structures of GPCR are not yet available, structural insight is accruing for the nonintegral membrane signaling proteins downstream of these receptors. The introduction of radioligand-binding assays for most receptors in the late 1970s/early 1980s revolutionized the analysis of ligand/receptor interaction and the drug discovery process, and remain indispensable in these processes today. The molecular cloning of many GPCR and the concomitant increase in understanding of the signaling pathways utilized by these GPCR has led more recently to opportunities in biochemical assessment of drug/receptor interaction that were not readily conceivable with the intact tissue preparations that historically were the cornerstone of drug development. Technology associated with both physical and biochemical quantitation of the functional readout of ligand/receptor interactions also is developing rapidly, and such measurements no longer need be considered too cumbersome or insensitive for primary screens of drugs for a particular GPCR or in first-step delineation of drug structure/activity relationships. The recently formalized concept of inverse agonism of molecules at GPCR evolved entirely from the types of biochemical assessments that have been considered here. Moreover, continued evolution of ideas related to inverse agonism will depend in large part on these biochemical approaches as will the identification of molecules that belong under this rubric of pharmacological agents. There is no reason to expect that the rapid and open interchange of basic ideas between the academic and industrial pharmacological/biochemical enterprises will not continue in the near future, and further and greater impact of biochemical readouts of drug/receptor interactions on drug discovery can be expected.

REFERENCES

1. Rall TW, Sutherland EW, Berthet J. The relationship of epinephrine and glucagon to liver phosphorylase. IV. Effect of epinephrine and glucagon on the reactivation of phosphorylase in liver homogenates. J Biol Chem 1957; 224:463–475.
2. Robison GA, Butcher RW, Sutherland EW. Cyclic AMP. New York: Academic Press, 1971.
3. Hardman JG, O'Malley BW. Methods in Enzymology, Vol. 38: Hormone Action. San Diego: Academic Press, 1974.
4. Conn PM, Means AR. Methods in Enzymology, Vol. 141: Cellular Regulators. San Diego: Academic Press, 1987.
5. Murphy RC, Fitzpatrick FA. Methods in Enzymology, Vol. 187: Arachidonate Related Lipid Mediators. San Diego: Academic Press, 1990.
6. Johnson RA, Corbin JD. Methods in Enzymology, Vol. 195: Adenylyl Cyclase, G Proteins, and Guanylyl Cyclase. San Diego: Academic Press, 1991.
7. Dennis EA. Methods in Enzymology, Vol. 197: Phospholipases. San Diego: Academic Press, 1991.
8. Iyengar R. Methods in Enzymology, Vol. 237: Heterotrimeric G Proteins. San Diego: Academic Press, 1994.
9. Iyengar R. Methods in Enzymology, Vol. 238: Heterotrimeric G Protein Effectors. San Diego: Academic Press, 1994.

10. Simon MI, Strathmann MP, Gautam N. Diversity of G proteins in signal transduction. Science 1991; 252:802–808.
11. Hepler JR, Gilman AG. G proteins. Trends Biochem Sci 1992; 17:383–387.
12. Nathans J, Hogness DS. Isolation, sequence analysis, and intron-exon arrangement of the gene encoding bovine rhodopsin. Cell 1983; 34:807–814.
13. Dixon RAF, Kobilka BK, Strader DJ, Benovic JL, Dohlman HG, Frielle T, et al. Cloning of the gene and cDNA for mammalian beta-adrenoceptor and homology with rhodopsin. Nature 1986; 321:75–79.
14. Yarden Y, Rodriguez H, Wong SK-F, Brandt DR, May DC, Burnier J, et al. The avian β-adrenoceptor: primary structure and membrane topology. Proc Natl Acad Sci USA 1986; 83:6795–6799.
15. Strader CD, Fong TM, Graziano MP, Tota MR. The family of G-protein-coupled receptors. FASEB J 1995; 9:745–754.
16. van Rhee AM, Jacobson KA. Molecular architecture of G protein-coupled receptors. Drug Dev Res 1996; 37:1–38.
17. Coughlin SR. Expanding horizons for receptors coupled to G proteins: diversity and disease. Curr Opin Cell Biol 1994; 6:191–197.
18. Henderson R, Baldwin JM, Cesca TA, Zemlin F, Beckman E, Downing KH. Model for the structure of bacteriorhodopsin based on high-resolution electron cryo-microscopy. J Mol Biol 1990; 213:899–929.
19. Schertler GF, Villa C, Henderson R. Projection structure of rhodoopsin. Nature 1993; 362:770–772.
20. Gilman AG. G proteins and regulation of adenylate cyclase. Angew Chem Int Ed Engl 1995; 34:1406–1419.
21. Biddlecome GH, Bernstein G, Ross EM. Regulation of phospholipase C-β1 by Gq and m1 muscarinic cholinergic receptor. Steady-state balance of receptor-mediated activation and GTPase-activating protein-promoted deactivation. J Biol Chem 1996; 271:7999–8007.
22. Clapham DE, Neer EJ. New roles for G-protein $\beta\gamma$-dimers in transmembrane signalling. Nature 1993; 365:403–406.
23. Boyer JL, Paterson A, Harden TK. G-protein-mediated regulation of phospholipase C: involvement of $\beta\gamma$ subunits. Trends Cardiovascular Med 1994; 4:88–95.
24. Noel JP, Hamm HE, Sigler PB. The 2.2 A crystal structure of transducin-α complexed with GTPγS. Nature 1993; 366:654–663.
25. Lambright DG, Noel JP, Hamm HE, Sigler PB. Structural determinants for activation of the α-subunit of a heterotrimeric G protein. Nature 1994; 369:621–628.
26. Coleman DE, Berghuis AM, Lee E, Linder ME, Gilman AG, Sprang SR. Structures of active conformations of $G_i\alpha1$ and the mechanism of GTP hydrolysis. Science 1994; 269:1405–1412.
27. Sondek J, Bohm A, Lambright DG, Hamm HE, Sigler PB. Crystal structure of a G-protein beta gamma dimer at 2.1A resolution. Nature 1996; 379:369–374.
28. Offermanns S, Simon MI. $G\alpha_{15}$ and $G\alpha_{16}$ couple a wide variety of receptors to phospholipase C. J Biol Chem 1995; 270:15175–15180.

29. Camps M, Hou C, Sidiropoulos D, Stock JB, Jakobs KH, Gierschik P. Stimulation of phospholipase C by guanine-nucleotide-binding protein $\beta\gamma$ subunits. Eur J Biochem 1992; 206:821–831.
30. Boyer JL, Waldo GL, Harden TK. $\beta\gamma$-Subunit activation of G-protein-regulated phospholipase C. J. Biol Chem 1992; 267:25451–25456.
31. Blank JL, Brittain KA, Exton JH. Activation of cytosolic phosphoinositide phospholipase C by G protein $\beta\gamma$-subunits. J Biol Chem 1992; 267:23069–23075.
32. Boyer JL, Graber SG, Waldo GL, Harden TK, Garrison JC. Selective activation of phospholipase C by recombinant G-protein α- and $\beta\gamma$-subunits. J Biol Chem 1994; 269:2814–2819.
33. Kleuss C, Scherübl H, Hescheler J, Schultz G, Wittig B. Selectivity in signal transduction determined by γ subunits of heterotrimeric G proteins. Science 1993; 259:832–834.
34. Iyengar R. Molecular and functional diversity of mammalian G_s-stimulated adenylyl cyclases. FASEB J 1993; 7:768–775.
35. Taussig R, Gilman AG. Mammalian membrane-bound adenylyl cyclases. J Biol Chem 1995; 270:1–4.
36. Exton JH. Phosphoinositide phospholipases and G proteins in hormone action. Annu Rev Physiol 1994; 56:349–369.
37. Neer EJ. Heterotrimeric G proteins: organizers of transmembrane signals. Cell 1995; 80:249–257.
38. Brunton LL, Mayer SE. Extrusion of cyclic AMP from pigeon erythrocytes. J Biol Chem 1979; 254:9714–9720.
39. Berridge MJ, Irvine RF. Inositol trisphosphate and diacylglycerol: two interacting second messengers. Annu Rev Biochem 1987; 56:159–193.
40. Michell RH, Drummond AH, Downes CP. Inositol Lipids in Cell Signalling. London: Academic Press, 1989.
41. Divecha N, Irvine RF. Phospholipid signaling. Cell 1995; 80:269–278.
42. Putney JW. Advances in Second Messenger and Phosphoprotein Research, Vol. 26: Inositol Phosphates and Calcium Signaling. New York: Raven Press, 1992.
43. Dennis EA. Diversity of group types, regulation, and function of phospholipase A_2. J Biol Chem 1994; 269:13057–13060.
44. Mosior M, McLaughlin S. Peptides that mimic the pseudosubstrate region of protein kinase C bind to acidic lipids in membranes. Biophys J 1991; 60:149–159.
45. Morris AJ, Engebrecht J, Frohman MA. Structure and regulation of phospholipase D. Trends Pharmacol Sci 1996; 17:182–185.
46. Exton JH. Phosphatidylcholine breakdown and signal transduction. Biochim Biophys Acta 1994; 1212:26–42.
47. English D. Phosphatidic acid: a lipid messenger involved in intracellular and extracellular signaling. Cell Signaling 1996; 8:341–347.
48. Singer WD, Brown HA, Jiang X, Sternweis PC. Regulation of phospholipase D by protein kinase C is synergistic with ADP-ribosylation factor and independent of protein kinase activity. J Biol Chem 1996; 271:4504–4510.

49. Cockcroft S, Thomas GMH, Fensome A, Geny B, Cunningham E, Gout I, et al. Phospholipase D: a downstream effector of ARF in granulocytes. Science 1994; 263:523–526.

50. Brown HA, Gutowski S, Moomaw CR, Slaughter C, Sternweis PC. ADP-ribosylation factor, a small GTP-dependent regulatory protein, stimulates phospholipase D activity. Cell 1993; 75:1137–1144.

51. Ohguchi K, Banno Y, Nakashima S, Nozawa Y. Regulation of membrane-bound phospholipase D by protein kinase C in HL60 cells. J Biol Chem 1996; 271:4366–4372.

52. Singer WD, Brown HA, Bokoch GM, Sternweis PC. Resolved phospholipase D activity is modulated by cytosolic factors other than ARF. J Biol Chem 1995; 270:14944–14950.

53. Cockcroft S. Ca^{2+}-dependent conversion of phosphatidylinositol to phosphatidate in neutrophils stimulated with fMet-Leu-Phe or ionophore A23187. Biochim Biophys Acta 1984; 795:37–46.

54. Brown AM. Membrane-delimited cell signaling complexes: direct ion channel regulation by G proteins. J Membr Biol 1993; 131:93–104.

55. Brini M, Marsault R, Bastianutto C, Alvarez J, Pozzan T, Rizzuto R. Transfected aequorin in the measurement of cytosolic Ca^{2+} concentration ($[Ca^{2+}_c]$). J Biol Chem 1995; 270:9896–9903.

56. Schroeder KS, Neagle BD. FLIPR: a new instrument for accurate, high throughput optical screening. J Biomol Screening 1996; 1:75–80.

57. Hill CS, Treisman R. Transcriptional regulation by extracellular signals: mechanisms and specificity. Cell 1995; 80:199–211.

58. Sista P, Edmiston S, Darges JW, Robinson S, Burns DJ. A cell-based reporter assay for the identification of protein kinase C activators and inhibitors. Mol Cell Biochem 1994; 141:129–134.

59. Chen W, Shields TS, Stork PJS, Cone RD. A colorimetric assay for measuring activation of Gs- and Gq-coupled signaling pathways. Anal Biochem 1995; 226:349–354.

60. Dhundale A, Goddard C. Reporter assays in the high throughput screening laboratory: a rapid and robust first look? J Biomol Screening 1996; 1:115–118.

61. Bronstein I, Fortin J, Stanley PE, Stewart GSAB, Kricka LJ. Chemiluminescent and bioluminescent reporter gene assays. Anal Biochem 1994; 219:169–181.

62. Jayawickreme CK, Graminski GF, Quillan JM, Lerner MR. Creation and functional screening of a multi-use peptide library. Proc Natl Acad Sci USA 1994; 91:1614–1618.

63. Lerner MR. Tools for investigating functional interactions between ligands and G-protein-coupled receptors. Trends Neurosci 1994; 17:142–146.

64. McConnell HM, Owicki JC, Parce JW, Miller DL, Baxter GT, Wada HG, et al. The cytosensor microphysiometer: biological applications of silicon technology. Science 1992; 257:1906–1912.

65. Boyer JL, Zohn I, Jacobson KA, Harden TK. Differential effects of putative P_2-purinergic receptor antagonists on adenylyl cyclase- and phospholipase C-coupled P_{2Y}-purinergic receptors. Br J Pharmacol 1994; 113:614–620.

66. Boyer JL, Schachter JB, Romero T, Harden TK. Identification of competitive antagonists of the $P2Y_1$-purinergic receptor. Mol Pharmacol 1996; 50:1323–1329.
67. Bond RA, Leff P, Johnson TD, Milano CA, Rockman HA, McMinn TR, et al. Physiological effects of inverse agonists in transgenic mice with myocardial overexpression of the β_2-adrenoceptor. Nature 1995; 374:272–276.
68. Milligan G, Bond RA, Lee M. Inverse agonism: pharmacological curiosity or potential therapeutic strategy? Trends Pharmacol Sci 1995; 16:10–13.
69. Leff P. The two-state model of receptor activation. Trends Pharmacol Sci 1995; 16:259–260.
70. Costa T, Herz A. Antagonists with negative intrinsic activity at δ opioid receptors coupled to GTP-binding proteins. Proc Natl Acad Sci USA 1989; 86:7321–7325.

The Use of Electrophysiology to Improve Understanding of Drug-Receptor Interactions

John G. Connolly and Charles Kennedy
University of Strathclyde, Glasgow, Scotland

I. THE CENTRAL IMPORTANCE OF LIGAND-RECEPTOR INTERACTIONS IN DRUG DESIGN

The major steps in designing a drug to treat a particular disease state are deceptively clear. First, find a therapeutic target that can be manipulated to ameliorate the condition, and then find a specific drug that will alter the function of the target appropriately. Ideally, the drug will not provoke deleterious side effects and will work as well in vivo as it does in vitro. Alternatively, one may already have a partially useful drug, but need to know its target and mechanism of action so as to improve its efficacy. A fundamental requirement for making progress in any of these steps is the need to be able to define the exact nature of the drug-receptor interaction, and the effect it has on receptor function.

Electrophysiological techniques are especially helpful in meeting these needs. They allow us to directly relate the nature of drug-receptor interactions to a functional outcome. This can be done for an individual receptor molecule, a cell, a synapse, a neuronal circuit, a particular region of the brain, or for the entire brain. Electrophysiological techniques can also be used to assess functional consequences of longer-term adaptive effects such as changes in gene expression.

These approaches have been used for many years to study the function of naturally occurring receptors in their native tissue. However, the molecular-genetic revolution in neuroscience has created a new set of tasks for electrophysiology. Many new receptor-like clones, discovered by the genome project, need to be functionally identified and characterized pharmacologically. In some cases the natural ligands and modulators for these new receptor subtypes still need to be discovered. In other cases, peptides that might be ligands have been found, but not the receptors that they act upon. Electrophysiology is a very powerful tool for solving these problems.

This review will examine some of the ways in which electrophysiology can contribute to the process of finding new drugs and therapeutic targets, and optimizing their interactions. Since other types of receptors are dealt with elsewhere in this volume, we will focus on ligand-gated ion channels.

II. RECEPTOR DIVERSITY

Historically, receptors were classified and characterized according to the relative potencies and estimated affinities of groups of ligands at binding sites on the receptors. Discrepancies between the responses obtained in different tissues often led to the discovery of a new subtype of receptor. This classic approach to ligand-receptor interaction has led to the identification of many new receptor subtypes, but is limited by the availability of specific chemical ligands.

In contrast, molecular biology can identify new receptors without using a selective ligand. The application of the molecular-genetic approach to neurobiology has provided an explosive increase in our knowledge about the variety, structure, and distribution of receptor subunits (1), each of which can be uniquely identified by its amino acid sequence. The description of a receptor has now expanded far beyond the chemical selectivity of its ligand-binding site. Also included are its subunit composition, structural features, ion selectivity, single-channel conductance, developmental expression, anatomical distribution, and the ways in which its activity is regulated. When combined with knowledge of its pharmacological profile, these can assist in predicting the physiological outcome of receptor-ligand interactions. However, another facet of the molecular approach is that it allows us to understand how different receptor subtypes are structurally related to one another. Among the ligand-gated ion channels this has led to the recognition of families and superfamilies of genes.

Gene Families and Superfamilies

The relationship between nicotinic acetylcholine (ACh) receptor subunits illustrates this point. In 1984, functional expression of the electroplax nicotinic ACh receptor of *Torpedo californica* was first achieved (2). Shortly afterward, neuronal nicotinic receptor subunits were also cloned (3–7) and functionally expressed (8). Examination of the nucleotide and deduced amino acid sequences of these subunits revealed many regions of sequence identity, suggesting that their genes may have arisen by duplication of a single ancestor gene, which subsequently diversified structurally and functionally. Receptor subunits of a particular kind that show a high degree of sequence identity (usually >45%) are considered members of a *gene family*. Sometimes, closely related members of a gene family are found clustered together on chromosomes. For example, the genes encoding the neuronal nicotinic $\alpha3$, $\beta4$, and $\alpha5$ subunits are clustered together in rat (9), chick (10), and human (11) chromosomes.

Shortly after neuronal nicotinic ACh receptor subunits were discovered, $GABA_A$ (12) and glycine (13) receptor subunits were also isolated. They shared several common amino acid motifs with nicotinic receptors and their gene families had a low level of evolutionary relatedness. 5HT3 receptors also share these properties (14). Together they are therefore said to constitute a *gene superfamily*.

Hydropathy analysis (15) of the predicted amino acid sequences revealed that the various subunits in this gene superfamily each contain four hydrophobic regions of 20–22 amino acids, which are predicted to lie within the cell membrane itself (see Refs. 16,17 for reviews). Thus, each subunit is a single polypeptide that spans the membrane four times. Five of these subunits come together to form the complete receptor complex with its central ion pore.

Both the carboxy- and amino-terminal ends of the individual peptides are thought to lie outside the cell membrane. Starting at the amino-terminal, the subunit polypeptide begins with the first extracellular domain, which continues for about 200 amino acids and contains the agonist binding site. The polypeptide then traverses the membrane (transmembrane domain 1, TMD1), loops around in the cytoplasm, and then traverses back through the membrane (TMD2) to the outside of the cell. TMD2 is α-helical in structure and contributes one of the five walls that make up the central ion pore in the complete receptor complex. The peptide now crosses back through the membrane (TMD3) to the inside of the cell. The next section of the peptide that lies within the cytoplasm varies considerably in length and sequence between different subunits. The final transmembrane domain (TMD4) takes the polypeptide chain back out of the cell and leaves the carboxy-terminus in an extracellular position. Thus, the protein domains to which the agonist binds and those that form the ion pore are all part of the same protein complex. Receptor molecules that have this particular structure-function relationship are therefore termed *ligand-gated ion channels*.

NMDA receptors (18) and AMPA-kainate receptors (19) are ligand-gated ion channels that respond to glutamate, but belong to separate gene families. Together they comprise a second gene superfamily that is structurally distinct from ligand-gated ion channels of the nicotinic type. They have a very large extracellular amino-terminal and the carboxy-terminal is located intracellularly (20,21). Also, instead of having an α-helical TMD2 region to line the wall of the ion channel region, they are thought to have a hairpin loop or "P-loop," which dips into the lipid bilayer to span the membrane, but exits on the same side at which it entered. Thus, these receptors follow the general rule of three membrane-spanning domains plus an ion channel domain. In contrast, a third gene superfamily, the more recently discovered ATP receptor (P2 receptor) subunits, have only two strongly hydrophobic transmembrane regions, while the nature of the ion channel domain is not yet clear (22–24). A FMRFamide-gated ion channel has been identified, which has a similar structure to P2 receptors, but they have no significant sequence homology (25). The general structural properties of these ligand-gated ion channel superfamilies are summarized in Table 1. A further ligand-gated ion channel, at which cyclicGMP is the agonist, has also been identified (26), but as the agonist binding site is intracellular, it will not be discussed further here.

III. ELECTROPHYSIOLOGICAL RECORDING METHODS

Structural introspection alone is of limited usefulness in deducing the nature of drug-receptor interactions. A means of measuring activity is essential in order to study structure-function relationships within subunits, ligand specificity, and the mechanism of action of drugs. Binding studies alone cannot do this. Nor can they indicate whether or not a ligand is an agonist or an antagonist at the receptor site. The great advantage of electrophysiological recording methods is that they can measure receptor activity directly and in real time. A detailed description of such methods is out of the scope of this work, but a number of excellent texts on the subject are available (27–29). However, the essential features of the most commonly used recording methods are described below and in Table 2.

A. Two-Electrode Voltage Clamp

The two-electrode voltage clamp technique is commonly used for monitoring the activity of cloned ligand-gated ion channels when expressed in *Xenopus* oocytes. Two low-resistance micropipettes are inserted into the oocyte. One monitors the transmembrane voltage and the second is used to inject current into the cell and so maintain the membrane potential at

Table 1 Structural Properties of the Ligand-Gated Ion Channel Superfamilies

LGIC superfamily	Nicotinic, glycine, GABA$_A$, 5HT3	Kainate, AMPA, NMDA	ATP	FMRF amide
Number of subunits in receptor complex	5	5?	5?	?
N-terminal location	Extracellular	Extracellular	Intracellular	Intracellular
C-terminal location	Extracellular	Intracellular	Intracellular	Intracellular
TMDs per subunit	4	3 + P-Loop	2	2
Ligand-binding site	1st extracellular domain	1st and 2nd extracellular	Extracellular	Extracellular
Channel pore region	TM2, alpha helix	TM2, P-Loop	?	?

Table 2   Recording Methods

Technique	Configuration	Resolution	Measurement
Patch clamp	On-cell	Single-channel	Current, pA
	Inside-out	Single-channel	Current, pA
	Outside-out	Single-channel	Current, pA
	Whole-cell	Whole-cell	Current, pA
Voltage-clamp	Switch clamp	Whole-cell	Current, pA
	2 electrode	Whole-cell	Current, mA
Current clamp	1 or 2 electrode	Whole-cell	Voltage, mV
Extracellular recording	Electrode close to cell surface	Single-cell, unit recordings	Firing rate (Hz), action potentials (mV)
	Electrode near a nerve tract	Populations of cells	Population spikes, field potentials (mV)

a constant value. Agonist application causes ligand-gated ion channels to change conformation, allowing ions to pass through the central ion pore into the cell. If unopposed, this influx of ions would alter the membrane potential of the cell. However, a feedback recording amplifier detects this change and injects an equal and opposite amount of current into the cell to maintain the voltage at its command level. The injected current is a direct measurement of the total amount of current that has entered the cells through the ion channels.

Alternatively, the receptor subunit cDNAs may be inserted into suitable vectors and transiently or stably transfected into cell lines such as COS cells or fibroblasts. These cells are too small for two-electrode voltage clamp and instead are studied by the patch clamp technique.

B.   Patch Clamp Recording

The elegance of the patch clamp technique lies in its ability to monitor the activity of a single receptor complex (30,31). A blunt, heat-polished micropipette is pushed up against the surface of the target cell, and with the aid of slight suction, a physicochemical bond is made between the pipette glass and the cell membrane. This creates an electrically tight seal that isolates a small patch of membrane in the mouth of the pipette. The resistance to current passing through the pipette and membrane patch is extremely high (tens of $G\Omega$) and as a result it is possible to detect currents of less than 1 pA, i.e., the size of the currents that are produced by the opening of individual receptor complexes in neurons.

C. Patch Clamp Configurations

There are four basic patch clamp configurations. That described above is the *on-cell* configuration, in which the cell membrane remains intact and the external domains of the receptors face into the mouth of the pipette. This configuration is of limited use as it is difficult to exchange solutions within the pipette and to control the cell membrane potential. Rapid withdrawal of the pipette produces the *inside-out* configuration, where the extracellular membrane surface still faces the interior of the pipette, but the cytoplasmic side of the membrane now faces the bath solution. This can be used for studying the effects of altering the conditions at the cytoplasmic face of the receptor, but again it is very difficult to exchange solutions inside the pipette. In both configurations, single-channel activity is recorded.

The *whole-cell* configuration is formed by breaking through the patch of membrane encircled by the pipette tip in the cell-attached mode. The interior of the pipette then communicates with the interior of the cell. Now, it is possible to clamp the membrane potential of the entire cell, and the current produced by the sum of activity of hundreds of individual ion channels is recorded. The last commonly used configuration is the *outside-out* patch. From the whole-cell clamp mode the pipette is withdrawn from the cell, forming a tube of membrane that eventually collapses to form a patch. The external side of the cell membrane faces the bath solution and the internal side faces the patch pipette. An advantage of this mode is that it is possible to apply a variety of concentrations of drugs to the receptors, while fully controlling the transmembrane potential. Single-channel activity is recorded in this configuration.

IV. GETTING STARTED

At the start of a drug discovery campaign looking for a modulator of ligand-gated ion channels, it is likely that no potent, selective ligands are available. Therefore, the first step will probably be a high-through-put screening (HTS) process, using nonelectrophysiological methods. Often, this produces a number of relatively low-potency ($IC_{50} = 1$–10 mM) "hits." Electrophysiology can then be used to discover how these compounds act. This is obviously a critical exercise, since it is important to make sure that only quality "hits" are pursued to avoid unnecessary use of time and resources.

Distinguishing Competitive and Noncompetitive Antagonism

There are a number of different ways in which a hit could inhibit the activity of a ligand-gated ion channel. The ideal, and easiest to study, is competitive antagonism, seen when the hit competes in a reversible manner for the agonist binding site on the receptor and so prevents the agonist binding to, and activating, the receptor. In this case, it is relatively simple to determine the effect of each hit on a concentration-effect curve to the agonist, with competitive antagonism indicated by a parallel shift of the curve to the right, with no change in the maximum response obtained. Compounds displaying such behavior can be considered as potential chemical leads with some degree of confidence.

Competitive antagonism is not the only way by which drugs may inhibit ligand-gated ion channels, and many are known to interact with sites other than the agonist binding site to depress channel activity in a potentially useful manner. Therefore, it may be worthwhile to carry out a further series of experiments to determine the mechanism of action of noncompetitive hits. A check for specificity could be to look at the activity of a hit versus another channel type, an ideal choice being a channel, which, if blocked, could produce a false positive in HTS. For example, given the Ca^{2+} permeability of many ligand-gated ion channels, a common strategy at the HTS phase is to use a Ca^{2+}-sensitive dye to assay Ca^{2+} influx resulting from agonist-induced channel activation. This approach can, however, produce false hits, some of which turn out to be decreasing the signal by blocking the potassium channels involved in setting the cell's membrane potential. The cell therefore depolarizes, the driving force for Ca^{2+} entry decreases, and the agonist-induced signal gets smaller. Electrophysiological studies on potassium channels in the cell type used by HTS, therefore, represent a sensible control system in such a case. Other channels may be more appropriate in other situations.

Compounds that pass this test can then be progressed to more complex electrophysiological analysis. For ligand-gated ion channels that do not desensitize too rapidly, it is possible to generate a current-voltage relationship and look for changes in this relationship in the presence of the hit: for example, does the compound affect the current in a voltage-dependent manner? If the hit is charged, the degree of channel block will change as the membrane potential is made more positive or negative. Such compounds may also produce apparent "flickering" as the hit rapidly blocks and unblocks the open channel. Although not totally definitive, some evidence of a voltage-sensitive block or a change in channel gating may provide enough optimism to pursue a compound further.

V. CHARACTERIZATION OF DRUG-RECEPTOR INTERACTIONS IN CLONED RECEPTOR SUBUNITS

Perhaps one of the dreams of those involved in rational drug design is that one day they will have enough information about the molecular basis of specific drug-receptor interactions to be able to design new drugs entirely *in silico* (on computer), independent of laboratory experiments. Such accurate models of native receptors are still a way off, but electrophysiology can help us progress toward this by identifying cDNA sequences, defining the structure-function relationships underlying ligand-receptor interactions, providing a pharmacological profile of defined receptor subunit combinations, and expediting the drug-discovery process.

A.   Revealing the Pharmacological Identity of DNA Sequences

The first task required by cloning studies of electrophysiology is matching a pharmacological identity to a given DNA sequence. Functional expression can confirm such an identity and establish the existence of gene families. It is also absolutely necessary, for even if sequence identity strongly suggests that a cDNA encodes a subunit of a particular receptor gene family, the pharmacological identity of the clone cannot be automatically assumed. For example, the first cloned neuronal ligand-gated ion channels to be functionally expressed were nicotinic receptors (8). Several years later in the same laboratory, an additional rat neuronal clone was isolated that had a higher sequence identity with the neuronal nicotinic receptor subunits than did some of the muscle nicotinic subunits. However, comparison with published work (14) showed that it encoded the 5HT3 receptor. Functional expression also allows the subunit requirements for functional receptors to be defined and the degree to which subunits are interchangeable to be determined (5). If different subunit combinations are not pharmacologically identical, then the potential to develop subtype specific drugs is also indicated (32).

When a cDNA library is screened to identify a receptor subunit clone, the oligonucleotide probe is often based on a known nucleotide sequence of a related subunit, or on a "guessmer" deduced from the amino acid sequence of the purified subunit protein. If the hunted subunit belongs to a new class, or there is no high-affinity ligand with which to purify receptor protein, then these approaches cannot be used. Electrophysiology can provide a solution to this problem. First, a cDNA expression library is made from tissue in which the receptor is abundant. The library is usually enriched for receptor subunits by selecting a cDNA fraction between 1 and 6 kb. Then RNA is transcribed from the library and injected into *Xenopus* oocytes. Among the millions of different species of RNA injected, some will encode subunits of the desired receptor and a positive electrophysiological response will be evoked by an appropriate agonist. The parent expression library that contains these cDNAs is then divided into pools, transcribed into RNA, and each pool separately injected into *Xenopus* oocytes. Such serial dilution is continued until a single cDNA encoding a subunit of the receptor is isolated. This powerful approach was used to identify the first kainate (19), 5HT3 (14), NMDA (18), and ATP (22,23) receptor subunits.

B. Probing Structural Determinants of Ligand Receptor Interactions

To completely model receptor-ligand interactions requires not only a clear structural knowledge of the ligand, but also knowledge of the amino acid domains on the subunits to which the ligand binds, the relative positions of these domains, and how they functionally interact. Electrophysiology can be combined with molecular approaches to obtain this knowledge at several structural levels, from whole subunits to individual amino acids.

1. Subunit Swapping

The first structural level for defining receptor-ligand interactions is to determine which subunit(s) contain the ligand-binding site. This can be achieved by adding or substituting species of subunits in a multimeric receptor complex. For example, when the $\alpha3\beta2$ nicotinic receptor combination is expressed, nicotine is a poor agonist and neuronal bungarotoxin a very potent antagonist (33). In contrast, the $\alpha3\beta4$ combination is very sensitive to nicotine, but relatively insensitive to neuronal bungarotoxin. The agonist order of potency at the $\alpha3\beta2$ combination is dimethylphenylpiperazinium iodide (DMPP) > ACh > lobeline > carbachol > nicotine > cystisine, whereas in the $\alpha3\beta4$ combination, cytisine > nicotine $\approx$ ACh > DMPP > carbachol > lobeline (34–36). There are also some additional striking differences in antagonist sensitivity. Curiously, while (+)-tubocurarine antagonizes the $\alpha3\beta2$ combination at all concentrations, at 1–10 μM, it enhances the response of the $\alpha3\beta4$ combination to agonist (37). Similarly, the agonist response of the $\alpha3\beta2$ combination is inhibited by niflumic and flufenamic acid, but the agonist responses of the $\alpha3\beta4$ combination are potentiated by these drugs (38). Specific pharmacological properties are therefore associated with particular subunit combinations.

Subunit exchange has also been applied to other ligand-gated ion channels. Loreclezole is an anticonvulsant that can strongly potentiate $GABA_A$ receptor responses. Expression of combinations of human $GABA_A$ subunits showed that strong sensitivity to loreclezole was associated with $\beta2$ and $\beta3$ subunits, rather than $\beta1$ subunits (39). Furthermore, its site of action was not shared with barbiturates, benzodiazepines, or steroids. Thus, electrophysiology had revealed a new therapeutic target.

Some receptor subunits, e.g., the neuronal nicotinic $\alpha5$ subunit, cannot form functional channels when expressed singly or as one of a pair of subunits. Therefore, it can be difficult to determine whether or not it is assembled into the receptor complex. This problem was solved by comparing the electrophysiological properties of the $\alpha4\beta2$ combination with those of $\alpha4\alpha5\beta2$ (40). The combination containing $\alpha5$ had a larger single-channel conductance than $\alpha4\beta2$, but was more rapidly desensitized by nicotine. The EC_{50} for ACh at the $\alpha4\alpha5\beta2$ combination was estimated to be shifted 125-fold to the right of that of the $\alpha4\beta2$ combination. Therefore, this demonstrated that $\alpha5$ subunits could functionally participate in nicotinic receptor complexes.

2. Chimeric Receptors

Although the above studies assist in understanding subunit function, they do not provide much detail about the amino acid domains within the subunits that are responsible for the differences in function that have been detected. A first step toward localizing the amino acids responsible for such distinctions is to construct chimeric receptor subunits. This technique allows sections of related subunits to be exchanged with each other and has the advantage that the amino acid sequence surrounding the site of the exchange is not altered, ensuring minimal structural disturbance of the subunit at that point.

This approach was used to localize the residues responsible for the difference in sensitivity to neuronal bungarotoxin of the $\alpha3\beta2$ (sensitive) and $\alpha2\beta2$ (insensitive) nicotinic receptor combinations (41). The region of the first extracellular domain up to amino acid 84 was found to be important for ACh and nicotine sensitivity, but not for neuronal bungarotoxin sensitivity. In contrast, the regions of amino acids 84–121 and 121–181 contained important determinants of differential sensitivity to neuronal bungarotoxin. The region from amino acid 195 to 215 was important for differences in both agonist and antagonist sensitivity. Since the $\alpha3$ and $\alpha2$ subunits are highly homologous, it was then possible to compare the amino acid sequences within these domains and identify individual nonconserved amino acids that might be critical for determining the pharmacological profile of the combination (see above).

The chimeric approach was taken one stage further in a study designed to identify the agonist binding site in glutamate receptors (42). Functional expression of GluR3 subunits yields receptors with properties similar to those of native AMPA receptors (kainate EC_{50}, 73 μM; AMPA EC_{50}, 16 μM), whereas GluR6 subunits produce a receptor that more closely resembles native kainate receptors (kainate EC_{50}, 1 μM; AMPA, no agonist response). Twenty-six chimeric exchanges were made and it was found that 150 amino acid segments of both the first extracellular domain (before TMD1) and the second extracellular domain, (between TMD2 and TMD3) needed to be exchanged in order to confer the reciprocal set of pharmacological properties upon the chimeric receptor. This demonstrates that the two discontinuous regions must functionally interact in the ligand-binding process.

Further analysis of the sequences within the identified domains showed a strong degree of homology between these domains and bacterial binding proteins for amino acids, the exact structures of which are known from X-ray crystallography studies. Consequently, it was proposed that the structure of the amino acid binding sites in GluR6 and GluR3 might be similar to those of the bacterial binding proteins. Thus, this chimeric study provided a highly detailed model for the ligand-binding site.

The chimeric approach was extended beyond the gene family to the level of superfamilies (43) by taking advantage of the fact that nicotinic $\alpha7$ and 5HT3 receptor subunits both form monomeric receptors when expressed in *Xenopus* oocytes. $\alpha7$ receptors are activated by ACh and nicotine and antagonized by α-bungarotoxin, whereas 5HT3 receptors are completely insensitive to these agents and are instead activated by serotonin. Also, the ion pore in $\alpha7$ receptors is permeable to Ca^{2+}, but that of the 5HT3 receptor is blocked.

Several chimeric receptors were constructed that contained an $\alpha7$ amino-terminus and 5HT3 carboxy-terminus. One of these exchanged domains around valine 201 of the first extracellular domain. The chimeric receptor was activated by ACh and nicotine, blocked by α-bungarotoxin, but was also blocked by Ca^{2+} ions. Therefore, it possessed the pharmacological properties of the nicotinic $\alpha7$ receptor, but the ion channel properties of 5HT3 receptors. Thus, this study localized many of the pharmacological properties of the monomeric receptors to their first extracellular domain. More remarkably, it suggested that ligand-gated ion channels might be composed of exchangeable functional units and that their mechanisms of activation might be common.

3. Site-Directed Mutagenesis

Although the chimeric approach can suggest the involvement of particular amino acids in determining pharmacological specificity, it cannot confirm their involvement. Nor can it distinguish whether it is the size or charge of the candidate amino acid that is important in ligand binding. Information at this level can be obtained by combining electrophysiology with site-directed mutagenesis.

The $\alpha2$ nicotinic subunit contains a proline at position 198, whereas $\alpha3$ contains a glutamine residue. When site-directed mutagenesis was used to substitute the proline residue in $\alpha2$ by glutamine, the resultant mutant $\alpha2\beta2$ combination showed an increased sensitivity to neuronal bungarotoxin, but a reduced relative potency for nicotine, properties of the normal $\alpha3\beta2$ (41).

Thus, it was suggested that a difference in the abilities of the two amino acids to form hydrogen bonds with ligands might be responsible for the pharmacological differences observed between $\alpha2$ and $\alpha3$ subunits, providing an atomic level hypothesis about competitive ligand-receptor interactions.

Site-directed mutagenesis can also alter the effect of antagonists. An unusual example of this was seen when leucine 247 of the nicotinic $\alpha7$ subunit was mutated to threonine (44). Leucine 247 is located in TMD2 and so is part of the ion channel pore of the assembled receptor. Antagonists of the rapidly desensitizing wild-type $\alpha7$ receptor, such as dihydro-β-erythroidine, hexamethonium, and (+)-tubocurarine, become agonists in the mutant receptor. The receptors also appear to lose the property of desensitization. Thus, leucine 247 is essential for antagonist activity at this receptor.

Modulatory sites of ligand-gated ion channels have also been explored using electrophysiology-based approaches. Full activation of the NMDA receptor by glutamate requires the coagonist glycine (45). (This action of glycine is unrelated to its action at glycine-gated ion channels.) Mutagenesis was used to locate the glycine-binding site on the NMDAR1 subunit (46). Substitution of phenylalanine 466 alone caused a reduction of glycine efficacy by more than 1000-fold. However, the EC_{50} of glutamate was not strongly affected, showing that the glycine modulatory site was clearly distinct from the glutamate binding site. Substitution of valine 666 and serine 669 in the region between TMD3 and TMD4 also reduced glycine efficacy. Substitution of the aromatic residues phenylalanine 390, tyrosine 392, and phenylalanine 466 in the first extracellular domain markedly reduced inhibition of NMDAR1 by the glycine binding site antagonist 7-chloro-kynurenate. Therefore, more than one group of amino acids is involved in the binding of glycine to the NMDAR1 subunit.

It was noted that the *phenylalanine-X-tyrosine* motif is also found in the agonist-binding site of the glycine receptor that is a member of the nicotinic superfamily (13). As with the glutamate receptors discussed above, the regions that determined glycine efficacy were structurally similar to those found in bacterial amino acid–binding proteins. This enabled a detailed model of the glycine modulatory site in the NMDAR1 subunit to be constructed.

A mutation in the benzodiazepine modulatory site of GABA$_A$ receptors explains the enhanced postural reflex activity found in an alcohol-nontolerant (ANT) strain of rat (47). The wild-type GABA$_A$ $\alpha6$ subunit in the receptor located in cerebellar granule cells in these rats is diazepam-insensitive and normally contains an arginine residue at position 100. However, in ANT rats the $\alpha6$ subunits contain a glutamine residue at this site. When the wild-type and mutant $\alpha6$ subunits were coexpressed with $GABA_A$ $\beta2$ and $\gamma2$ subunits in *Xenopus* oocytes, the former combination was found to be insensitive to diazepam, whereas the agonist responses of the mutant receptor were potentiated. Thus, an electrophysiological study of a ligand-receptor interaction led to an explanation of a behavioral phenomenon.

A more common problem of abnormal ligand-receptor interactions in $GABA_A$ receptors is pesticide resistance in agriculture. Perhaps 60% of these cases can be tracked down to a single mutation in TMD2 of insect $GABA_A$ subunits (48). In the Rdl mutant of *Drosophila*, mutation of alanine 302 to the larger, polar residue serine resulted in a loss of sensitivity to the noncompetitive antagonist picrotoxin and the insecticide dieldrin.

A case of developmental drug resistance has long been known, whereby fetal rats are not affected by strychnine, but it is lethal to adults. Strychnine is an antagonist at the glycine-gated ion channel. A variant of the glycine $\alpha2$ subunit was found in fetal rats ($\alpha2^*$) that was 99% homologous to the adult, strychnine-sensitive subunit (49). When the glutamate at position 167 in the first extracellular domain of $\alpha2^*$ was mutated to glycine, as in $\alpha2$, then the affinity of the mutant receptor for glycine increased by 35-fold and the sensitivity to strychnine increased 500-fold. Thus, the presence of a larger, negatively charged amino acid in $\alpha2^*$ is sufficient to provide resistance to strychnine.

The structure-function studies applied here to receptors can equally be applied to ligands that act at receptors (50). Slowly a detailed picture of the critical interactions between agonists and their binding sites on ligand-gated ion channels is being assembled. Such studies should greatly accelerate the process of drug design.

C. Pharmacological Profiles of Receptors

The above discussion has focused on structure-function relationships within receptor subunits, but equally important for designing specific drugs is comparison of ligand actions between different receptor subtypes. Electrophysiological techniques can easily be used to obtain the relative order of potency of a series of agonists and antagonists at functionally expressed receptors. This approach has several advantages, the first of which is accessibility. It is much easier to study the properties of a receptor subtype in an expression system than, for example, on a human presynaptic terminal! A distinctive pharmacological profile can be a useful diagnostic tool for identifying the presence of particular receptor subunits in native receptors. The use of human receptor clones may avoid problems arising from inter-species differences and be a good indicator of in vivo activity. Studies on receptors expressed in cell lines may also reduce some of the need for animal experimentation.

Obtaining a profile of the action of a particular drug against a variety of cloned receptor subtypes is also useful for predicting and eliminating possible side effects. For example, amitriptyline is a tricyclic antidepressant that has strong "anticholinergic" properties. Although the term "anticholinergic" is often used to mean "antimuscarinic," amitriptyline is also an effective antagonist at nicotinic ACh receptors. Antipsychotic drugs such as trifluoperazine also have antinicotinic activity. This may be associated with long-term side effects of antipsychotic drug treatment, such as tardive dyskinesia in schizophrenic patients, as well as diminishing cognitive function (51). As many human muscarinic and nicotinic receptor subtypes have already been cloned and expressed, the problems of anticholinergic specificity could be detected at a very early stage of drug development, using a combination of electrophysiological and molecular-genetic approaches. Understanding the complete range of action of drugs may also reveal that part of the therapeutic action of a ligand is mediated by an unsuspected receptor system, thus providing a new drug target for treating a disease.

VI. CHARACTERIZATION OF DRUG-RECEPTOR INTERACTIONS AT NATIVE RECEPTORS

A. Comparison of the Properties of Cloned Subunit Combinations with Native Receptors

While studies on cloned receptor subunits have provided a wealth of information, it is essential to realize that, in many cases, the exact subunit composition of native receptors is unknown. That a subunit combination can be functionally expressed does not mean that the same combination occurs in nature. This is clearly demonstrated by the promiscuity of the rat neuronal nicotinic $\beta2$ subunit, which can functionally substitute for the mouse muscle $\beta1$ subunit in forming a receptor (5). However, this combination is unlikely to occur naturally. Furthermore, it has recently been shown that the chick neuronal nicotinic $\beta2$ subunit can coassemble with an insect α subunit (52). Therefore, to extrapolate data obtained from cloned receptors to native receptors, it is necessary to compare the properties of native and cloned combinations to see how well they match.

This matching process has had mixed success. The mouse muscle nicotinic receptor combination expressed in *Xenopus* oocytes has similar single-channel properties to the native receptor in BC3H1 cells (53,54). Also, comparison of the chick homomeric neuronal nicotinic $\alpha7$ clone with the purified receptor revealed an almost exact match in pharmacological binding properties (55). However, the kinetic and conductance properties of native neuronal nicotinic receptors have not yet been reproduced by studying cloned subunit combinations (56). This emphasizes the need to study the properties of native, as well as cloned, receptors.

B.   Recording from Native Receptors in Isolated Cells

Just as electrophysiological techniques have been used to provide the pharmacological profile of cloned receptor subtypes, so they can also be used to study ligand-receptor interactions on native receptors. Often, tissues are enzymatically dissociated to produce single cells, on which it is generally technically easier to apply the patch clamp technique. Using single cells also has the advantage that drugs can be applied rapidly and under concentration-clamp conditions and thus inactivation of ligands by uptake or breakdown can be minimized (57,58). Under these conditions, the relative order of potency of a series of agonists and antagonists can usually be obtained and compared with that of cloned receptors.

In some cases, evidence for the presence of a particular cloned subunit in the native receptor under study may be obtained, if highly specific ligands are available. Methyllycaconitine (MLA) and α-bungarotoxin are both selective antagonists at the neuronal nicotinic $\alpha7$ homomers with $\geq$nM affinity, whereas their affinity for other mammalian neuronal nicotinic receptor subtypes (apart from $\alpha9$) is more than 1000 times less (59). They were used to characterize nicotinic receptors in cultured fetal rat hippocampal neurons (60), where there is no expression of the $\alpha9$ subunit. The whole-cell currents in response to ACh or anatoxin A, a specific nicotinic agonist, were abolished by 0.6 nM MLA. This action of MLA was voltage-independent, consistent with a competitive antagonism. Also, binding studies showed that the competitive antagonist α-bungarotoxin was displaced by MLA with an IC_{50} of about 10 nM. Thus, the electrophysiological response in the presence of distinguishing ligands indicated the probable presence of one species of cloned receptor subunit ($\alpha7$) in a native receptor.

In a reciprocal situation, distinctive electrophysiological responses to a single ligand (ATP) have suggested the presence of different receptor subtypes in different cells (61). The cell bodies of nociceptive and stretch-sensing sensory neurons were identified following retrograde axonal transport of a dye that had been injected into the tissues they innervate (tooth pulp and masseter muscle, respectively). Application of ATP to the nociceptive neurons typically caused a train of action potentials, the frequency of which rapidly diminished. A second application of ATP evoked little response. However, in stretch-sensing neurons, ATP evoked a train of action potentials that did not adapt. Also, the nociceptive neurons expressed two types of current, a fast, transient, rapidly desensitizing current and a slower, persistent current. In contrast, the stretch-sensing neurons only responded to ATP with a slow, persistent current. The nociceptor currents were similar to those carried by cloned P2X3 receptors, suggesting a role for these subunits in nociception (62). This study illustrates a general principle that the molecular diversity of receptors underlies their functional specialization in the nervous system.

C. Recording from Native Receptors In Situ

A disadvantage of studying native receptors in dissociated or cultured cells is that the cells are divorced from their normal neuronal circuitry. Culturing of cells may also alter the nature of the receptors that are expressed. One approach that avoids these problems is to record from neurons within brain slices. Intracellular recording techniques have successfully been used for many years to record from brain slices 400–600 μm thick. A recent development (63) has been to use brain slices that are thin enough (200 μm) to allow visible or infrared light to pass through. This has enabled patch clamp recordings to be obtained from cells in their native circuitry and allows anatomical studies to be correlated with electrophysiological investigations. It is also possible to use outside-out patches to detect neurotransmitter release from synapses in these visualized neuronal circuits (64).

1. Postsynaptic Receptors

An example of the usefulness of the thin brain slice technique is shown in a recent study on native nicotinic receptors in the medial habenula (65). The medial habenula contains high-affinity binding sites for nicotine (66), and excitatory nicotinic responses have been recorded in this brain region using intracellular recording in the intact slice (67) and patch clamp techniques in dissociated cells (68–71). Which nicotinic receptor(s) mediate these responses is unclear as in situ hybridization studies show that numerous nicotinic receptor subunits are expressed (5,33,72,73). The neuronal $\alpha 7$, $\alpha 4$, $\beta 2$, and $\beta 3$ genes are transcribed throughout the rat medial habenula, but the $\alpha 3$ and $\beta 4$ genes are transcribed only in the ventral areas. Both $\alpha 3$ and $\alpha 4$, combined with either $\beta 2$ or $\beta 4$ subunits (but not $\beta 3$), have been functionally expressed (in pairs) in *Xenopus* oocytes, while $\alpha 7$ expresses as a homo-oligomer (73,74). Thus, in expression systems, several different functional combinations of nicotinic receptor were possible.

Patch clamp recordings in thin slices of the ventral medial habenula revealed a heterogeneity of neuronal nicotinic acetylcholine receptors (67). There was not a close match between the properties observed for these channels and those of cloned subunit combinations. However, the work did demonstrate that functional heterogeneity in expression studies on cloned subunit combinations is mirrored by functional heterogeneity of native receptors in the brain.

2. Presynaptic Receptors

Such molecular and functional diversity reflects the specialized functions that each receptor subtype must perform within the brain. Electrophysiological experiments on cells that have their synaptic connections intact can help identify these functions. For example, in whole-cell patch clamp recordings from CA3 pyramidal cells in hippocampal slices, application of low concentrations of nicotine onto the neuronal dendrites increased the frequency, but not the amplitude, of miniature excitatory postsynaptic currents (mEPSCs) (75). Further experiments on cultured hippocampal neurons in which functional synapses had reformed showed that the mEPSCs were blocked by the glutamate receptor antagonists CNQX and kynurenate. Thus, nicotine appeared to be acting presynaptically to facilitate glutamate release. The effects of nicotine were blockable by MLA and α-bungarotoxin, suggesting that the presynaptic receptor contained the $\alpha7$ subunit. The $\alpha7$ subunit is highly permeable to Ca^{2+}, and fluorescence measurements showed that nicotine enhanced Ca^{2+} influx into the presynaptic terminal. From these studies a model was proposed whereby activation of the presynaptic nicotinic receptors just before the arrival of the action potential in the nerve terminal would greatly increase the probability of glutamate release and so explain the increase in mEPSCs frequency.

Presynaptic facilitation of glutamate release involving the nicotinic $\alpha7$ subunit and Ca^{2+} influx has also been seen in synapses formed between chick habenular and interpeduncular nucleus neurons in culture (76). Nicotinic presynaptic receptors also facilitate release of ACh in the visceral motor nucleus of Ternii (77) and the lumbar sympathetic ganglion (78). These studies show how electrophysiology can be used to elucidate the physiological consequences of ligand-receptor interactions. Such consequences depend not only on the species of receptor, but also on their location within cells and neuronal circuits.

D. Extracellular Recording

In some instances it is desirable to record nerve cell activity without disturbing the cell membrane integrity. This can be achieved with extracellular recording. For example, this can be used to monitor the rate at which a cell fires action potentials. In the rat substantia nigra reticulata and ventral palladium, nicotine-induced increases in neuronal action potential firing rate were enhanced by alcohol (79). However, in the rat locus coeruleus, alcohol inhibited the excitatory effects of nicotine, kainate, and NMDA (80). Thus, the effects of ethanol on nicotinic receptors are not uniform in the brain, but may be subtype specific, or depend upon the intracellular signaling status of the cell studied.

Extracellular recording can also monitor field potentials from an entire group of neurons. In an electroencephalogram, electrodes placed on the surface of the scalp can detect transient coordinated potential differences as waves of activity of thousands of neurons sweep through pathways in the brain. If a control individual is presented with an unexpected auditory stimulus, the wave of activity in response to the stimulus peaks after about 50 msec (P50 wave). However, if the stimulus is repeated, the peak of the wave is diminished, or "gated." Schizophrenics and their first-degree relatives lack the ability to gate the P50 wave and have lower levels of expression of the neuronal nicotinic $\alpha7$ receptor. An animal model of this wave has been developed in which α-bungarotoxin, MLA, and $\alpha7$ antisense oligonucleotides can block gating, whereas nicotine can restore it (81,82). The gating deficit in schizophrenic patients can also be temporarily restored by smoking tobacco.

In this example, electrophysiological monitoring of the activity of a population of neurons has been used to address the contribution of a specific ligand-receptor interaction to a behavioral phenomenon. The study has also shown how a combination of electrophysiological and other techniques can lead to the discovery of a new therapeutic target and an explanation of a symptom in a disease process.

VII. LIGAND-RECEPTOR INTERACTIONS IN DISEASE STATES

A. Rasmussen's Disease

An unusual example of how studying a ligand-receptor interaction can lead to the understanding of a disease process occurred with Rasmussen's disease, a progressive and severe form of epilepsy that does not respond to conventional drug treatment. The discovery of the cause of the disease began with an experiment in rabbits in which antibodies were raised against fragments of the N-terminal of the GluR3 subunit (which contained the putative glutamate-binding domain). As the rabbits developed high titers of antibody, they also exhibited seizure-like behavior. The pathology of the rabbits' brains was similar to that seen post mortem in patients with Rasmussen's disease. Also, the blood of patients with Rasmussen's disease was found to have high titers of antibody against the GluR3 receptor and exchange plasmaphoresis brought about an immediate reduction in the severity of symptoms (83).

Subsequently, the effects of sera from rabbits and patients were tested on cultured fetal mouse cortical neurons using the whole-cell patch clamp technique. Both sera activated glutamate receptors present on the neurons (84). Thus, the antibody against the GluR3 receptor appears to act as an ever-present agonist of the excitatory glutamate receptor. In vivo, this would trigger a wave of neuronal activity that would result in an epileptic seizure. Interestingly, activation of the GluR3 receptor by the sera was blocked by a peptide that incorporated the amino acid residues 372–395 of the GluR3 receptor subunit, suggesting that this domain may contribute to the ligand binding site of the GluR3 receptor. This study introduced a new class of highly specific ligand-receptor interactions, explained a confounding illness, suggested a new way forward for the development of a therapeutic treatment, and provided an excellent model of a disease process.

B. Transgenic Animal Models

New models of neurological disorders are constantly being developed in transgenic animals. To fully interpret the results in such models, it is essential to know how synaptic transmission has been affected by the expression, or lack of it, of the transgene in the experimental animals. For example, long-term potentiation (LTP) has long been suggested as a model of learning and memory processes (85). A short burst of high-frequency stimulation can lead to a sustained potentiation of synaptic transmission in CA1 pyramidal cells. Conversely, prolonged low-frequency stimulation can cause a decrease in the synaptic response (long-term depression, LTD). In a transgenic model of Alzheimer's disease (86) LTD was normal, but LTP was diminished with respect to control. These mice overexpressed the 104 carboxy-terminus amino acids of the Alzheimer amyloid precursor protein and developed an increase in extracellular β-amyloid immunoreactivity and the associated inflammatory response as they aged. They exhibited spatial learning deficits in the Morris water maze and also had a reduced number of CA1 pyramidal cells, consistent with the observed decrease in LTP.

VIII. THE CHALLENGE AHEAD

The problem stated at the outset of this review was how electrophysiology can help in finding a therapeutic receptor target and a specific, appropriate ligand to modulate its activity. Some of these ways have been outlined and are summarized in Table 3. However, electrophysiology still has much more to contribute to drug discovery, especially when considering the vast number of putative receptors and ligands that the genome project presents to us for characterization. Much of this work will be carried out on isolated cells expressing cloned subunit combinations or identified native receptor subtypes. The awareness of all these different receptors is especially relevant

Table 3 Uses of Electrophysiology to Study Receptor-Ligand Interactions

Procedure	Information
Functional expression of cloned receptor subunits	Proof of functional identity of subunits
	Expression cloning of novel receptor subtypes
	Profiling of ligand selectivity of individual receptors
	Profile of receptor selectivity of individual ligands
Structure-function studies in cloned receptor subunits	Explaining developmental changes in receptor properties
	Determining functional relationship between different domains within subunits
	Determining structural relationships between different receptor gene families
	Defining amino acids responsible for pharmacological distinctions between subunits
	Modeling of ligand-binding sites for drug design
	Creating constructs for transgenic models of disease states
Studies on native receptors	Identifying subunit composition of native receptors
	Understanding physiological role of receptors
	Understanding complete action of therapeutic ligands
	Evaluating disease models
	Understanding disease states

to the determination of their specific tasks in the healthy and dysfunctional nervous systems. As their involvement becomes better understood, it then becomes possible to create more accurate transgenic models of disease states, which also have to be investigated, particularly in brain slices. The complexity of the situation is further complicated by the fact that the activity of ion channels can be modulated by intracellular signaling pathways. Thus, the demand for electrophysiological recording of ligand-receptor interactions seems destined to increase.

ACKNOWLEDGMENTS

This work was supported by grants from Astra Charnwood, the Medical Research Council, the Royal Society, Strathclyde University Research and Development Fund, and the Wellcome Trust. We are also grateful to Alasdair Gibb for helpful discussions and to the Scottish Hospital Endowments Research Trust for refurbishing the laboratories in which some of these experiments were performed.

REFERENCES

1. Alexander SPH, Peters JA. 1997 receptor and ion channel nomenclature supplement. Trends Pharmacol Sci 1997; 1–84.
2. Mishina M, Kurosaki T, Tobimatsu T, Morimoto Y, Noda M, Yamamoto T, Terao M, Lindstrom J, Takahashi T, Kuno M, Numa S. Expression of functional acetylcholine receptor from cloned cDNAs. Nature 1984; 307:604–608.
3. Boulter J, Evans K, Goldman D, Martin G, Treco D, Heinemann S, Patrick J. Isolation of a cDNA clone coding for a possible neural nicotinic acetylcholine receptor alpha-subunit. Nature 1986; 319:368–374.
4. Goldman D, Deneris E, Kochhar A., Patrick J, Heinemann S. Members of a nicotinic acetylcholine receptor gene family are expressed in different regions of the mammalian central nervous system. Cell 1987; 48:965–973.
5. Deneris E, Connolly J, Boulter J, Wada K, Wada E, Swanson L, Patrick J, Heinemann S. Identification of a cDNA coding for a subunit common to distinct acetylcholine receptors. Neuron 1988; 1:45–54.
6. Nef P, Onesyer C, Alliod C, Couturier S, Ballivet M. Genes expressed in the brain define three distinct neuronal nicotinic acetylcholine receptors. EMBO J 1988; 7:595–601.
7. Schoepfer R, Whiting P, Esch F, Blacher R, Shimasaki S, Lindstrom J. cDNA clones coding for the structural subunit of a chicken brain nicotinic acetylcholine receptor. Neuron 1988; 1:241–248.
8. Boulter J, Connolly JG, Deneris E, Goldman D, Heinemann S, Patrick J. Functional expression of two neuronal nicotinic acetylcholine receptors from cDNA clones identifies a gene family. Proc Natl Acad Sci USA 1987; 84:7763–7767.
9. Boulter J, O'Shea-Greenfield A, Duvoisin RM, Connolly JG, Wada E, Jensen A., McKinnon D, Ballivet M, Deneris ES, Heinemann S, Patrick J. Three members of the rat neuronal nicotinic acetylcholine receptor gene family form a gene cluster. J Biol Chem 1990; 265:4472–4482.
10. Couturier S, Erkmann L, Valera S, Rungger D, Bertrand S, Boulter J, Ballivet M, Betrand D. Alpha 5, alpha 3 and non-alpha3. Three clustered avian genes encoding neuronal nicotinic acetylcholine receptor-related subunits. J Biol Chem 1990; 265:17560–17567.
11. Raimondi E, Rubbiloi F, Moralli D, Chini B, Fornasari D, Tarroni P, De Carli L, Clementi F. Chromosomal localisation and physical linkage of the genes the human alpha 3, alpha 5 and beta 4 neuronal nicotinic receptor subunits. Genomics 1992; 12:849–850.
12. Schofield PR, Darlison MG, Fujita N, Burt DR, Stephenson FA, Rodriguez H, Rhee LM, Ramachandran J, Reale V, Glencorse TA., Seeburg PH, Barnard, EA. Sequence and functional expression of the $GABA_A$ receptor shows a ligand-gated receptor super-family. Nature 1987; 328:221–227.
13. Grenningloh G, Rienitz A., Schmitt B, Methfessel C, Zensen M, Beyreuther K, Gundelfinger ED, Betz H. The strychnine-binding subunit of the glycine receptor shows homology with nicotinic acetylcholine receptors. Nature 1987; 328:221–227.

14. Maricq AV, Peterson AS, Brake AJ, Myers RM, Julius D. Primary structure and functional expression of the 5HT3 receptor, a serotonin gated ion channel. Science 1991; 254:432–436.
15. Kyte J, Doolittle RF. A simple method for displaying the hydropathic character of a protein. J Mol Biol 1982; 157:105–132.
16. Browning MD, Rogers SW. Ligand-gated ion channels and functional regulation by phosphorylation. In: Sibley DR, Housley MD, eds. Regulation of Cellular Signal Transduction Pathways by Desensitization and Amplification. New York: Wiley, 1994:167–201.
17. Karlin A, Akabas MH. Toward a structural basis for the function of nicotinic acetylcholine receptors and their cousins. Neuron 1995; 15:1231–1244.
18. Moriyoshi K, Masayuki M, Ishii T, Shigemoto R, Mizuno N, Nakanishi S. Molecular cloning and characterisation of the rat NMDA receptor Nature 1991; 354:31–37.
19. Hollman M, O'Shea-Greenfield A, Rogers S, Heinemann S. Cloning by functional expression of a member of the glutamate receptor family. Nature 1989; 342:643–648.
20. Hollman M, Maron C, Heinemann S. N-Glycosylation site tagging suggests a three transmembrane domain topology for the glutamate receptor GluR1 subunit. Neuron 1994; 13:1331–1343.
21. Wyszynski M, Lin J, Rao A, Nigh E, Beggs A, Craig A-M, Sheng M. Competitive binding of α-actinin and calmodulin to the NMDA receptor. Nature 1997; 385:439–442.
22. Valera S, Hussy N, Evans RJ, Adami N, North RA, Suprenant AM, Buell G. A new class of ligand-gated ion channel defined by P_{2x} receptor for extracellular ATP. Nature 1994; 371:516–519.
23. Brake AJ, Wagenbach MJ, Julius D. New structural motif for ligand-gated ion channel defined by an ionotropic ATP receptor. Nature 1994; 371:519–523.
24. North RA. P2X purinoceptor plethora. Seminars in the Neurosciences 1996; 8:187–194.
25. Lingueglia E, Champigny G, Lazdunski M, Barby P. Cloning of the amiloride-sensitive FMRFamide peptide-gated sodium channel, Nature 1995; 378: 730–733.
26. Kaupp UB, Niidome T, Tanabe T, Terada S, Bonigk W, Stuhmer W, Cook N, Kangawa K, Matsuo H., Hirose T, Miyata T, Numa S. Primary structure and functional expression from complementary cDNA of the rod photoreceptor cyclic GMP-gated channel. Nature 1989; 342:762–766.
27. Ogden D, ed. Microelectrode Techniques: The Plymouth Workshop Handbook, 2nd ed. Cambridge: The Company of Biologists Limited, 1987.
28. Sakmann B, Neher E, eds. Single Channel Recording. New York: Plenum Press, 1983.
29. Wallis DI, ed. Electrophysiology: A Practical Approach. Oxford: Oxford University Press, 1993.
30. Neher E, Sakmann B. Single-channel currents recorded from membrane of denervated frog muscle fibres. Nature 1976; 260:799–802.

31. Hamill OP, Marty A, Neher E, Sakmann B. Improved patch clamp techniques for high resolution current recording from cells and cell free patches. Pflugers Arch 1981; 391:85–100.

32. Wada K, Ballivet M, Boulter J, Connolly J, Wada E, Deneris E, Swanson LW, Heinemann S, Patrick J. Primary structure and expression of beta-2, a novel subunit of neuronal nicotinic acetylcholine receptors. Science 1988; 240:330–334.

33. Duvoisin RM, Deneris ES, Patrick J, Heinemann SF. The functional diversity of neuronal nicotinic acetylcholine receptors is increased by a novel subunit b4. Neuron 1989; 3:589–596.

34. Leutje CW, Patrick J. Both α and β subunits contribute to the agonist sensitivity of neuronal nicotinic acetylcholine receptor subunit combinations. J Neurochem 1991; 55:632–640.

35. Hussy N, Ballivet M, Bertrand D. Agonist and antagonist effects of nicotine on chick neuronal nicotinic receptors are defined by a and b subunits. J Neurophysiol 1994; 72:1317–1326.

36. Covernton PJO, Kojima H, Sivilotti LG, Gibb AJ, Colquhoun D. Comparison of neuronal nicotinic receptors in rat sympathetic neurones with subunit pairs expressed in Xenopus oocytes. J Physiol 1994; 481:27–34.

37. Cachelin AB, Rust G. Unusual pharmacology of (+)-tubocurarine with rat neuronal nicotinic acetylcholine receptors containing b4 subunits. Mol Pharmacol 1994; 46:1168–1174.

38. Zwart R, Oortgiesen M, Vijverberg HPM. Differential modulation of a3b2 and a3b4 neuronal nicotinic receptors expressed in Xenopus oocytes by flufenamic acid and niflumic acid. J Neurosci 1995; 15(3):2168–2178.

39. Wafford KA, Bain CJ, Quirk K, McKernan RM, Wingrove PB, Whiting PJ, Kemp JA. A novel allosteric modulatory site on the GABA$_A$ receptor b subunit Neuron 1994; 12:775–782.

40. Ramirez-Latorre J, Yu CR, Perin F, Karlin A, Role L. Functional contributions of a5 subunit to neuronal acetylcholine receptor channels. Nature 1996; 380:347–351.

41. Leutje CW, Piattoni M, Patrick J. Mapping of ligand-binding sites of neuronal nicotinic acetylcholine receptor subunits using chimeric α subunits. Mol Pharmacol 1993; 44:657–666.

42. Stern-Bach Y, Bettler B, Hartley M, Sheppard PO, O'Hara PJ, Heinemann SF. Agonist selectivity of glutamate receptors is specified by two domains structurally related to bacterial amino acid-binding proteins. Neuron 1994; 13:1345–1357.

43. Eisele J-L, Bertrand S, Galzi J-L, Devillers-Thiery A, Changeux J-P, Bertrand D. Chimeric nicotinic serotonergic receptor combines distinct ligand-binding and channel specificities. Nature 1993; 366:479–483.

44. Bertrand D, Devillers-Thiery A, Revah F, Galzi J-L, Hussy N, Mulle C, Bertrand S, Mallivet M, Changeux J-P. Unconventional pharmacology of a neuronal nicotinic receptor mutated in the channel domain. Proc Natl Acad Sci USA 1992; 89:1261–1265.

45. Johnson JW, Ascher P. Glycine potentiates the NMDA response in cultured mouse brain neurones. Nature 1987; 325:529–531.

46. Kuryatov A, Laube B, Betz H, Kuhse J. Mutational analysis of the glycine-binding site of the NMDA receptor: structural similarity with bacterial amino acid-binding proteins. Neuron 1994; 12:1291–1300.

47. Korpi ER, Kleingoor C, Helmut K, Seeburg PH. Benzodiazepine-induced motor impairment linked to point mutation in cerebellar GABA$_A$ receptor. Nature 1993; 361:356–359.

48. ffrench-Constant RH, Rocheleau TA, Steichen JC, Chalmers AE. A point mutation in a *Drosophila* GABA receptor confers insecticide resistance. Nature 1993; 363:449–451.

49. Kuhse J, Schmieden V, Betz H. A single amino acid exchange alters the pharmacology of neonatal rat glycine receptor subunit. Neuron 1990; 5:867–873.

50. Pillet L, Tremeau O, Ducancel F, Drevet P, Zinn-Justin S, Pinkasfield S, Boulain JC, Menez A. Genetic engineering of snake toxins. Role of invariant residues in the structural and functional properties of a curaremimetic toxin, as probed by site-directed mutagenesis. J Biol Chem 1993; 268:909–916.

51. Connolly JG, Boulter J, Heinemann S. a4-2β2 and other nicotinic acetylcholine receptor subtypes as targets of psychoactive and addictive drugs. Br J Pharmacol 1992; 105:657–666.

52. Bertrand D, Ballivet M, Gomez M, Bertrand S, Phannavong B, Gundelfinger ED. Physiological properties of neuronal nicotinic receptors reconstituted from the vertebrate β2 subunit and *Drosophila* α subunits. Eur J Neurosci 1994; 6:869–875.

53. Gibb AJ, Kojima H, Carr JA, Colquhoun D. Expression of cloned receptor subunits produces multiple receptors. Proc R Soc Lond B 1990; 242:108–112.

54. Sine SM, Steinbach JH. Activation of acetylcholine receptors on clonal mammalian BC3H1 cells by high concentrations of agonist. J Physiol 1987; 385:325–359.

55. Anand R, Peng X, Lindstrom J. Homomeric and native α7 acetylcholine receptors exhibit remarkably similar but non-identical pharmacological properties, suggesting that the native receptor is a heteromeric protein complex. Fed Eur Biochem Soc 1993; 327(2):241–246.

56. Sivilotti LG, McNeil DK, Lewis TM, Nassar MA, Schoepfer R, Colquhoun D. Recombinant nicotinic receptors, expressed in *Xenopus* oocytes, do not resemble native rat sympathetic ganglion receptors in single-channel behaviour. J Physiol 1997; 500:123–138.

57. Evans, RJ, Kennedy C. Characterisation of P2-purinoceptors in the smooth muscle of the rat tail artery: a comparison between contractile and electrophysiological responses. Br J Pharmacol 1993; 113:853–860.

58. Kennedy C, Leff P. How should P2x-purinoceptors be characterised pharmacologically? Trends Pharmacol Sci 1995; 16:168–174.

59. Drasdo A, Caulfield M, Bertrand D, Bertrand S, Wonnacott S. Methyllycaconitine: a novel nicotinic antagonist. Mol Cell Neurosci 1992; 3(3):237–243.

60. Alkondon M, Pereira EFR, Wonnacott S, Albuquerque EX. Blockade of nicotinic currents in hippocampal neurons defines methyllcaconitine as a potent and specific receptor antagonist. Mol Pharmacol 1992; 41:802–808.
61. Cook SP, Vulchanova L, Hargreaves KM, Elde R, McCleskey EW. Distinct ATP receptors on pain-sensing and stretch-sensing neurons. Nature 1997; 387:505–508.
62. Kennedy C, Leff P. Painful connection for ATP. Nature 1995; 377:385–386.
63. Edwards F.A., Konnerth A., Sakmann B, Takahashi T. A thin slice preparation for patch clamp recordings from synaptically connected neurones of the mammalian central nervous system. Pflugers Arch 1989; 414:600–612.
64. Allen TGJ. The sniffer-patch technique for detection of neurotransmitter release. Trends Neurosci 1997; 20:192–198.
65. Connolly JG, Gibb A, Colquhoun D. Multiple subtypes of large conductance neuronal nicotinic acetylcholine receptors studied in thin slices of rat medial habenula. J Physiol 1995; 484(1):87–105.
66. Clarke PBS, Schwarz RD, Paul SM, Pert CB, Pert A. Nicotinic binding in the rat brain. Autoradiographic comparison of [^{3}H]-acetylcholine, [^{3}H]-nicotine and [^{125}I]-alpha-bungarotoxin. J Neurosci 1985; 5:1307–1315.
67. McCormick DA, Prince DA. Acetylcholine causes rapid nicotinic excitation in the medial habenular nucleus of guinea pig, in vitro. J Neurosci 1987; 7:742–752.
68. Mulle C, Changeux J.-P. A novel type of nicotinic receptor characterised in the rat central nervous system by patch clamp techniques. J Neurosci 1990; 10:169–175.
69. Mulle C., Vidal C., Benoit P, Changeux J-P. Existence of different subtypes of nicotinic acetylcholine receptors in the rat habenulo-interpeduncular system. J Neurosci 1991; 11:2588–2597.
70. Mulle C, Choquet D, Korn H, Changeux J-P. Calcium influx through nicotinic receptor rat central neurons: its relevance to cellular regulation. Neuron 1992; 8:135–143.
71. Mulle C., Lena C, Changeux J-P. Potentiation of nicotinic receptor response by external calcium in rat central neurons. Neuron 1992; 8:937–945.
72. Wada E, Wada K, Boulter J, Deneris E, Heinemann S, Patrick J, Swanson LW. The distribution of alpha2, alpha3, alpha4 and beta2 neuronal nicotinic acetylcholine receptor subunit mRNAs in the central nervous system: a hybridisation histochemical study in the rat. J Comp Neurol 1988; 284:314–335.
73. Seguela P, Wadiche J, Dinely-Miller K, Dani JA, Patrick JW. Molecular cloning, function properties, and distribution of rat brain $\alpha 7$: a nicotinic cation channel highly permeable to Ca^{++}. J Neurosci 1993; 13:596–604.
74. Deneris ES, Connolly JG, Rogers SW, Duvoisin R. Pharmacological and functional diversity of neuronal nicotinic acetylcholine receptors. Trends Pharmacol Sci 1991; 12:32–40.
75. Gray R, Rajamn AS, Radcliffe KA, Yakehiro M, Dani JA. Hippocampal synaptic transmission enhanced by low concentrations of nicotine. Nature 1996; 383:713–716.
76. McGehee DS, Heath MJS, Gelber S, Devay P, Role LW. Nicotine enhancement of fast excitatory synaptic transmission in CNS by presynaptic receptors. Science 1995; 269:1692–1696.

77. Ullian EM, Sargent P. Pronounced cellular diversity and extrasynaptic location of nicotinic acetylcholine receptor subunit immunoreactivities in the chicken pretectum. J Neurosci 1995; 15:7012–7023.
78. Zhang Z-W, Coggan JS, Berg DK. Synaptic currents generated by neuronal acetylcholine receptors sensitive to α-bungarotoxin. Neuron 1996; 17:1231–1240.
79. Criswell HE, Simson PE, Duncan GE, McCown TJ, Herbert JS, Morrow AL, Breese GL. Molecular basis for regionally specific action of ethanol on gamma-amino butyric acid (A) receptors: generalisation to other ligand-gated ion channels. J Pharmacol Exp Ther 1993; 267:552–537.
80. Frolich R, Patzelt R, Illes P. Inhibition by ethanol of excitatory amino acid receptors and nicotinic acetylcholine receptors in rat locus coeruleus neurons. Naun-Schmied Arch Pharmacol 1994; 350:626–631.
81. Leonard S, Adams C, Breese CR, Adler LE, Bickfort P, Byerley HC, Griffith JM, Miller C, Myles-Worsley M, Nagamoto T, Collins Y, Stevens KE, Waldo M, Freedman R. Nicotinic receptor function in schizophrenia. Schizophrenia Bull 1996; 22:431–445.
82. Freedman R, Coon H, Myles Worsley M, OrrUrtreger A, Olincy A, Davis A, Polymeropoulos M, Holik J, Hopkins J, Hoff M, Rosenthal J, Waldo MC, Reimherr F, Wender P, Yaw J, Young DA, Breese CR, Adams C, Patterson D, Adler LE, Kruglyak L, Leonard S, Byerley W. Linkage of a neurophysiological deficit in schizophrenia to a chromosome 15 locus. Proc Natl Acad Sci USA 1997; 94:587–592.
83. Rogers SW, Andrews PI, Gahring LC, Whisenand T, Cauley K, Crain B, Hughes TE, Heinemann SF, McNamara JO. Autoantibodies to glutamate receptor GluR3 in Rasmussen's encephalitis. Science 1994; 265:648–651.
84. Twyman RE, Gahring LC, Spiess J, Rogers SW. Glutamate receptor antibodies activate a subset of receptors and reveal an agonist binding site. Neuron 1995; 14:755–762.
85. Nalbantoglu J, Tirado-Santiago G, Lahsaini A, Poirier J, Goncalves O, Verge G, Momoli F, Weiner SA, Massicotte G, Julien J-P, Shapiro, ML. Impaired learning and LTP in mice expressing the carboxy terminus of the Alzheimer amyloid precursor protein. Nature 1997; 387:500–505.
86. Bliss TVP, Collingridge GLA. A synaptic model of memory: long term potentiation in the hippocampus. Nature 1993; 361:31–39.

6

Selective Adrenergic and Glucocorticoid Treatments for Asthma

David Jack
Glaxo Holdings plc, Wheathampstead, Hertfordshire, England

I. INTRODUCTION

A. Objectives of the Chapter

This chapter gives a brief account of how selectively acting β-adrenoceptor agonists and glucocorticoid steroids were developed in Glaxo-Allenburys Laboratories for use by inhalation in asthma and of their therapeutic use. It is written in tribute to the colleagues and clinical investigators whose efforts have transformed the treatment of asthma during the past 30 years and in the hope that it may lead to better understanding and use of these drugs. The views expressed are, however, personal and may not accord with those of former colleagues or with current medical opinion.

B. The Therapeutic Problem

Asthma is characterized by variable obstruction of the airways that is a consequence of inflammation in the lungs. There are two distinct causes of airways obstruction. The most obvious is contraction of bronchial muscle, which responds rapidly to epinephrine and other β-adrenergic bronchodilators. The most dangerous is physical occlusion of the airways caused by osmotic swelling and other derangements of bronchial mucosal structure and function. It is not relieved by bronchodilators and is reliably prevented or reversed only by cortisol and other glucocorticoid steroids. Bronchodilators are effective in less severe asthma because bronchoconstriction is the dominant cause of airway obstruction at this stage. As asthma worsens,

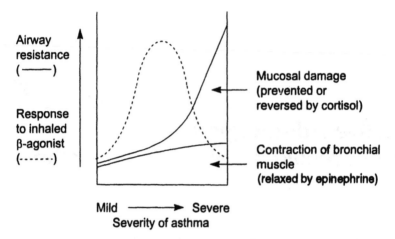

Figure 1 Causes of airway obstruction in asthma, effects of epinephrine and cortisol on them, and relationship between the severity of asthma and the bronchodilating effect of inhaled β-adrenergic bronchodilators.

however, they become progressively less effective because, with increasing physical occlusion, they are unable to create open airways. A waning response to bronchodilator treatment is, therefore, a prime indicator of worsening asthma and of the need for anti-inflammatory steroid treatment. These relationships are illustrated in Figure 1, which is considerably based on Hume and Gandevia's observations (1).

If this analysis was sound, the rational research objectives were new medicines with only the desirable effects of epinephrine and cortisol selectively in the lungs. The effects of such agents were expected to be complementary and, if exerted concurrently, to provide an unusually effective way to treat asthma by controlling inflammation and its consequences in the lungs of patients. This chapter gives a brief account of how appropriately selective β-agonists and glucocorticoids for use by inhalation were discovered and an explanation of why they should be used concurrently for best results in asthma.

II. THE EVOLUTION OF HIGHLY SELECTIVE β₂-ADRENOCEPTOR AGONISTS FROM EPINEPHRINE

A. Why β-Agonism Is a Desirable Bronchodilating Mechanism

Apart from acute bronchoconstriction that occurs during attacks of asthma, basal bronchial tone is greatly enhanced in all asthmatics. This condition is undesirable because, by impeding inspiration, it is a prime cause of the hyperresponsiveness (2) that is a cardinal feature of the disease.

The cause of enhanced bronchial tone may be deduced from the response of patients to β-blockade and the attenuating action of inhaled antimuscarinic agents on it. β-Blockade causes marked bronchoconstriction, even in patients who appear well, thus revealing both an inhibitory β-adrenergic action on bronchial muscle and an ongoing inflammatory process that generates endogenous spasmogens. Inhaled ipratropium greatly attenuates the bronchoconstricting effect of β-blockade, so the enhanced bronchial tone is largely the result of muscarinic cholinergic stimulation and β-adrenergic inhibition of muscle contraction. As shown in Figure 2, the cause of the former is enhanced vagal activity and of the latter, high circulating levels of epinephrine especially during the day (3). Falling epinephrine levels during the night are accompanied by bronchoconstriction, which is relieved by inhalation of a β-agonist. Epinephrine, by virtue of its β-adrenergic action, is, therefore, a natural protective hormone in asthma to which there is no evidence of tolerance; β-blockade still causes bronchoconstriction in long-standing disease.

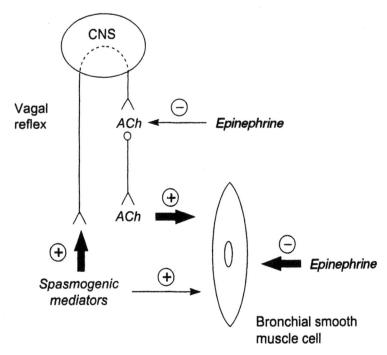

Figure 2 Origins of bronchoconstriction and enhanced bronchial tone in asthma.

The main endogenous spasmogens in asthma are leukotrienes C_4 and D_4 and histamine, which has a lesser effect. The other candidates, prostaglandin D_2 (which acts at thromboxane A_2 receptors) and the platelet-activating factor (PAF), make much smaller contributions because specific antagonists of them have little, if any, acute bronchodilating activity. These observations and the finding that the intensity of the bronchodilating effect of zafirlukast, a potent specific antagonist of LTC_4 and D_4, is only about 60% of that of inhaled salbutamol, 200 μg (4), indicate that another potent spasmogen [possibly bradykinin (5)] may be involved in asthma.

The cellular origins of the endogenous spasmogens are indicated in Table 1. Their prime action, and that of exogenous spasmogens such as atmospheric pollutants and cold air, is to activate vagal reflexes and thus release of acetylcholine, which activates muscarinic receptors on bronchial muscle cells. Ipratropium is a bronchodilator because it blocks these receptors but its maximum effect is only about 80% of that of β-adrenergic bronchodilators (6). The reason for the difference is that the endogenous spasmogens have additional lesser direct spasmogenic actions on bronchial muscle that are unaffected by muscarinic blockade but are sensitive to β-agonists. Accordingly, continuous augmentation of the action of endogenous epinephrine by an exogenous β-agonist is a desirable rational therapeutic objective in asthma because of the superior bronchodilating effect of β-agonism and of the resultant decreases in bronchial tone and hyperresponsiveness (2).

B. Ahlquist's Classification of Adrenoceptors and Isoprenaline

Modern drug research on asthma remedies began in 1948 when Ahlquist (7) subdivided adrenoceptors in α- and β-types on the basis of the relative activities of a few close analogs of epinephrine in a range of pharmacological tests. His evidence for the new classification is summarized in Table 2. Both

Table 1 Cellular Origins of the Principal Spasmogenic Mediators in Asthma

Spasmogen	Mast cell	Basophil	Eosinophil	Macrophage
Leukotriene C_4	+	+	+	
Leukotriene D_4[a]	−	−	−	−
Histamine	+	+		
Prostaglandin D_2	+			+
Platelet-activating factor	+		+	+

[a] Leukotriene D_4 is formed by extracellular enzymic hydrolysis of leukotriene C_4.

Table 2 Summary of Ahlquist's Evidence for the Classification of Adrenoceptors into α and β Types

Type of receptor	Biological responses	Order of potency of catecholamines	Effect of dibenamine
α-Adrenoceptor	Contraction of smooth muscle in the arteries of the skin and viscera, and in the uterus, ureter iris, and nictitating membrane; relaxation of smooth muscle in the gut	Epinephrine > norepinephrine > α-methylnorepinephrine > α-methylepinephrine > isoprenaline	Blockade
β-Adrenoceptor	Relaxation of smooth muscle in the bronchi, uterus, and in the arteries supplying the heart and skeletal muscles Stimulation of the heart	Isoprenaline > epinephrine > α-methylepinephrine > α-methylnorepinephrine > norepinephrine	No blockade

the desired bronchodilating action of epinephrine and the unwanted cardiac stimulant activity are mediated by β-receptors.

Because of its selectivity for β-receptors isoprenaline steadily replaced epinephrine as an inhaled bronchodilator during the 1950s. Therapeutic doses of the drug induce prompt intense bronchodilation, which lasts for about 1–1.5 hr and is accompanied by obvious cardiac stimulation. Its parenteral use is precluded by its intense inotropic and chronotropic actions on the heart. A longer-acting, more selectively acting bronchodilator was needed to replace isoprenaline, and around 1960, different approaches were made to the problem in the laboratories of Winthrop in the United States and of Boehringer-Ingelheim in Germany.

The profound significance of Ahlquist's work for drug research was not recognized at the time. Indeed, even now it is not fully realized that, apart from clarifying adrenergic physiology and pharmacology, his use of close analogs of a physiological mediator as biological probes provided the first general method for detecting chemical differences between the receptor proteins for that mediator in different kinds of cells. Such differences are, of course, an absolute condition for selective activity by a generally distributed drug. Not surprisingly, Ahlquist's approach has been used to classify receptors for other mediators and has contributed to the discovery of many important medicines.

C. Lands' Classification of β-Adrenoceptors and Selective β_2-Adrenergic Bronchodilatation

Lands and his colleagues in Winthrop extended Ahlquist's work to catecholamines with larger nonpolar N-substituents and found some of them to be about 10–20 times more active on bronchial than on heart muscle. Typical results for selected catecholamines are given in Table 4. These and similar results for other responses led them to propose the subdivision of β-receptors into β_1 and β_2 subtypes (8,9). This classification is summarized in Table 3.

The important finding for asthma research was that the desired bronchodilating action is mediated by β_2-receptors and unwanted cardiac stimulation by β_1-receptors, so that they were separable. The new classification also gave an accurate forecast of the main side effects found with systemic use of β_2-agonists, namely tremor of skeletal muscles and hypokalemia, which are caused by accelerated repolarization and hyperpolarization of the muscle cells, and tachycardia, which is secondary to dilatation of arteries in skeletal muscle beds.

The practical outcome of Lands' work was isoetharine, the first β_2-selective agonist to be used to treat asthma. Its bronchodilating action after

Table 3 Lands' Classification of β-Adrenoceptors in Mammalian Tissues

Responses mediated by β_1-adrenoceptors	Responses mediated by β_2-adrenoceptors
Increased force and rate of contraction of cardiac muscle	Relaxation of smooth muscle in bronchi, uterus, and arteries that supply skeletal muscles
Relaxation of smooth muscle in the alimentary tract	Decreased twitch tension in skeletal muscle
Lipolysis	Glycolysis
	Glycogenolysis

inhalation is essentially free of side effects, but since it too is a catechol, it is not much longer-acting than isoprenaline. Nevertheless, its β_2-selectivity made it a substantial advance on isoprenaline.

D. Orciprenaline and Prolonged β-Adrenergic Bronchodilatation

The Boehringer group achieved relatively prolonged bronchodilatation with orciprenaline, the resorcinol analog of isoprenaline, because it is not a substrate for the catechol-specific mechanisms that inactivate catecholamines. However, like isoprenaline, orciprenaline activates all β-receptors. By inhalation, its bronchodilating action persists for 3–4 hr, mainly because it is slowly absorbed from the bronchi and, for the same reason, its cardiovascular effects are less intense than those of isoprenaline. It was, therefore, a clear advance on isoprenaline by this route. Unlike isoprenaline, orciprenaline is well absorbed from the gut but its bronchodilating action is accompanied by use-limiting cardiac stimulation.

E. Salbutamol and Highly Selective β_2-Adrenergic Bronchodilatation

Our starting objective in Glaxo Allenburys in 1963 was simply a long-acting β-adrenergic bronchodilator to replace isoprenaline, and work soon centered on Lunts' proposal (10) to make noncatechol analogs of isoprenaline since they might, like orciprenaline, be longer acting. In 1966, however, our understanding of what was achievable was transformed when AH3021, the saligenin analog of isoprenaline, was found to be hundreds of times more active on bronchial than on heart muscle and AH3365 (salbutamol), its t-butyl homolog, proved to be more potent and even more selectively active. These and other relevant results for selected catecholamines and their saligenin analogs are summarized in Table 4. The great selectivity of

Table 4 Relative Activities and Affinities of Selected Catecholamines and Their Saligenin Analogs at Preparations Containing β1- or β2-Adrenoceptors

Catechols R^1 = OH
Saligenins R^1 = CH_2OH

HO, $R1$ — [ring] — CH(OH)—CH_2—NH—$R2$

Compound	R_2	Relative activity Equipotent concentration: (−)-isoprenaline = 1		Relative affinity (−)-isoprenaline = 1	
		Guinea pig tracheal muscle[a] (β2-adrenoceptors)	Guinea pig left atria[b] (β1-adrenoceptors)	Rat lung membrane[c] (β2-adrenoceptors)	Rat left atrial membrane[c] (β1-adrenoceptors)
Catechols					
(−) Norepinephrine	H−	56	10	80	14
(−) Epinephrine	CH_3−	5	12	6	10
(−) Isoprenaline	$(CH_3)_2CH$−	1 (pEC_{50} = 7.7)[d]	1 (pEC_{50} = 7.9)[d]	1 (pK_1 = 6.71)[a]	1 (pK_1 = 6.52)[a]
(±)t-Butylnorepinephrine	$(CH_3)_3C$−	0.4	3.6		
(±)p-Hydroxyphenyl-isopropylnorepinephrine	HO—[ring]—CH_2—CH(CH_3)−	0.1	0.5		
Saligenins					
(±) AH3364	H	580	10,000		
(±) AH4053	CH_3−	430	3500		
(±) AH3021	$(CH_3)_2CH$−	6	>1000[e]		
(±) Salbutamol (AH3365)	$(CH_3)_3C$−	3.7	>2000[e]	10	26
(±) AH4553	HO—[ring]—CH_2—CH(CH_3)−	1.3	>2000[e]		

[a] Inhibition of induced tone (11).
[b] Electrically stimulated inotropic response (12).
[c] Comparative binding vs 1^{125} -(−) iodopindolol.
[d] pEC_{50} and pK_1 are the negative logarithms of the concentration producing half-maximal response and of the equilibrium constant, respectively.
[e] Partial agonist.

action of the saligenin derivatives, which was quite unexpected, was strong evidence in favor of Lands' subdivision of β-receptors into β_1 and β_2 subtypes and his classification was generally accepted from about 1970 onward.

The cause of the highly selective action of salbutamol at β_2-receptors is obvious from the data in Table 4. The ratio of its potencies at β_2- and β_1-receptors is at least 500 whereas the corresponding ratio for its affinities at these receptors is 3 at most. Accordingly, the main cause of its selectivity is greater efficacy at β_2-receptors. The same is probably true for other highly β_2-selective agonists because the highest β_2/β_1 affinity ratio reported is only about 25 (13).

The data in Table 4 also show that more potent agonists than epinephrine at β_2-receptors are obtained simply by increasing the size of the nonpolar N-subsitutent. The reason for this is that the fit of the larger molecules within the β_2-receptor protein involves additional interactions and, therefore, increasing affinities of binding. In consequence, the resultant agonist receptor protein complexes are more stable and process more substrate G_s-protein molecules than epinephrine during their lifetimes. The cause of the parallel increases in potency in the saligenin series is the same but they are less active than the corresponding catecholamines because their efficacies are lower. These and other theoretical considerations are dealt with later.

Salbutamol, the first highly selective β_2-adrenergic bronchodilator, came to be widely used after its introduction in 1969. It is at its best by inhalation because its intrinsic β_2-selectivity is reinforced by local application within the bronchi. Near-maximal bronchodilation is achieved within minutes with 200-μg doses with no significant side effects. The duration of action is about 3–4 hr. The bronchodilating effect of a 4-mg oral dose is less intense, its onset of action is slower, and its duration no longer. Larger oral doses are not well tolerated because of generalized activation of β_2-receptors in the body. Tremor of skeletal muscles and tachycardia are the most obvious side effects.

Fortunately the side effects of salbutamol are not intrinsically dangerous, since, as shown in Table 5, its actions on the cardiovascular system in humans are significantly different from those of isoprenaline (14). The most important difference is that salbutamol lacks the biochemically wasteful inotropic action of epinephrine, so increased cardiac output is achieved without disproportionate oxygen consumption. The increase in heart rate with salbutamol is mainly of reflex origin being secondary to the decreased peripheral resistance that follows dilatation of arteries in the skeletal muscle beds.

Apart from its bronchodilating action, salbutamol inhibits antigen-induced activation of IgE-sensitized guinea pig and human mast cells (15),

Table 5 Effects of Intravenous Isoprenaline and Salbutamol in Patients with Mitral Valve Disease

Response measured	Isoprenaline (μg/kg)		
	Control	0.05	0.1
Heart rate (beats/min)	70.0	+8.0**	+10.6**
Cardiac output (L/min)	2.8	+0.7**	+1.1**
Aortic pressure (mmHg)	98.0	−9.5*	−10.0
Mean ejection rate (ml/sec)	135.0	+42.0**	+62.0**
Oxygen uptake (ml/min)	170.0	+26.7**	+47.5**

	Salbutamol (μg/kg)			
	Control	0.4	1	2
Heart rate (beats/min)	78.0	+5.7*	+17.5**	+25.5**
Cardiac output (L/min)	3.0	+0.2	+0.6**	+1.1**
Aortic pressure (mmHg)	102.0	+3.0	−6.7*	−16.0*
Mean ejection rate (ml/sec)	160.0	+1.0	+4.0	+8.0
Oxygen uptake (ml/min)	185.0	0	+11.0	+21.0*

*$p < 0.05$; **$p < 0.01$.
Source: From Ref. 14.

and such an action would be expected to be beneficial in asthma if it were maintained by repeated doses. The clinical data in Section IV show that inhaled salbutamol, 200 μg four times daily, provides effective β_2-agonism during the day but that it is too short-acting to prevent worsening of asthma during the night. The drug should, therefore, be regarded primarily as a bronchodilator.

F.   Salmeterol and Continuous β_2-Agonism in the Lungs

1.   The Rationale for Salmeterol

Despite considerable effort during the 1970s the Glaxo-Allenburys team failed to discover an orally active β-adrenergic bronchodilator without use-limiting side effects or to solve the problem of selective bronchodilatation with a systemic agent by other means. In 1981, therefore, I had to think about the problem again and concluded that the desired drug would have to be given by inhalation if side effects were to be avoided and would have to be much longer acting than salbutamol and similar β_2-agonists to prevent nocturnal attacks of asthma. Such a drug would also be expected to inhibit mast cell activation and, as found later, the extravasation of plasma proteins into the bronchial mucosa that follows contraction of endothelial cells in the postcapillary venules.

A possibility that occurred to me in 1991 was that the required prolongation of action might be achieved with an inhaled β_2-agonist if it formed a stable active complex with the β_2-receptor protein (16). Its duration of action would have to be at least 8 hr to prevent nocturnal asthma and for convenient chronic dosage. The general hypothesis is illustrated and explained in Figure 3.

The extensive literature on desensitization of β-receptors exposed to β-agonists and reports of tolerance to β-adrenergic bronchodilators in patients made some of my colleagues doubtful about this approach. Clearly, if the prevailing expert opinion was right, similar desensitization would occur even more rapidly with the new kind of agonist I envisaged. I have already explained why I believe that a waning response to regular inhalations of a β_2-agonist indicates worsening asthma and not desensitization of β_2-receptors in the lungs. More fundamentally, however, I found it hard to believe that ready desensitization would occur to any hormone whose physiological role requires its continuous presence in the extracellular fluid, and especially to epinephrine and glucocorticoid steroids because of their physiological roles in stress. Indeed, ready desensitization of their receptors would be a catastrophic design fault in mammalian biology! I had already characterized the actions of these hormones as type 1 agonism in which the affected cell is capable of responding continuously to a continuing stimulus. Its essential characteristics are listed in Table 6. [Type 2 agonism

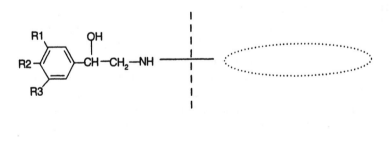

A   B

Figure 3 A hypothesis for a long-acting β-adrenoceptor protein complex. (A) A β-active phenylethanolamine head preferably with high affinity for the epinephrine-binding area in the β-receptor protein. (B) A large, flexible, nonpolar side chain to ensure β_2-selectivity and to anchor the agonist by interacting with nonpolar areas of the receptor protein accessible to the epinephrine-binding site.

Table 6   Essential Characteristics of Type 1 Agonism

The affected cell is capable of responding continuously to a continuing stimulus.
Pharmacological tolerance does not occur with stimuli of physiological
    dimensions or, as a rule, with tolerated doses in humans.
Increasing efficacy and increasing receptor affinity enhance the potency of
    agonists.
Prolonged agonism is achievable by formation of stable agonist-receptor protein
    complexes.
Antagonists form inactive receptor-protein complexes and their potency is
    determined solely by their affinities for their receptors.

is different in that the affected cell, having responded to an effective stimulus, must recover before it can respond to another (16–18).] The only way to resolve the question was to find a truly long-acting β_2-agonist and test it.

Lunts, who was responsible for the chemistry program, decided to test the hypothesis by making appropriately N-substituted analogs of salbutamol. Their rates of onset and offset of action were measured using superfused, electrically stimulated guinea pig tracheal muscle in which the contractions elicited are of cholinergic origin (12). Interest in the project rose sharply when it was found that this preparation is not easily desensitized by continuous superfusion with isoprenaline or other β-agonists and that the rates of onset and offset of action of β-agonists vary considerably. Eventually, by painstaking optimization of the N-substituent in the saligenin series, salmeterol emerged (19).

2.   The Pharmacology of Salmeterol

Like salbutamol, salmeterol is a highly selective β_2-agonist but the kinetics of its action are quite different (16,19,20). As shown in Figure 4, its onset of action on electrically stimulated guinea pig tracheal muscle is slower than that of isoprenaline or salbutamol and it is much longer-acting. Indeed, its inhibitory effect persists for hours even during superfusion with drug-free Krebs' solution and, under the same conditions, submaximal responses to the drug are stable and similarly persistent. This outcome was, of course, gratifying but hardly suprising since that is what was intended. That its action was found to be reversibly antagonized by sotalol and other β-blockers was, however, quite unexpected and needed to be explained.

The results of more recent studies of the effects of isoprenaline, salbutamol, formoterol, and salmeterol on spontaneous tone in superfused human bronchial muscle are summarized in Table 7 (22,23). Again the action of salmeterol was slow in onset, persistent, and reversible by ICI 118551, a selective β_2-blocker. Isoprenaline and salbutamol also behaved in the ex-

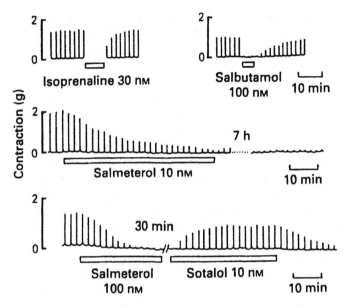

Figure 4 Effects of isoprenaline, salbutamol, and salmeterol on superfused, electrically stimulated, isolated guinea pig trachea and of sotalol on the response to salmeterol. (From Ref. 14.)

pected ways, but the rapid recovery with formoterol was suprising because it is known to be long-acting after inhalation by asthmatic patients, another result that requires explanation. Details of the mechanism of action of salmeterol and other β-agonists are dealt with later.

The persistent action of salmeterol on respiratory muscle is also evident in whole animals and humans. In the guinea pig, for example, inhaled

Table 7 Relative Potencies and Rates of Onset and Offset of Action of Isoprenaline, Salbutamol, Formoterol, and Salmeterol on Spontaneous Tone of Isolated Superfused Human Bronchial Muscle

Agonist	Equieffective concentration $(-)$-isoprenaline = 1	Onset time (Ot_{50} min)	Offset time (Ot_{50} min)
$(-)$ Isoprenaline	1 (EC_{50} = 24.4 nM)	1.5	2.2
$(\pm)$ Salbutamol	9.3	3.3	6.8
$(\pm)$ Formoterol	0.08	5.3	6.6
$(\pm)$ Salmeterol	0.15	35	>275

salmeterol inhibits histamine-induced bronchoconstriction for much longer than salbutamol. More importantly, 50- or 100-μg single inhaled doses of the drug induce near-maximal bronchodilatation in asthmatic patients, which is greatest after 1–2 hr and still marked after 12 hr, without causing side effects other than tremor with the higher dose (24). Furthermore, when inhalations are repeated at 12-hr intervals, effective bronchodilatation is maintained both night and day after prolonged use of the drug.

The inhibitory actions of salmeterol on mast cells and on endothelial cells may also be important for patients because they are potentially anti-inflammatory. The action of salmeterol or human mast cells is evident in its inhibition of release of histamine, leukotrienes C_4 and D_4, and prostaglandin D_2 from fragments of IgE-sensitized human lung after challenge by allergen. Unlike the corresponding inhibition by isoprenaline, salbutamol, or formoterol, that of salmeterol cannot be reversed by washing out the agonist, so this inhibitory action is similar to that on bronchial muscle (25). The prolonged relaxant action of salmeterol on endothelial cells is manifested by its ability to inhibit exudation of plasma proteins into the airways of guinea pigs exposed to a histamine aerosol, and to inhibit infiltration of eosinophils and neutrophils into the airways induced by PAF and lipopolysacharide, respectively, in the same species (26).

These results and those obtained in early clinical pharmacological and trial studies fostered hope that salmeterol would prove to have a useful anti-inflammatory action in patients, different from and perhaps complementary to that of glucocorticoids. The encouraging results were (1) that single 50-μg inhaled doses abolished both the early and late asthmatic responses and prevented bronchial hyperresponsiveness for at least 24 hr in allergen-challenged patients (27), and that 100-μg doses also prevented expected increases in serum eosinophil cationic protein and serum eosinophil protein X (28), and (z) that salmeterol, given 50 μg twice daily for 4 weeks to 12 young asthmatics, decreased bronchial hyperresponsiveness to histamine, improved the macroscopic appearance of the bronchial mucosa, and decreased the concentrations of eosinophil cationic protein and of oxygen free radicals derived from macrophages in bronchial alveolar washings (29).

Another study, in which 23 atopic asthmatics were treated with salmeterol twice daily for 6 weeks, had a less encouraging outcome. The treatment improved respiratory performance but had no significant effect on the levels of mediators (histamine, PGD_2, and tryptase), which are markers for mast cell activity, or on plasma protein extravasation (30). This uncertainty has been resolved by recent clinical trial results that strongly indicate that the effects of inhaled salmeterol and beclomethasone diproprionate in asthma are synergistic. These studies are dealt with below.

G. Mode of Action of Epinephrine and Unnatural β-Adrenoceptor Agonists

1. Epinephrine

Epinephrine is a physiological activator of β-adrenoceptor proteins whose function is to catalyze the formation of α_s-GTP from G-proteins (31). α_s-GTP, in turn, activates adenylyl cyclase to generate cyclic AMP from ATP. These events are illustrated in Figure 5.

The β_2-adrenoceptor protein is one of the superfamily of rhodopsin-like seven-transmembrane (7TM) domain receptors. Polar amino acids predominate on its extracellular and intracellular surfaces and nonpolar amino acids in its hydrophobic core. Like other catalytic proteins it is highly flexible and can adopt many conformations in the ground state at 37°C. One or a few of these are catalytically active and may make a small contribution to the basal level of intracellular cyclic AMP (32). This effect is small compared with that of epinephrine, which greatly increases the proportion of activated receptors and, therefore, the rate of formation of cyclic AMP.

The epinephrine binding site is situated at the base of the deep pocket on the outer surface of the receptor protein and is easily accessible from the extracellular fluid. Binding of epinephrine is known to involve an ionic interaction between its protonated amino group and an aspartate carboxylate group in TM-III (Asp III:08), hydrogen bonding of its catechol hydroxyl groups with serine residues in TM-V (Ser V:09 and SerV:12), and a nonpolar interaction between its benzene ring and that of a phenylalanine residue in TM-VI (Phe VI:17) (33).

Binding of epinephrine induces a major conformational change in the protein with creation of a catalytic site that processes G-protein molecules on its inner surface. A primitive illustration of these events is given in Figure 6 in which epinephrine is presumed to be "engulfed" within the deep pocket during the activation process. The onset and offset of action of epinephrine are rapid because it associates with and dissociates from its binding site quickly and the activated protein rapidly reverts to the ground state.

2. Unnatural β-Adrenoceptor Agonists (16)

General Principles. Unnatural β-agonists interact with all or part of the epinephrine binding site and with other chemical groups in the receptor protein that are determined by the structures of the agonist and the protein. The binding site for each agonist and the nature of its active complex are, therefore, unique. These interactions involve formation of low-energy ionic and/or hydrogen bonds, which contribute much to the specificity of binding, and of hydrophobic (nonpolar) binding, which contributes to the specificity

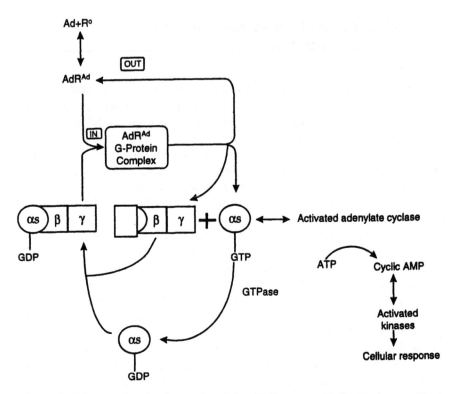

Figure 5 Diagram showing how epinephrine (Ad) reacts with the β-adrenoceptor protein ($R°$) to form the activated receptor protein (AdR^{Ad}), which catalyzes the formation of αs-GTP from a trimeric G_s-protein ($\alpha_s \beta \gamma$). α_s-GTP activates adenylate cyclase and the cellular response is continued. The effect of epinephrine terminates when ADR^{Ad} dissociates, and that of α_s-GTP when it disengages from adenylate cyclase and/or is hydrolyzed to α_s-GDP by its intrinsic GTPase activity. While they exist each molecule of ADR^{Ad} (and of the α_s-GTP-adenylate cyclase complex) processes numerous substrate molecules. Accordingly, a very low concentration of epinephrine in the extracellular fluid gives rise to a much greater concentration of cyclic-AMP inside the cell. (From Ref. 14.)

and, especially, to the intensity of binding. The latter process involves (1) a decrease in free energy because of the replacement of high-energy lipid-water interfaces within the receptor protein (and/or between the agonist and its aqueous environment) with lipid-lipid interfaces of lower energy and (2) an increase in entropy because "ordered" water molecules are released from the high-energy interfaces. These events cause activating

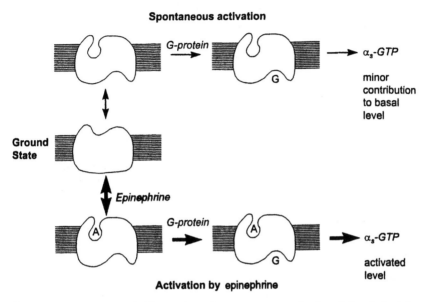

Figure 6 Diagrammatic illustration of spontaneous and epinephrine-induced activation of the β-adrenoceptor.

conformational changes in the protein and the resultant structures have greater or lesser catalytic activity and stability.

The potency of a β-agonist is determined solely by its capacity to generate α_s-GTP and this is determined by its efficacy, which is a function of the average rate at which each G-protein molecule is bound and processed, and by the life of the agonist-receptor protein complex, which determines the average number of G-protein cycles each complex catalyzes. An unnatural agonist is more potent than epinephrine if it forms a more stable receptor-protein complex and/or a complex of greater catalytic activity. Thus, as shown in Table 4, epinephrine analogs with larger nonpolar N-substituents are more potent than the natural hormone because their greater affinities result in more stable receptor-protein complexes and norepinephrine is less active because of its lower affinity. The activities of a great variety of other β-agonists have been found to correlate well with their affinities for β_1-receptors (34) and for β_2-receptors (13). Unnatural β_2-agonists more potent than epinephrine are, therefore, commonplace, but few, if any, noncatechol analogs have greater efficacy than corresponding catechols. Possible exceptions are 3-formamido analogs, such as formoterol, and carbostyrils, such as procaterol (35), whose structures are shown in Table 8.

Table 8 Activities of Variously Substituted Carbostyrils Derivatives at β_1-Adrenoceptors (cardiac stimulation) and β_2-Adrenoceptors (tracheal muscle)

| | Relative equipotent concentration $(-)$ - Isoprenaline = 1 | |
R1	Relaxation of tracheal muscle[a]	Cardiac stimulation (rate)[b]
H	4	1827
-CH(CH$_3$)$_2$ (Procaterol)	0.00005	810
-CH$_2$CH(CH$_3$)$_2$	0.0003	16
-C(CH$_3$)$_3$	0.0004	23
-CH$_2$Ph	0.07	3468
-C(CH$_3$)$_2$Ph	0.001	0.001

[a]Isolated guinea pig tracheal chain.
[b]Isolated guinea pig atria.
Source: Ref. 35.

3. Receptor Mechanisms and the Duration of Action of Inhaled β-Adrenoceptor Agonists (16)

The structures of the β-agonists whose mechanisms of action are considered are shown in Figure 7. Isoprenaline was chosen as the archetypal selective β-agonist and salbutamol as the first noncatechol highly selective β_2-agonist. Salmeterol and formeterol are distinguished by their prolonged actions, and clenbuterol and quinterenol by their ability to create drug-free, catalytically active β_2-receptor proteins.

The relevant experimental evidence, which is summarized in Tables 7 and 9, was generated by Coleman and Sumner and their colleagues at Glaxo Research Ltd (21–23). It consists of (1) relative potencies of the agonists and estimates of their rates of onset and offset of action on super-fused, isolated human bronchial muscle and guinea pig tracheal muscle and (2) relative affinities for and rates of dissociation from the rat lung β_2-adrenoceptor protein. The latter was assessed from the rates of uptake of ^{125}I-(-)iodoprindolol (^{125}I-PIN) by the receptor protein preloaded with the relevant agonist to at least 90% of its capacity.

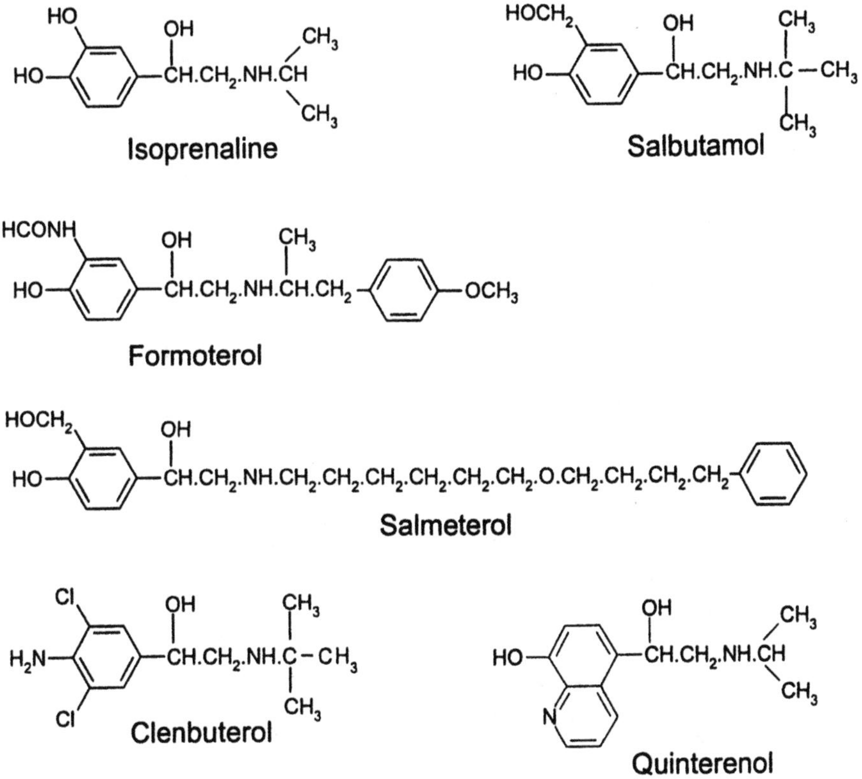

Figure 7 Structures of β-adrenoceptor agonists used in onset and offset of action studies.

    a. Isoprenaline and Salbutamol. Like epinephrine, isoprenaline rapidly associates with and dissociates from the receptor protein and is a substrate for the uptake-2 and intracellular inactivating processes which are catechol-specific. Its onset and offset of action are, therefore, rapid in both the human and guinea pig preparations. The essential reactions are summarized in Figure 8 together with appropriate mechanisms for other agonists that are qualitatively different. The slower onset of action of salbutamol is mainly a consequence of its lower efficacy and its slower offset of the fact that it is not a substrate for the uptake (2) or inactivating processes for catecholamines.

    b. Salmeterol. The kinetics of action of salmeterol on tracheal muscle are quite different from those of isoprenaline or salbutamol and its slow onset and persistence of action and its reversibility by β-blockers require

Table 9 Effects of Selected β-Adrenoceptor Agonists on Electrically Stimulated Superfused Isolated Guinea Pig Tracheal Muscle and on β_2-Adrenoceptors in Rat Lung Membrane

Agonist	Equipotent concentrations (−) isoprenaline = 1	Onset time (Ot$_{50}$ min)	Offset time (Ot$_{50}$ min)	Rat lung β_2-adrenoceptors Affinity (−) isoprenaline = 1	Time for 50% dissociation min
(−) Isoprenaline	1 (pEC$_{50}$) = 7.7	>1<3	>2<4	1 (pK$_i$ = 6.7)	≤4
(±) Salbutamol	3.0	3.2 (2.0–5.4)	11 (8–15)	10	≤4
(±) Formoterol	0.03	8.3 (3.9–12.6)	19.6 (7.7–31.6)	0.10	About 5
(±) Salmeterol	0.4	29 (24–36)	No recovery	0.07	No significant dissociation
(±) Clenbuterol	0.6	6.7 (4.5–8.9)	48 (39–57)	0.28	≤4
(±) Quinterenol	27	About 5	No recovery	43	>4<10
(^{125}I)-Iodopindolol	—	—	—	—	Time for 50% association—about 4

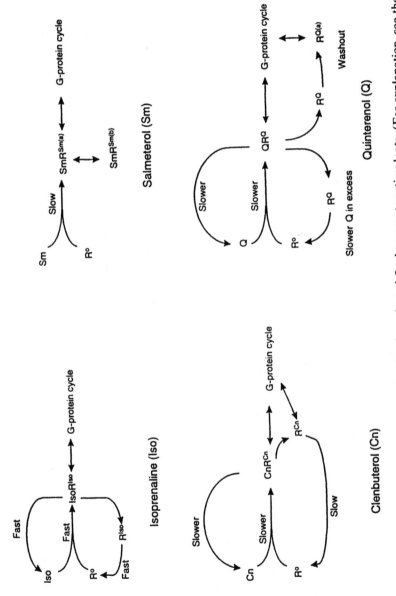

Figure 8 Postulated mechanisms of action for selected β-adrenoceptor stimulants. (For explanation, see the text. The term "slower" means slower than the corresponding events with isoprenaline.) (From Ref. 16.)

explanation. Salmeterol is a much larger, less polar molecule than isoprenaline, so it associates with the β_2-receptor protein slowly but once bound it does not dissociate from it; this persistence of binding is evidenced by nonuptake of ^{125}I-PIN by salmeterol-loaded β_2-receptor protein. The consequences of these interactions and the low efficacy of the saligenin head are an unusually slow onset of action and uniquely persistent β_2-agonism. The reversal of this action by pindolol was at first puzzling. Why should uptake of ^{125}I-PIN be prevented in salmeterol-loaded β_2-receptors but not that of β-blockers which reverse its action? The explanation is that ^{125}I-PIN, by virtue of its 2-iodo-substituent, is a larger molecule than the others and its binding to the receptor protein is sterically hindered when salmeterol is in place.*

The overall mechanism of action of salmeterol shown in Figure 8 accounts for all its pharmacological actions (16). It binds irreversibly to the β_2-receptor protein to form an active complex in which its phenylethanolamine head activates the protein by engaging the epinephrine binding areas. This active conformation alternates with an inactive conformation[†] in which the epinephrine-binding areas are free to accept β-blocking molecules if they are small enough to gain access to them. When the β-blocker disengages from its binding site, salmeterol is still in place and β_2-agonism resumes.

Binding of salmeterol is irreversible because its long, flexible, nonpolar N-substituent forms low-energy lipid-lipid interfaces with nonpolar amino acid residues in the hydrophobic core of the protein. The ether oxygen in the side chain plays an essential directive role in this process probably by hydrogen bonding with an amino acid residue such as tryptophan. A probable site of these interactions first proposed by Dave Timms, of ICI (now Zeneca) Pharmaceuticals, and me (16) using bacteriorhodopsin as a model for the β_2-receptor protein is similar to that recently deduced from binding of the drug by the β_2-receptor protein and analogs of it obtained by site-directed mutagenesis (38).

The potency of a "captive" or "anchored" agonist such as salmeterol is determined by the efficacy of its phenylethanolamine head and by the

---

* Similar but more extreme steric hindrance has been reported for in vitro interactions between carbostyril derivatives with large N-substituents and the β-adrenoceptor protein in rat reticulocytes. The agonists are inactive on receptor proteins preloaded with β-blocking drugs and the latter are unable to reverse established β-agonism (36). A critical factor in this system is the relatively large size of the carbostyril grouping compared with that in phenylethanolamines.
[†] Similar oscillations between inactive and activated conformations of ground-state β-adrenoceptor proteins were reported in 1992 (32,37).

affinity of that head for the area of the epinephrine-binding site, since this determines the proportion of time that the drug-receptor complex spends in the active conformation. Its duration of action is determined by the average life of the occupied β_2-receptor protein, which in the case of salmeterol is long enough for twice-daily inhalations of 50 or 100 μg to maintain continuous β_2-agonism in the airways with effective bronchodilation both night and day.

c. Formoterol. In its general pharmacology, formoterol is essentially a more potent, more efficacious salbutamol since it is relatively short-acting by systemic routes of administration. After inhalation, however, its prolonged bronchodilating action in asthmatics is similar to that of salmeterol and, as is obvious from the data in Tables 7 and 9, this effect is not a consequence of persistent binding of the drug within the receptor protein. Why is it long-acting?

In 1981, two general ways to achieve prolonged bronchodilatation with an inhaled β_2-agonist occurred to me and I have already outlined how one of them led to salmeterol. The other arose from the knowledge that salbutamol, which is a base, is relatively slowly absorbed from the respiratory tract compared, for example, with cromoglycate, which is an acid, perhaps because diffusion of the base from the airways is impeded by reversible binding to acidic groups in mucoprotein. If this were true, a β-agonist that binds more strongly to mucoprotein would be absorbed more slowly than salbutamol and, if sufficiently potent, would be longer acting (16,39). Formoterol has the required potency but its affinity for mucoprotein is unknown.

Anderson (40) explains the prolonged bronchodilating action of formoterol by supposing that it is concentrated in the bronchial muscle cell membrane and maintains β_2-agonism by slowly leaching from this depot. The main evidence in favor of his hypothesis is that loading human or especially guinea pig respiratory muscle by superfusion with increasingly high concentrations of the drug leads to progressive increases in the recovery times during continuous superfusion with drug-free Krebs' solution (22,23). There is, however, no reason for supposing that inhaled formoterol is selectively concentrated in the bronchial muscle membrane. If Anderson's hypothesis is right, the effective depot for inhaled drug must involve all the cell membranes on its diffusion pathway from the bronchial lumen to its target β_2-receptor protein. The end result would be an increase in the capacity and "depth" of the drug depot, which would further impede its diffusion. In addition, I would not entirely discount the possibility that reversible binding of the drug to bronchial mucoprotein also contributes to the prolonged action of inhaled formoterol. Whatever mechanism(s) are involved, however, it is clear that the basic cause of the prolonged action of inhaled formoterol is impeded diffusion of the drug in the lungs.

d. Clenbuterol and Quinterenol. One surprising result shown in Table 9 is that the action of clenbuterol on the tracheal muscle clearly exceeds the time required for its disengagement from the β_2-receptor protein. Accordingly, its prolonged action is not caused by its persistence at its site of action but by the receptor protein continuing to be catalytically active after disengagement of the agonist (16). This was the first report of β-agonism being maintained by an agonist-free receptor protein. The mechanisms involved are illustrated in Figure 5.

Quinterenol proved to be the ultimate in continuing β_2-agonism by an agonist-free receptor protein; the tracheal muscle does not recover from its inhibitory action, which may even intensify, when the drug is washed out (16). It is not easy to rationalize this unexpected phenomenon. The drug behaves as a modestly active β-agonist as long as it is present in active concentration in the extracellular fluid and the permanently activated unnatural receptor protein is formed only when it is washed out. That change must, therefore, be thermodynamically favored compared with a return to the physiological ground state, an explanation that implies that the modestly active complex formed in the presence of the drug is unusual in being stabilized by it. These events are illustrated in Figure 8.

III. SELECTIVE ANTI-INFLAMMATORY GLUCOCORTICOID ACTIVITY IN THE LUNGS

The following account of the development of the use of anti-inflammatory steroids by inhalation is the minimum needed to rationalize their clinical use.

A. Rationale

Another major starting point for drug research in asthma was the 1949 report by Hench and co-workers (41) of the beneficial effects of cortisone in rheumatoid arthritis, which initiated a new, uniquely effective way to treat inflammatory diseases. Cortisol, its active metabolite, was soon found to be effective in asthma, but enthusiasm for steroids turned to caution as the harmful effects of prolonged treatment became obvious.

Cortisol is active at mineralocorticoid receptors as well as glucocorticoid receptors which mediate its anti-inflammatory action and, since these receptors are different, selective glucocorticoid activity was soon achieved. Prednisolone was followed by the more potent, more selective dexamethasone and betamethasone. These remain lifesaving medications for asthmatics but they must be used cautiously in chronic asthma because the receptors responsible for the anti-inflammatory activity also mediate the adverse effects. Selective systemic glucocorticoid activity is, therefore, unattainable with steroids of this kind.

A very significant advance in steroid medication was made in the late 1950s when fluocinolone acetonide was found to have a potent anti-inflammatory action after local application to the skin, an effect that is achieved without significant systemic glucocorticoid actions in most patients. Betamethasone valerate soon followed, and steroids of this kind transformed the treatment of eczema and other common skin diseases. The use of such a steroid by inhalation in asthma was first proposed by Wilfred Simpson in Allen and Hanburys and by Phillipps and Snell in Glaxo. The concept was brilliantly simple. If the steroids were effective anti-inflammatory agents on the skin, they might be in the airways, which are simply another body surface.

A distinguishing feature of potent topical steroids is their ability to cause local vasoconstriction when applied to the skin under occlusive conditions, a phenomenon first described by McKenzie and Stoughton (42) but never adequately explained. It may simply be a consequence of continuous intense local glucocorticoid activity, which, inter alia, inhibits nitric oxide formation and thus removes its continuous physiological vasodilating action. The McKenzie skin-blanching test played a key role in the selection of glucocorticoids for use on the skin because the intensity, area, and duration of skin blanching were found to give accurate forecasts of clinical potency (and, incidentally, to correlate well with the affinities of glucocorticoids for their receptor protein.) Skin-blanching activity also proved to be a reliable forecaster of anti-inflammatory activity in the lungs with inhaled glucocorticoids, but this activity had to be free from harmful local or systemic effects.

The general criteria that emerged for selective glucocorticoid activity in the lungs are outlined in Figure 9; they do not require differentiation of glucocorticoid receptors.

B. Beclomethasone Dipropionate

Beclomethasone dipropionate (BDP) was chosen for clinical development because (1) it has potent skin-blanching activity, (2) its systemic glucocorticoid activity was found to be less after oral than after intravenous administration in humans (indicating poor absorption from the gut or substantial first-pass metabolism), and (3) it was not toxic to the lungs of rat or dog in prolonged inhalation studies. The relevant topical and systemic glucocorticoid activities of BDP and other steroids in humans are summarized in Table 10 in which its favorable therapeutic ratio is obvious. Significant first-

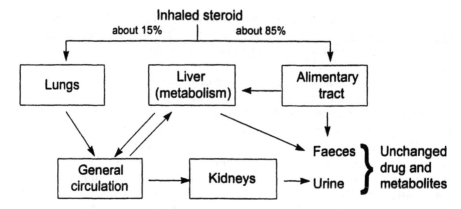

Figure 9 Fate of inhaled steroids in the body. For a clinically useful selective effect in asthma, the steroid must: (1) have a persistent action in the lungs, and (2) be poorly absorbed from the gut, and/or (3) be substantially inactivated by first-pass metabolism.

pass metabolism was confirmed later when BDP was found to be inactivated by oxidation in the liver.

Daily doses of inhaled BDP up to 1 mg proved to be effective in chronic asthma without causing significant systemic glucocorticoid actions. The greater doses used to control more severe asthma do, however, cause modest, but significant suppression of hypophysial-pituitary-adrenal function and perhaps a degree of osteoporosis after prolonged use (43). This is why a more selectively acting topical steroid was sought and found in fluticasone propionate, and also why the effectiveness of concurrent treatment with low-dose BDP and salmeterol is important (see Section IV).

Beclomethasone dipropionate inhaler was first marketed in 1972 and became one of the mainstay treatments of chronic asthma during the 1980s as concern about possible long-term toxic effects was overcome by experience. Ironically, as confidence in inhaled steroids grew, concern grew, especially in Europe, about possible adverse consequences of regular inhalations of β_2-adrenergic bronchodilators. The latter is dealt with in Section IV.

C. Fluticasone Propionate

Fluticasone propionate was chosen for development from the Glaxo library of steroids because it was known to be twice as active as BDP in the McKenzie and Stoughton skin-blanching test in humans and to have little glucocorticoid activity after oral administration in the mouse (44). It was later found to be similarly inactive by mouth in humans because drug absorbed from the gut is quantitatively hydrolyzed in the liver to inactive

Table 10 Relative Systemic and Topical Activities of Selected Steroids in Humans

Biological activity	Cortisol	Betamethasone	Beclomethasone dipropionate	Fluticasone propionate
Systemic glucocorticoid[a]	1	25	4 (oral) 20 (i.v.)	~0 (oral)
Topical anti-inflammatory[b]	0.1 (cortisol acetate = 1)	0.8	500	950
Ratio Anti-inflammatory / Oral glucocorticoid	0.1	0.03	125	Very high
Oral bioavailability	High	High	About 20%	~0

[a] Suppression of early-morning plasma cortisol levels.
[b] McKenzie and Stoughton skin-blanching test.

products. In clinical trials it proved to be twice as active, on a weight basis, as BDP in asthmatic patients and to have less effect on HPA function (45). The exceptional anti-inflammatory activity of fluticasone propionate is a consequence of its persistent binding to the glucocorticoid receptor protein (46).

Fluticasone propionate (Flixotide) inhaler was marketed in 1994. It may have a special place for treating severe asthma.

IV. CLINICAL USE OF INHALED β_2-ADRENOCEPTOR AGONISTS AND TOPICAL GLUCOCORTICOID STEROIDS IN ASTHMA

A. Recent Concern About Inhaled β_2-Adrenoceptor Agonists

Inhaled β_2-agonists have been problematic for a long time. During the 1960s and the 1970s there were epidemics of unexpected deaths, especially of young asthmatics, associated with their use; Pearce and co-workers have provided a detailed analysis of these events (47). The main reason for these deaths is now accepted by most to be inadequate use of glucocorticoids to control the life-threatening physical occlusion of the airways that occurs as asthma worsens. As explained above, rational treatment of chronic asthma requires the concurrent use of glucocorticoids, to contain the inflammatory process, and inhaled β_2-adrenergic bronchodilators, to relieve acute bronchospasm and to reduce bronchial tone.

More recent concern is that regular multiple daily inhalations of β_2-agonists may cause asthma to worsen and thus be one of the causes of the gradual increase in mortality reported by Jackson and co-workers (48). The most damning indictment of this use of β_2-agonists was that of Sears and co-workers (49). Their conclusions are discussed separately in the next paragraph because of the profound influence they have had on asthma treatment and on clinical pharmacology in the United Kingdom and elsewhere in Europe. Succinct accounts of European medical reservations about β_2-agonists in asthma have been given by Chung (50) and by van Schayck and Herwaarden (51).

Following a double-blind, placebo-controlled, cross-over study of the effects of regular versus on-demand inhaled bronchodilator therapy in a total of 56 assessable subjects, Sears and co-workers (49) reported that respiratory performance of asthmatics treated with regular inhalations of fenoterol hydrobromide (400 μg four times daily) was significantly worse than that during the corresponding control period in which a matching placebo was given together with occasional inhalations of a β_2-agonist only when needed to relieve symptoms. On this meager evidence, they offered a general hypothesis that regular inhalations of any β-agonist, including the then new long-acting agents salmeterol and formoterol, are likely to increase asthma morbidity, and strongly recommended that inhaled β-agonists be used only when needed to relieve symptoms and that regular inhalations of them be avoided.

Sears' hypothesis and recommendations were clearly refuted when no evidence of worsening asthma attributable to β-agonist treatment was found in any of six long-term clinical trials (detailed later) in which many hundreds of patients received regular inhalations of salbutamol or salmeterol for 3, 6, or 12 months. On the contrary, the patients clearly benefited from β_2-agonist treatment whether or not they received additional inhaled steroid. Much of this data was available from 1990 onward, so it is not easy to understand why the British Thoracic Society (BTS) guidelines for asthma treatment (52a) issued in 1993 were considerably based on Sears' recommendations.

According to these guidelines, less severe chronic asthma should be treated with regular inhalations of an anti-inflammatory steroid supplemented when necessary by occasional inhalations of a short-acting β_2-agonist such as salbutamol. Regular inhalations of a long-acting β_2-agonist are reserved for patients with more severe disease that is still inadequately controlled after the dose of steroid is increased. These recommendations are clearly inconsistent in that regular β_2-agonist inhalations are avoided for fear of worsening less severe asthma and yet are expected to ameliorate more severe disease. Surprisingly, this general approach to therapy is maintained in revised BTS guidelines issued in 1997 (52b). The only significant change is that concurrent salmeterol and low-dose steroid treatment is allowed as an alternative to increasing the inhaled steroid intake.

B. Early Clinical Trials of Salmeterol

Salmeterol has performed to best expectations in clinical use. The key results of a large early trial* (53,54) are summarized in Figures 10 and 11. Evening and especially morning respiratory performance were obviously better with salmeterol than with salbutamol. That nocturnal asthma attacks were virtually abolished by salmeterol treatment was particularly gratifying since this was one of the starting objectives of the project. More importantly, however, no evidence was found of tolerance to the actions of either β_2-agonist; if anything, the incidence of exacerbations of disease fell during the 12 months' treatment with salmeterol. Salmeterol and salbutamol behaved similarly in all the other early clinical studies* (55–57). These results, and

---

* In all of these studies respiratory performance was worse in the morning than in the evening, which is an expectable consequence of low epinephrine levels during the night (3). There is, therefore, a good case for increasing the standard evening dose of salmeterol to 100 μg.

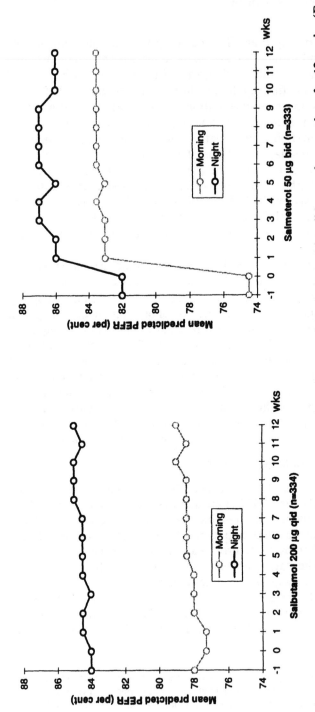

Figure 10 Comparison of the effects of treatment with inhaled salmeterol or salbutamol in mild to moderate asthma for 12 weeks. (From Ref. 53.)

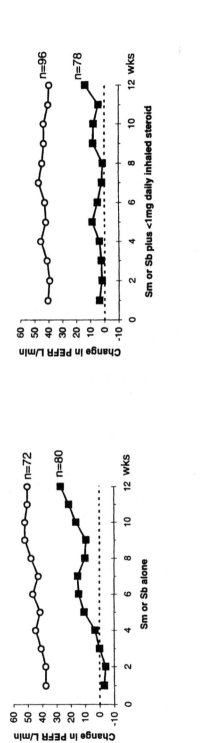

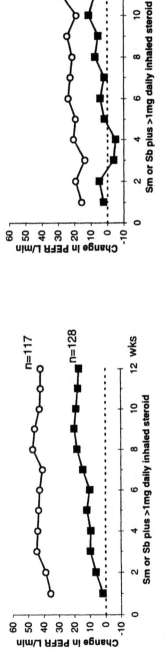

Figure 11 Effects of inhaled salmeterol (Sm), 50 μg bid, or salbutamol (Sb), 200 μg qid, alone or with added steroid, on morning PEFR in mild to moderate asthmatic patients. ——○—— = salmeterol treatment; ——■—— = salbutamol treatment. (From Ref. 54.)

the fact that serious unexpected adverse events were noticeably absent during postmarketing surveillance of 15,000 patients treated with salmeterol (58), show beyond reasonable doubt that Sears' hypothesis is false.

Salmeterol inhaler was introduced in the United Kingdom in 1992 into an environment that had become hostile to regular use of β_2-adrenergic bronchodilators. In consequence, the therapeutic potential of this major new medicine is only being slowly realized. Recent progress on its concurrent use with inhaled steroids is outlined next.

C. Concurrent Use of Inhaled Long-Acting β_2-Adrenoceptor Agonists and Topical Anti-inflammatory Glucocorticoids in Asthma

The reason why chronic asthma should be treated by concurrent inhalation of a β_2-agonist, such as salmeterol, and a topical anti-inflammatory steroid is that their inhibitory actions on the inflammatory cells involved are complementary and their therapeutic effects are synergistic. The cellular basis for this approach is summarized in Table 11. The effects of β_2-agonists on these cells have already been outlined in Section II. The complexity of the actions of glucocorticoids on them is well summarized in Taylor and Shaw's review (59).

Greening and co-workers (60) found a marked improvement in respiratory performance when salmeterol was added to the dosage regimen of patients previously stabilized on low-dose BDP, an improvement substantially greater than that obtained by increasing the steroid intake. The design and essential results of their trial are summarized in Figure 12. An obvious explanation of the great improvement in the salmeterol/BDP group is β_2-mediated bronchodilatation, and the reason for the lesser improvement in

Table 11 Complementary Actions of β_2-Adrenoceptor Stimulants and Anti-inflammatory Steroids on Cells Involved in the Pathology of Asthma

Activation inhibited by β-adrenoceptor stimulants	Activation inhibited by anti-inflammatory steroids[a]
Bronchial smooth muscle	Eosinophils
Endothelial cells in postcapillary venules	Basophils
Mast cells	Alveolar macrophages
	T lymphocytes

[a] Anti-inflammatory steroids also reduce the number of proinflammatory cells by inhibiting their replication and their migration into the bronchial mucosa and by inducing death by apoptosis.

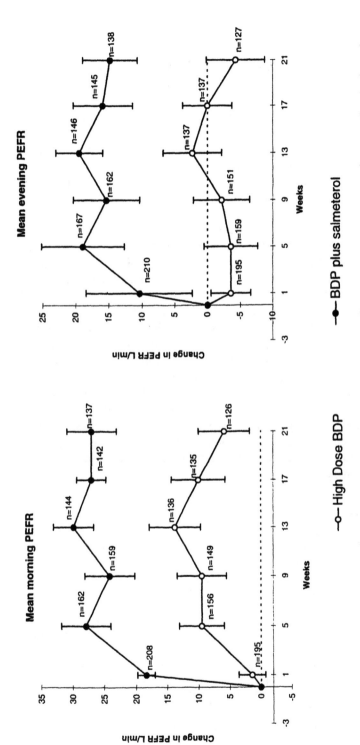

Figure 12 Comparative effects of treatment with beclomethasone dipropionate (200 μg) bid plus salmeterol (50 μg bid) or beclomethasone dipropionate (500 μg bid) in asthmatic patients stabilized on beclomethasone dipropionate (200 μg bid). (From Ref. 59.)

167

the high-dose BDP group is the relatively flat dose-response curve to steroids. Salmeterol must, however, do more than relax bronchial muscle in the former group because any contribution attributable to bronchodilatation would occur rapidly whereas the peak expiratory flow rates in the salmeterol/BDP group took more than a week to reach new therapeutic plateaus. Accordingly, additional β_2-adrenergic action(s), probably inhibition of mast cells and/or endothelial cell contraction, are involved.

Regardless of mechanism, Greening's results and similar findings by Woolcock and co-workers (61) provide compelling evidence that concurrent, continuous β_2-agonism and inflammatory steroid activity in the airways gives exceptional control of chronic asthma. Other advantages of this treatment are that unnecessarily high doses of inhaled steroids are avoided, bronchial hyperresponsiveness is greatly reduced, and given the right dosage regimen of β_2-agonist, inhalations of short-acting bronchodilators as rescue medication are seldom necessary.

In response to these outstanding results, the BTS guidelines were changed in 1997 to allow concurrent use of inhaled salmeterol in patients inadequately controlled by low-dose steroid and on-demand β_2-agonist treatment, as an alternative to increasing the steroid intake. Although this change is welcome, I am sure that combined β_2-agonist/steroid treatment would also greatly benefit patients with less severe disease and might even prevent or retard the irreversible structural changes in the airways that characterize long-standing disease since they are consequences of inadequately controlled chronic inflammation. New thinking of this kind is required if patients are to benefit maximally from the new long-acting β_2-agonists and topical steroids.

REFERENCES

1. Hume KM, Gandevia B. Forced expiratory volume before and after isoprenaline. Thorax 1956; 12:276.
2. Skloot G, Perutt S, Togias A. Airway hyper-responsiveness in asthma: a problem of limited smooth muscle relaxation with inspiration. J Clin Invest 1995; 96:2393–2403.
3. Barnes P, Fitzgerald G, Brown M, Dollery C. Nocturnal asthma and changes in circulating epinephrine, histamine, and cortisol. N Engl J Med 1980; 303:263–267.
4. Hui KP, Barnes NC. Lung function inprovement in asthma with a cysteinyl-leukotriene receptor antagonist. Lancet 1991; 337:1062–1063.
5. Regoli D, Gobeil F, Rhaleb NE. Receptors for bradykinin and related kinins: a critical analysis. Can J Phys Pharmacol 1993; 71:556–567.
6. Gross NJ. Ipratropium bromide. N Engl J Med 1988; 311:486–494.

7. Ahlquist RP. A study of adrenotropic receptors. Am J Physiol 1948; 153:586–600.
8. Lands AM, Arnold A, McAuliff JP, Luduena FP, Brown TG. Differentiation of receptor systems activated by sympathomimetic amines. Nature 1967; 214:597–598.
9. Lands AM, Luduena FP, Buzzo HJ. Differentiation of receptors responsive to isoproterenol. Life Sci 1967; 8:2241–2249.
10. Lunts LHC. Salbutamol: a selective β_2-stimulant bronchodilator. In: Roberts SM, Price BJ, eds. Medicinal Chemistry. The Role of Organic Chemistry in Drug Research. London: Academic Press, 1985:49–67.
11. Coleman RA, Farmer JB. The inducement of tone and its inhibition in isolated tracheal muscle. J Pharm Pharmacol 1971; 23:220–223.
12. Coleman RA, Nials AT. The characterisation and use of the electrically stimulated, superfused guinea-pig tracheal strip preparation. Br J Pharmacol 1986; 91:409P.
13. Minneman KP, Hegstrand LR, Molinoff PB. The pharmacological specificity of beta-1 and beta-2 adrenergic receptors in rat heart and lung in vitro. Mol Pharmacol 1979; 16:21–33.
14. Gibson DG, Coltart DJ. Haemodynamic effects of intravenous salbutamol in patients with mitral valve disease—comparison with isoprenaline and atropine. Post Grad Med J 1971; 46(Suppl):40–44.
15. Butchers PR, Fullarton JR, Skidmore IF, Thompson LE, Vardey CJ, and Wheeldon A. A comparison of the antianaphylactic activities of salbutamol and disodium cromoglycate in the rat, the rat mast cell, and in human lung tissue. Br J Pharmacol 1979; 84:23–32.
16. Jack D. A way of looking at agonism and antagonism: lessons from salbutamol, salmeterol and other β-adrenoceptor agonists. Br J Clin Pharmacol 1991; 31:501–514.
17. Jack D. Rational approaches to drug discovery. In: Shanks RG, ed. Advanced Medicine—Topics in Therapeutics, Vol 3. London: Pitman Medical, 1977:135–147.
18. Jack D. Safety by design. Drug Safety 1990; 5(Suppl):4–23.
19. Wallis CJ. Salbutamol, salmeterol—a chemist's perspective. Actual Chim Ther 1994; 20:265–292.
20. Ball DI, Brittain RT, Coleman RA, Denyer LH, Jack D, Johnson M, Lunts LHC, Nials AT, Sheldrick KE, Skidmore IF. Salmeterol, a novel, long-acting β_2 adrenoceptor agonist: characterisation of pharmacological activity in vitro and in vivo. Br J Pharmacol. 1991; 104:665–671.
21. Nials AT, Sumner MJ, Johnson M, Coleman RA. Investigations into factors determining the duration of action of the β_2-adrenoceptor agonist, salmeterol. Br J Pharmacol 1993; 108:507–515.
22. Nials AT, Coleman RA, Johnson M, Magnussen HJ, Rabe KF, Vardey CJ. The effects of β-adrenoceptor agonists in human bronchial smooth muscle. Br J Pharmacol 1993; 110:1112–1116.
23. Nials AT, Ball DI, Butchers PR, Coleman RA, Humbles AA, Johnson M, Vardey CJ. Formoterol on airway smooth muscle and human lung mast cells:

a comparison with salbutamol and salmeterol. Eur J Pharmacol 1994; 251:127–135.

24. Ullman A, Svedmyr N. Salmeterol, a new long-acting inhaled β_2-adrenoceptor agonist: comparison with salbutamol in adult asthmatic patients. Thorax 1988; 43:674–678.

25. Butchers PR, Vardey CJ, Johnson M. Salmeterol: a potent and long-acting inhibitor of inflammatory mediator release from human lung. Br J Pharmacol 1991; 104:672–676.

26. Whelan CJ, Johnson M. Inhibition of increased vascular permeability and granulocyte accumulation in guinea-pig lung and skin. Br J Pharmacol 1992; 105:831–838.

27. Twentyman OP, Finnerty JP, Harris A, Palmer J, Holgate ST. Protection against allergen-induced asthma by salmeterol. Lancet 1990.

28. Pedersen B, Dahl R, Larsen BB, Venge P. The effect of salmeterol on the early- and late- phase reaction to bronchial allergen and post challenge variation in bronchial allergen and post-challenge variation in bronchial reactivity, blood eosinophils, serum eosinophil cationic protein, and serum eosinophil protein X. Allergy 1993; 48:377–382.

29. Dahl R, Pedersen B, Venge P. Bronchoalveolar lavage studies. Eur Respir Rev 1991; 1:272–275.

30. Holgate ST. β_2-agonists and airways inflammation. In: Long-Acting β_2-Agonists: Issues and Challenges. Satellite symposium held at the Annual Congress of the European Respiratory Society, Barcelona, Spain, September 1995. Chairman F Manresa. Parthenon Publishing Group, 1996:17–21.

31. Gilman AG. G-proteins: transducers of receptor generated signals. Annu Rev Biochem 1987; 56:615–649.

32. Lefkowitz RJ, Cotecchia S, Samama PS, Costa. Constitutive activity of receptors coupled to guanine regulatory proteins. Trends Pharmacol Sci 1993; 14:303–307.

33. Schwarty TW, Rosenkilde MM. Is there a "lock" for all agonist "keys" in 7TM receptors? Trends Pharmacol Sci 1996; 17:213–216.

34. Bilezikian JP, Dornfeld AM, Gammon DE. Structure-binding-activity analysis of β adrenergic amines. I. Binding to the beta receptor and activation of adenylate cyclase. Biochem Pharmacol 1978; 27:1445–1454.

35. Yoshisaki S, Tanimura K, Tamada S, Yabuuchi Y, Nakagawa K. Sympathomimetic amines having a carbostyril nucleus. J Med Chem 1976; 19:1138–1142.

36. Standifer KM, Pitha J, Baker SP. Carbostyril-based beta-adrenergic agonists: evidence for long-lasting or apparent irreversible receptor binding and activation of adenylate cyclase activity in vitro. Naunyn-Schmiedeberg's Arch Pharmacol 1989; 339:129–137.

37. Costa T, Ogino Y, Hunson PJ, Onaran HO, Rodbarel D. Drug efficacy at guanine nucleotide-binding regularly protein-linked receptors: thermodynamic interpretation of negative antagonism and of receptor activity in the absence of ligand. Mol Pharmacol 1992; 41:549–560.

38. Green SA, Spasoff AP, Coleman RA, Johnson M, Liggett SB. Sustained activation of a G-protein-coupled receptor via "anchored" agonist binding.

Molecular localisation of the salmeterol exosite within the β_2-adrenergic receptor. J Biol Chem 1996; 271:24029–24035.

39. Jack D. Albuterol: are improved selective beta$_2$-stimulant brochodilators attainable? In: New Concepts in the Topical Treatment of Asthma and Related Disorders. Proceedings of a symposium sponsored by Brigham and Womens Hospital and Harvard Medical School. Cliggott Publishing Co, 1982:23–27.

40. Anderson GP. Formoterol: pharmacology, molecular basis of agonism, and mechanism of long duration of a highly potent and selective β2-adrenoceptor agonist bronchodilator. Life Sci 1993; 52:2145–2160.

41. Hench PS, Kendall EC, Slocumb CH, Polley HF. The effect of a hormone of the adrenal cortex (17-hydroxy-11-dehydro-corticosterone: compound E) and of pituitary adrenocorticotropic hormone on rheumatoid arthritis. Proc Staff Meet Mayo Clin 1949; 24:181–197.

42. McKenzie AW, Stoughton RB. Method for comparing percutaneous absorption of steroids. Arch Dermatol 1962; 86:608–610.

43. Utiger RD. Differences between inhaled and oral glucocorticoid therapy. N Engl J Med 1993; 329:1731–1733.

44. Phillipps GH. Structure-activity relationships of topically active steroids: the selection of fluticasone propionate. Respir Med 1990; 84(Suppl A):19–23.

45. Holliday SM, Faulds D, Sorbin EM. Inhaled fluticasone propionate. A review of its pharmacodynamic and pharmacokinetic properties, and therapeutic use in asthma. Drugs 1994; 47(2):318–331.

46. Högger P, Rohdewald P. Binding kinetics of fluticasone propionate to the human glucocorticoid receptor. Steroids 1994; 59:597–602.

47. Pearce N, Crane J, Burgess C, Jackson R, Beasley R. Beta agonists and asthma mortality: déja vu. Clin Exp Allergy 1991; 21:401–410.

48. Jackson R, Sears MR, Beaglehole R, Rea HH. International trends in asthma mortality: 1970–1985. Chest 1988; 94:14–18.

49. Sears MR, Taylor DR, Print CG, Lake DC, Li Q, Flannery EM, Yates DM, Lucas MK, Herbison CP. Regular inhaled beta-agonist treatment in bronchial asthma. Lancet 1990; 336:1391–1396.

50. Chung KF. The current debate concerning β-agonists in asthma: a review. J Roy Soc Med 1993; 86:96–100.

51. Van Schayck CP, Van Herwaarden CLA. Do bronchodilators adversely affect the prognosis of bronchial hyper-responsiveness? Thorax 1993; 48:470–473.

52a. British Thoracic Society. Guidelines on the management of asthma. Thorax 1993; 48(Suppl):S1–S24.

52b. The British Guidelines on Asthma Management. Thorax 1997; 52(Suppl 1):S1–S21.

53. Britton M. Salmeterol and salbutamol: large multicentre studies. Eur Respir Rev 1991; 1:288–292.

54. Britton MG, Earnshaw JS, Palmer JBD. A twelve month comparison of salmeterol with salbutamol in asthmatic patients. Eur Respir J 1992; 5:1062–1067.

55. Faurschou P. Chronic dose-ranging studies with salmeterol. Eur Respir Rev 1991; 1:282–287.

56. Pearlman DS, Chervinski P, Laforce C, Seltzer JM, Southern DL, Kemp JP, Dockhorn RJ, Grossman J, Liddle RF, Yancey SW, Cocchetto DM, Alexander WJ, van As A. A comparison of salmeterol with albuterol in the treatment of mild to moderate asthma. N Engl J Med 1992; 327:1420–1425.
57. Lötvall J, Lunde H, Ullman A, Törnqvist H, Svedmyr N. Twelve months treatment with inhaled salmeterol in asthmatic patients. Effects on beta$_2$-receptor function and inflammatory cells. Allergy 1992; 47:477–483.
58. Mann RD. Results of prescription event monitoring study of salmeterol. Br Med J 1994; 309:1018.
59. Taylor IK, Shaw RJ. The mechanism of action of corticosteroids in asthma. Respir Med 1993; 87:261–277.
60. Greening AP, Ind PW, Northfield M, Shaw G. Added salmeterol versus higher-dose corticosteroids in asthma patients in patients with symptoms on existing inhaled steroid. Lancet 1994; 344:219–224.
61. Woolcock A, Lundback B, Ringdal N, Jacques L. Comparison of addition of salmeterol to inhaled steroids with doubling of the dose of inhaled steroids. Am J Respir Crit Care Med 1996; 153:1481–1488.

7

Serotonin 5-HT$_{1B/D}$ Receptor Agonists

Graeme R. Martin
Roche Bioscience, Palo Alto, California

I. INTRODUCTION

Few areas of research have been as fruitful in the identification of new therapeutic products as the serotonin receptor field. At the beginning of 1996, at least 20 drugs with a primary therapeutic action mediated via one of five distinct serotonin receptors and nine drugs interacting specifically with the serotonin transporter were listed as prescription medicines. Between them they accounted for sales of around US$5.6 billion, about 2% of world drug sales for the year 1995 (1). In nearly every case, these drugs were developed using a receptor-targeted approach to mimic, block, or augment the actions of serotonin at specific receptor subtypes. Often, serotonin itself was used as the template for medicinal chemistry. With such an impressive precedent, it is perhaps understandable that serotonin receptors and gene products remain a focus for drug discovery programs. According to published information, at least six distinct receptor targets in addition to those alluded to above are currently the subject of intense research across a broad spectrum of therapeutic categories (see Table 1). Among the most important of these are the 5-HT$_{1B}$ and 5-HT$_{1D}$ receptors, which appear to be key targets for the acute treatment of migraine and associated vascular headaches.

II. SEROTONIN RECEPTOR DIVERSITY

The differential classification of receptors is based on primary structural information in conjunction with precise operational data obtained using

173

Table 1 Drugs with Actual or Proposed Therapeutic Actions Mediated by Serotonin Receptors or the Serotonin Transporter

Receptor	Ligand type	Actual (*proposed*) therapeutic utility	Novel targeted drugs	Historical drugs with activity
5-HT$_{1A}$	Agonist	Anxiety/depression/bulimia	Buspirone Ipsapirone,[a] Gepirone,[a] Tandospirone[a]	—
5-HT$_{1B}$	*Antagonist*	*Anxiety/schizophrenia*	*WAY100635*	—
	Agonist	Migraine/anxiety	Sumatriptan, Zolmitriptan[a] *CGS-12066, CP-93129*	Methysergide Ergot alkaloids
	Antagonist	*Motor disorders/vasospastic disease/depression*	*GR127935, GR55562*	—
5-HT$_{1D}$	Agonist	Migraine	Sumatriptan, Zolmitriptan[a]	Methysergide Ergot alkaloids
	Antagonist	*Motor disorders/depression*	*GR127935*	—
5-ht$_{1E}$	—	—	—	—
5-ht$_{1F}$	*Agonist*	*Migraine*	—	Methysergide Ergot alkaloids
5-HT$_{2A}$	Antagonist	Hypertension/vasospastic disease	Ketanserin	—
		Psychosis/schizophrenia/anxiety/dependence	Risperidone,[b] Ritanserin,[a] Seroquel,[a,b] Sertindole,[a,b] Olanzapine[a,b]	—
5-HT$_{2B}$	*Antagonist*	*Migraine prophylaxis/anxiety*	*SB200646*	Methysergide Pizotifen Cyproheptidine ($\pm$)Propranolol Metergoline

Receptor	Action	Proposed indication	Drugs	Clinical drugs
5-HT$_{2C}$	*Antagonist*	*Anxiety/epilepsy*	*Deramciclane*	Methysergide, Pizotifen, Cyproheptidine
5-HT$_3$	Antagonist	Chemotherapy/radiation-induced emesis Gastric motility disorder	Ondansetron, Granisetron, Tropisetron, Ramosetron,[a] Azasetron,[a] Dolasetron,[a] Mosapride[a]	Metoclopramide
5-HT$_4$	*Agonist*	*Cognition enhancement/anxiety*	*RS67333, FCE-29029, ML-10302*	—
5-HT$_4$	Antagonist	*Inflammatory bowel disease/GI motility disorder*	*GR125487, RS39604, SB204070*	Metoclopramide
5-ht$_{5A}$	—			—
5-ht$_{5B}$	—			—
5-ht$_6$	*Antagonist*	*Schizophrenia*	—	Traditional antipsychotics, e.g., Clozapine
5-HT$_7$	*Antagonist*	*Schizophrenia/circadian cycle modulation*	—	—
Serotonin transporter	Inhibitor	Depression/anorexia/compulsive disorders/pain	Sertraline, Fluoxetine, Paroxetine, Citalopram, Tianeptine, Venlafaxine[b]	MAOIs, Amitriptyline

Proposed indications and drugs under development and/or evaluation are shown in italics. For additional information, see Ref. 72.

[a] Drug with proven clinical efficacy, but not yet registered.

[b] Drug with pharmacological properties in addition to serotonin receptor actions.

selective agonist and antagonist ligands (see Chapter 1). For serotonin, 15 gene products encoding 14 receptor proteins and the serotonin transporter are currently recognized, although a physiological function for five of the receptor proteins has yet to be established (2) (see also Table 1).

Structural homologies among these gene products enable seven distinct serotonin receptor families to be defined: 5-HT_1 to 5-HT_7. With a single exception, all are members of the seven-transmembrane (TM) G-protein-coupled superfamily, the 5-HT_3 receptor being unique in belonging to the ligand-gated ion-channel superfamily. Within some of these families, receptor subtypes and/or allelic variants have been identified, increasing diversity further. In this respect the 5-HT_1 family exhibits the most marked heterogeneity. Recently, an important advance was made in the approach to serotonin receptor classification by aligning the nomenclature with the human genome, eliminating ambiguity and confusion resulting from structural and/or operational differences between species (3). This has resulted in an important nomenclature change for the receptor subtypes originally defined as $5\text{-HT}_{1D\alpha}$ and $5\text{-HT}_{1D\beta}$. These have now been renamed 5-HT_{1D} and 5-HT_{1B}, respectively, reflecting more accurately the fact that although they presently display near-identical operational profiles, they are nevertheless distinct gene products (4). In accomodating this change, the well-characterized "rodent" 5-HT_{1B} receptor is subsumed within the "human" 5-HT_{1B} class, with the appellation $r5\text{-HT}_{1B}$ (denoting rat 5-HT_{1B} receptor). However, it should be borne in mind that although these receptor homologs share a 99% sequence identity, a single amino acid difference in the seventh transmembrane domain results in completely different pharmacological profiles (5,6).

Although diverse in their pharmacology, serotonin receptor subtypes within each family share the same primary transduction system, presumably reflecting a common phylogenetic ancestry. Indeed, analysis of serotonin receptor evolution suggests that yet more serotonin receptor families remain to be discovered, although these receptors are likely to exhibit poor homology (<25%) with those currently identified (7).

III. SEROTONIN $5\text{-HT}_{1B/1D}$ AGONISTS FOR MIGRAINE

The possibility that acute relief from migraine might be achieved by selective activation of an atypical 5-HT receptor mediating cerebrovascular constriction was first proposed by Humphrey and colleagues toward the end of the 1970s (see Ref. 8). This concept arose from a careful appraisal of clinical and preclinical studies that suggested that (1) cranial blood vessels are in some way implicated in the manifestation of migrainous symptoms, (2) an alteration in serotonin biochemistry underlies attacks in many sufferers, (3) many effective antimigraine drugs interact with serotonin receptors as either agonists or antagonists, and (4) serotonin and other vasoconstrictors such as norepinephrine were reported to abort induced or spontaneous migraine, whereas vasodilator agents could provoke an attack. Coupled with earlier important experiments by Saxena (9,10) showing that many antimigraine drugs selectively constrict extracranial blood vessels in vivo, these data prompted Humphrey to conduct a systematic characterization of vascular 5-HT receptors. This led to the identification of a novel serotonin receptor mediating constriction of the carotid vascular bed (11–13). Initially characterized in isolated preparations of dog saphenous vein, this receptor was termed 5-HT$_1$-like to distinguish it from the 5-HT$_2$ (D) and 5-HT$_3$ (M) receptors recognized at that time (14). Using serotonin as a chemical template, novel, selective 5-HT$_1$-like receptor agonists were sought, resulting in the discovery of 5-carboxamidotryptamine (5-CT). Unfortunately, although selective for the saphenous vein 5-HT$_1$-like over 5-HT$_2$ and 5-HT$_3$ receptors, 5-CT also produced profound vasorelaxation, revealing the presence of a second 5-HT$_1$-like receptor mediating smooth muscle relaxation subtype now considered to be 5-HT$_7$ (2,15,16). Further chemical modification of 5-CT produced close analogs devoid of activity at the vasorelaxant 5-HT$_1$-like receptor, making available for the first time agonists with selectivity for the vasoconstrictor 5-HT$_1$-like receptor. An early compound, AH25086, was deemed suitable to test the hypothesis that such an agonist would be of value in aborting migraine, and to this end was infused intravenously into patients with an ongoing attack (17,18). Not only did this agent ameliorate or abolish headache in these patients, it was also shown to alleviate associated sensory disturbances and nausea, prompting the rapid development of a more robust drug candidate suitable for oral administration. This candidate, GR43175, is now widely known as sumatriptan, the most significant advance in migraine acute treatment in more than 60 years.

IV. SUMATRIPTAN: CURRENT VIEWS ON MECHANISM OF ACTION

An increasing body of evidence suggests that the vasoconstrictor 5-HT$_1$-like receptor that served as the target for sumatriptan is, in fact, the 5-HT$_{1B}$ receptor (19,20). However, it remains possible that the pharmacologically similar 5-HT$_{1D}$ receptor also contributes to the pharmacological and clinical actions of the drug (21,22). Like most of the so-called 5-HT$_1$-like agonists now under development, sumatriptan does not distinguish between these two receptor types; hence these agents are more properly referred to as 5-HT$_{1B/1D}$ agonists (see Fig. 1). With the continued discovery of novel serotonin receptors, sumatriptan has also been shown to display moderate to high

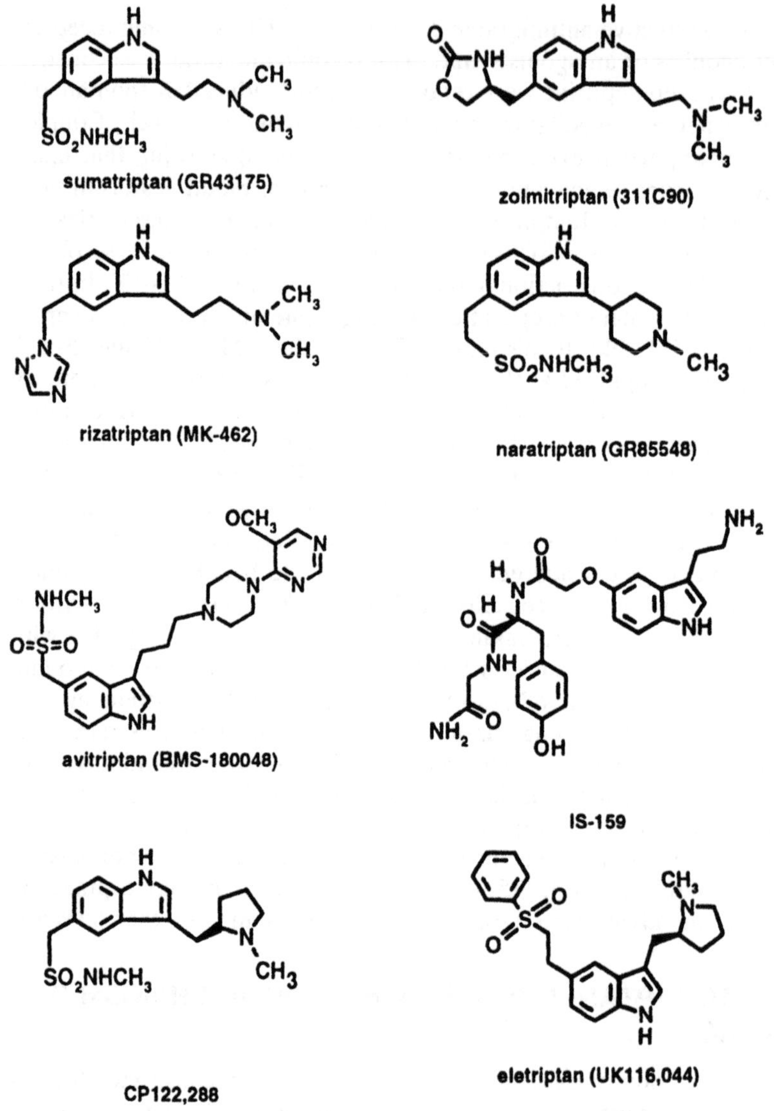

Figure 1 Chemical structures of some 5-HT$_{1B/1D}$ receptor agonists under development for the acute treatment of migraine. Published affinity estimates ($-\log K_I$ or $-\log [IC_{50}]$) at human recombinant 5-HT$_1$ receptor subtypes are shown in the table in Figure 2. (Modified from Ref. 73.)

Table 2 Affinities (-log K$_I$ or -log [IC$_{50}$]) at Human Recombinant 5-HT$_1$ Receptors of Antimigraine Drugs Known to Be or Putatively Effective in the Acute Treatment of Migraine

Drug	5-HT$_{1A}$	5-HT$_{1B}$	5-HT$_{1D}$	5-ht$_{1E}$	5-ht$_{1F}$
Sumatriptan	7.0	7.9	7.9	5.6	7.6
Zolmitriptan	6.5	8.2	9.2	<5.0	7.1
Rizatriptan	6.3	7.3	7.7	6.5	—
IS 159	6.0	8.5	8.8[a]	<5.0	<5.0
Naratriptan	7.1	8.7	8.3	—	—
BMS-180,048	6.7	7.7	8.3	<6.0	—
CP-122,288	—	7.5	8.1	—	—
Ergotamine	9.5	8.3	9.4	8.0	6.8
DHE	9.1	8.2	9.3	8.1	—

[a] Total 5-HT$_{1D}$ binding.
Source: From Ref. 24.

affinity for 5-HT$_{1A}$ and 5-HT$_{1F}$ receptors 23, but neither of these subtypes is thought to contribute to its antimigraine actions (24).

As with any innovative new therapeutic entity, the advent of sumatriptan stimulated intense clinical and preclinical research into migraine pathophysiology and opened vigorous debate about the drug's mechanism of action. At the outset, Humphrey and colleagues sought a 5-HT$_{1B/1D}$ receptor *full* agonist to ensure that the drug would exert a profound cranial vasoconstriction (8). Sumatriptan appeared to meet this criterion as judged by its intrinsic activity relative to 5-HT in a number of 5-HT$_{1B/1D}$ receptor-containing blood vessels, including mammalian cerebral vessels (25–27). It also quickly became clear that owing to its hydrophilic nature, sumatriptan penetrated the central nervous system very poorly (28). The idea was therefore promoted that sumatriptan acts simply as a cranial-selective vasoconstrictor, normalizing the status of meningeal and large intracranial vessels perceived to be dilated and inflamed during a migraine attack (29,30). The use of single positron emission tomography coupled with laser Doppler flowmetry to determine the cerebrovascular actions of sumatriptan in patients during and between migraine attacks suggests that the drug, at higher than therapeutic doses, may indeed produce constriction of the middle cerebral artery. However, no consistent evidence has been forthcoming to indicate that this represents a "normalization" of inappropriate vessel dilation (31–34). Whether or not vessel constriction per se represents the most important action of the drug therefore remains unproved. Indeed, the concept has been challenged repeatedly. On the basis of animal models of trigeminovascular activation, it has been proposed that the key activity of sumatriptan lies in its ability to inhibit directly the provoked release of vasodilator and proinflammatory neuropeptides (CGRP, substance P) from the trigeminal sensory nerves innervating dural and cerebral blood vessels (35,36). Significantly, it has been shown that the concentration of CGRP in jugular venous blood is elevated during migraine and that concentrations are reduced to normal after successful treatment of the attack with subcutaneous sumatriptan (35). This important result clearly implies that trigeminal activation is occurring during migraine and shows that sumatriptan can inhibit the process. However, whether this reflects a prejunctional neuroinhibitory effect of the drug or is secondary to vasoconstriction remains controversial (36). The debate has been fueled by in situ mRNA hybridization studies that have confirmed the presence of message for both 5-HT_{1B} and 5-HT_{1D} receptors in the human trigeminal ganglion as well as the smooth muscle of intracranial vessels (39). If the protein products of these mRNAs are expressed and functional in both the neuronal and vascular elements of the trigeminovascular system, then an action of sumatriptan at both of these sites would seem likely. In the light of these data, the most widely held view is that sumatriptan probably owes its clinical efficacy to a peripheral action on cephalic blood vessels, producing an inhibition of sensory neuroexcitability by directly constricting cerebral and meningeal vessels and blocking the release of CGRP and substance P from sensory terminals innervating these vessels.

V. DEVELOPMENT OF ZOLMITRIPTAN (311C90)

The research program that culminated in the development of zolmitriptan was started in April 1988, shortly after proof-of-concept studies with AH25086 and subsequently sumatriptan itself had confirmed the utility of $5\text{-HT}_{1B/1D}$ agonists in the acute treatment of migraine. The rationale for pursuing development of a second drug of this type was founded on scientific as well as commercial beliefs: first, that it was indeed possible to develop a drug with an improved pharmacodynamic and pharmacokinetic profile compared to sumatriptan that might offer therapeutic advantage, and second, that as an innovative new treatment for a debilitating illness with a poorly met therapeutic need, sumatriptan would increase awareness about the utility of $5\text{-HT}_{1B/1D}$ agonists and prime the market for the entry of a newer, potentially advantageous drug. Against a scientific and clinical background indicating that sumatriptan was a peripherally acting, $5\text{-HT}_{1B/1D}$ receptor full agonist with poor (14%) oral bioavailability in humans (40), the following characteristics were established as the basis for development of a new drug candidate:

The drug must be a selective 5-HT$_{1B/1D}$ receptor *partial* agonist to maximize the opportunity for organ-selective drug action.

The drug must exhibit reliable oral pharmacokinetics, meaning that ideal attributes would be rapid oral absorption, good bioavailability, and a plasma elimination half-life consistent with the symptomatic treatment of migraine.

Access to CNS components of the trigeminovascular system would not be ruled out, but evaluated early in the program for risk/benefit in terms of potential therapeutic advantage. The rationale for this approach was intuitive, but based on the fact that the bipolar organization of the trigeminal nerve offers an opportunity to inhibit neuroexcitation not only peripherally, but also at central synapses that might be perceived to be a "bottleneck" for cranial nociceptive transmission.

A. Identifying 5-HT$_{1B/1D}$ Receptor Partial Agonists

The decision to seek a partial agonist drug candidate was pragmatic and guided by reports suggesting that 5-HT$_{1B/1D}$ receptors are particularly prominent in intracranial blood vessels (26,27). A more objective reason concerned the propensity of 5-HT$_{1B/1D}$ receptors to exhibit pharmacological synergism such that in a tissue where 5-HT$_{1B/1D}$ receptor activation would normally produce no effect, or a barely discernible one, the presence of a second independently acting stimulant (at threshold concentrations and above) converts 5-HT$_{1B/1D}$ receptor occupancy into a significant response (41–43). The precise nature of this synergism is unknown, but available evidence suggests that it results from cross-talk between transducer components of G$_i$/G$_o$-coupled (e.g., 5-HT$_{1B}$, 5-HT$_{1D}$) and G$_q$/G$_{11}$ (e.g., AT$_1$, H$_1$, α_1, TxA$_2$) receptor systems.

To define the intrinsic efficacy required in novel 5-HT$_{1B/1D}$ agonists, the agonist activities of antimigraine agents known to be effective acutely were examined using ring preparations of the rabbit lateral saphenous vein as a bioassay (44,45). In addition to 5-HT itself, ergonovine, methylergonovine, ergotamine, dihydroergotamine, and methysergide were studied. Although methysergide is regarded as the prototypical migraine prophylactic, studies by Saxena (10) and Apperley et al. (11) implied that it behaved as a 5-HT$_1$-like receptor partial agonist. We were also aware of anecdotal evidence suggesting that the drug can be effective in acute treatment (46,47). The intrinsic efficacy of methysergide [τ relative to 5-HT = 0.3 (45)] was therefore set as the lower limit for novel agonist development. This decision, again pragmatic, carried two caveats. First, in humans methysergide undergoes rapid metabolism to methylergonovine, which appears in the plasma at 10 times higher concentrations than the parent and with an elimination half-life nearly four times longer (48). Second, our studies showed that relative to methysergide, methylergonovine is 10 times more potent as a 5-HT$_{1B/1D}$ receptor agonist with a higher intrinsic efficacy [τ relative to 5-HT = 0.7 (45)]. These data raise the possibility that methysergide acts effectively as a prodrug, releasing methylergonovine as a long-acting, though poorly selective, 5-HT$_{1B/1D}$ agonist. The possibility therefore existed that the intrinsic efficacy of methysergide might be too low to produce a therapeutic effect.

Having established bioassay conditions that produced reliable measures of agonist potency and intrinsic activity, a chemically diverse set of ligands was assessed to determine the key functional groups responsible for 5-HT$_{1B/1D}$ receptor affinity and selectivity. In this respect both methysergide and 5-CT were important early leads. In the absence of any knowledge about the receptor tertiary structure, the active analog approach described by Marshall et al. (49) was used to determine likely receptor-bound conformations of compounds in the set. Distances between functionally important groups were then computed with the aim of obtaining an overlay of these groups constrained by one set of distance ranges for all the active molecules. The synthesis of novel ligands based upon this preliminary template resulted, after several iterations, in the development of a more refined pharmacophore model, which, although only qualitative, nevertheless served as a framework to assess the "fit" of proposed chemical structures with the principal receptor recognition sites (see Ref. 50).

Initially, medicinal chemistry was guided by potency (p[A$_{50}$]) and intrinsic activity (relative to 5-HT = 1.0) measures for novel agonists. Experience with a number of serotonergic receptor systems in intact tissues had revealed that 5-HT itself invariably behaves as a partial agonist (51), providing confidence that p[A$_{50}$]s would be good approximations of the dissociation equilibrium constant, K_A. Moreover, early experience with the saphenous vein bioassay indicated that intrinsic activity values were highly reproducible, with a variance of ≤8% of the mean value (see Fig. 2). However, with more widespread use of the bioassay, estimates of agonist potency and, especially, intrinsic activity became increasingly variable and "operator-dependent." This problem compromised rigorous definition of the structure-activity relationship for intrinsic efficacy. The subsequent discovery that benextramine tetrahydrochloride could be used to inactivate the target receptors in saphenous vein provided a solution to the dilemma, allowing agonist affinity and relative efficacy estimates to be made that were largely free of operator-related variance (see table in Fig. 2). Indeed, when this approach was used in defining the parameters of agonism for sumatriptan, it was established that contrary to published reports and our own early data suggesting it was a full agonist (Fig. 2), the drug was unequiv-

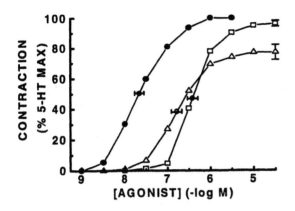

Operator	5-HT (p[A₅₀])	5-HT (α)	Test (p[A₅₀])	Test (α)	5-HT (pK_A)	Test (pK_A)	Test τrel.
A	8.40	100	7.17	0.89	7.30	6.40	0.25
A	8.29	100	6.85	0.75	7.28	6.45	0.27
A	7.91	100	6.77	0.60	7.33	6.41	0.22
B	8.06	100	6.66	0.78	7.32	6.37	0.29
B	7.54	100	6.51	0.61	7.30	6.42	0.30
Mean±s.e.m	8.04±0.15	100	6.79±0.11	0.73±0.06	7.31±0.01	6.41±0.01	0.27±0.01

Figure 2 5-HT₁ᵦ/₁ᴅ receptor-mediated constrictor effects of 5-HT (●), sumatriptan (□), and 311C90 (△) obtained in ring segments of rabbit saphenous vein. Data are the averages of four to six cumulative concentration-effect curves in each case. Vertical bars show the standard error of the mean and illustrate the excellent reproducibility of this assay (SE ≤ 8% of the mean) in the hands of a single operator. The table shows that variance is considerably increased when data from different operators (A and B) are pooled: measures of potency (p[A₅₀]) and intrinsic activity release to 5-HT (α) are given for 5-HT and a novel test agonist obtained using a paired-curve study design, which also enabled estimation of agonist affinities (pK_A) and the intrinsic efficacy of the test agonist relative to 5-HT (τrel).

ocally shown to be a partial agonist with an intrinsic efficacy (τ) value of 0.6 relative to 5-HT (45,52).

B. Identifying Orally Bioavailable 5-HT₁ᵦ/₁ᴅ Agonists

Using the approach outlined above, novel potent (pK_A ~ 8) indolamine agonists with selectivity for 5-HT₁ᵦ/₁ᴅ receptors and the requisite degree of partial agonism were rapidly identified. Unfortunately, as is often the case in drug discovery, the physicochemical attributes of these molecules were not compatible with oral bioavailability. Distribution studies revealed that the problem was poor absorption from the gastrointestinal tract. With the exception of active transport mechanisms, intestinal absorption depends either on partitioning of drug substance across cell membranes (transcellular absorption) or on the ability of the molecule to pass between the cells via aqueous pores (paracellular absorption). Thirty chemically diverse drugs known to be orally bioavailable were therefore surveyed and their distribution coefficients ($\log D_{pH7.4}$) and calculated molar refractivity (CMR) determined. These attributes represent measures of lipid solubility and molecular volume respectively and are shown plotted for each compound in Figure 3A.

Empirically, the plot can be divided into four regions:

Region 1 : comprising relatively small hydrophobic molecules presumed to cross the intestinal wall by either transcellular or paracellular absorption

Region 2 : comprising larger hydrophobic molecules, presumed to be dependent upon transcellular transport alone

Region 3 : comprising relatively small hydrophilic molecules, presumed to be dependent upon paracellular transport alone

Region 4 : an "exclusion zone," defining molecules too large and too hydrophobic to cross the intestinal wall via either route.

When these attributes were also measured for an array of novel 5-HT$_{1B/1d}$ agonists that were not absorbed, it became immediately obvious that these drugs occupied, or fell close to, the "exclusion zone." This dictated the need for smaller, more hydrophilic analogs. The pharmacophore model alluded to above suggested that this could be achieved by relaxing the criterion of high affinity, a maneuver that resulted in a series of partial agonists with oral bioavailability in rats ranging between 46 and 100% (Fig. 3B). These ligands retained selectivity for 5-HT$_{1B/1D}$ receptors, making the initial choice of development candidate dependent upon the balance between oral bioavailability and intolerable CNS penetration.

C. Optimizing Oral Bioavailability and CNS Penetration

A battery of behavioral tests were conducted on three candidate compounds (311C90, 324C91, and 1532W91) with similar affinities ($pK_A \sim 7$) and intrinsic efficacies (τ relative to 5-HT $= 0.5-0.6$) at the 5-HT$_{1B/1D}$ receptor, but with oral bioavailabilities in the rat of 46%, 70%, and 100%. Studies were performed in conscious mice, rats, beagle dogs, and cynomolgus monkeys. Central effects of 5-HT$_{1B/1D}$ agonists exhibit a marked species dependence

PANEL A

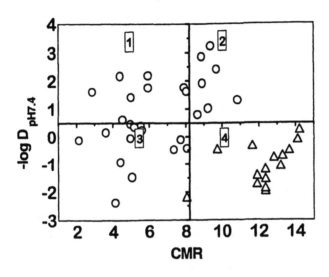

PANEL B

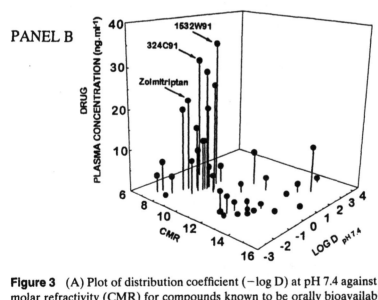

Figure 3 (A) Plot of distribution coefficient ($-\log D$) at pH 7.4 against calculated molar refractivity (CMR) for compounds known to be orally bioavailable (○) and for early nonorally bioavailable 5-HT$_{1D}$ agonist leads (△) identified at Wellcome. (B) Plasma concentrations (ng/ml) achieved 2 hr after oral administration against log D$_{pH7.4}$ and CMR for various novel 5-HT$_{1D}$ agonist drugs. The resulting three-dimensional surface was used to guide medicinal chemistry toward molecules with the appropriate attributes for oral bioavailability. See text for details.

with low oral doses of each compound tested producing a stereotypical set of behaviors in dogs, identical to those described for sumatriptan (28). Evidently, in this species, the blood-brain barrier is reasonably permeable even to sumatriptan, with approximately 20% of the administered dose appearing in the cerebrospinal fluid (28). In the other species tested, all three drugs were well tolerated following intravenous or oral administration, but the "tolerability index" (defined as the ratio: minimum drug dose to produce centrally mediated behaviors/ED_{50} for 5-$HT_{1B/1D}$ receptor-mediated effects) clearly followed the order: 311C90 > 324C91 $\gg$ 1532W91. Confirmation that this order reflected an increasing ability of these drugs to enter the brain of rats was obtained using GC-mass spectroscopy to determine the whole-brain concentration of each drug following intravenous steady-state infusion (Fig. 4).

As a result of the above studies, 311C90 was subjected to a comprehensive evaluation of its actions on the trigeminovascular system. As expected from its selectivity as a 5-$HT_{1B/1D}$ receptor partial agonist (311C90 pK_A = 6.7: $\tau_{rel. 5-HT}$ = 0.6), 311C90 constricted mammalian cerebral arterial vessels, reversed trigeminal-evoked increases in cerebral blood flow, and inhibited plasma protein extravasation into the dura provoked by electrical stimulation of the trigeminal ganglion (52–54). In anesthetized guinea pigs, the 5-$HT_{1B/1D}$ receptor-mediated constriction of extracranial vessels (presumed

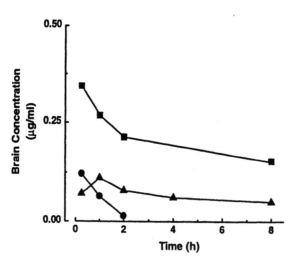

Figure 4 Time course for whole-brain concentrations of zolmitriptan ($\bullet$), 324C91 ($\blacktriangle$), and 1532W91 ($\blacksquare$) determined following steady-state intravenous infusion of each drug for 15 min into anesthetized rats (n = 3). Total administered dose was 10 mg/kg in each case.

to mimic intracranial vessel reactivity) was produced by 311C90 at the same intravenous doses required to inhibit extravasation. These effects are essentially similar to those described for sumatriptan (55). However, unlike sumatriptan, systemic administration of 311C90 also inhibited trigeminal excitation within the brainstem. This was first implied by studies showing that the drug not only blocked the trigeminal-evoked release of CGRP and substance P into jugular venous blood, it also blocked the release of vasointestinal polypeptide (VIP), a marker for VIIth cranial nerve excitation mediated by a brainstem reflex with trigeminal inputs (56). Subsequent elegant electrophysiological studies in anesthetized cats likewise showed that *intravenously* administered 311C90 inhibited both evoked potentials and unit activity of cells of the caudal (C2) trigeminal nucleus caudalis (TNC) during stimulation of the superior sagittal sinus, reaffirming that systematically administered 311C90 accesses central components of the trigeminovascular system to inhibit trigeminal neuroexcitation (57). This ability of 311C90 to produce a dual central and peripheral inhibition of trigeminal excitation was regarded as a potentially crucial distinction from sumatriptan, which cannot access the TNC unless the blood-brain barrier is first disrupted (58).

Confirmation that 311C90 recognizes and binds to specific central structures involved in cranial pain processing was provided by in vitro and ex vivo autoradiographic studies using [^{3}H]-311C90 (59–62). Incubation of the label with transverse sections of cat brainstem and cervical spinal cord revealed highly specific binding within the TNC, the nucleus tractus solitarius (NTS), and the dorsal horns of the spinal cord. In all other structures, binding was either absent or diffuse and at low density. The pharmacology of this [^{3}H]-311C90 binding was entirely consistent with the labeling of 5-HT$_{1B/1D}$ receptors (59,61). Furthermore, displacement studies with a 5-HT$_{1D}$ receptor–selective ligand, ketanserin, failed to reveal any regional differences in binding density inferring that the binding sites are homogeneously distributed (59,61). However, whether or not binding is to 5-HT$_{1B}$ or 5-HT$_{1D}$ receptors, or both, awaits the development of subtype-selective receptor probes. Ex vivo studies revealed an identical distribution of [^{3}H]-311C90, consistent with the ability of the drug to access these structures after systemic administration (60,62).

D. Choice of 311C90 as Development Candidate

The attributes identified at the outset as ideal for a 5-HT$_{1B/1D}$ agonist development candidate made 311C90 an obvious choice. Not only did the compound meet the necessary criteria of selectivity and 5-HT$_{1B/1D}$ receptor partial agonism, it emerged as a compound with perhaps the most appropriate balance between oral bioavailability and CNS penetration. Although the selection of a compound with access to the brain clearly represented a significant risk, the ability of such a compound to inhibit cranial nociceptive processing at a point of convergence within the brainstem was highly attractive and offered a real possibility of improving on the therapeutic profile of sumatriptan. This possibility was reinforced by the subsequent publication of two studies showing that sumatriptan is completely ineffective if administered subcutaneously 20–40 min before the anticipated onset of headache in cluster and migraine-with-aura patients (63,64). This result not only implies that patients must await the development of headache before they take the drug, it calls into doubt the relevance of peripheral actions, such as cerebral vasoconstriction, in the therapeutic actions of drugs in this class. It also implies that sumatriptan may indeed act at a central site of action, accessible to the drug only when central access is made possible by the onset of headache. This result is entirely consistent with the preclinical studies of Goadsby and colleagues showing that sumatriptan inhibits trigeminal neuroexcitation in the brainstem only after blood-brain barrier disruption (58). In turn, this raises the intriguing possibility that an increased permeability of this important barrier also accompanies the onset of migraine headache in humans.

Clinical studies with 311C90 have so far confirmed that the orally administered drug is a safe, highly effective therapy for the acute treatment of migraine, with an improved oral bioavailability relative to sumatriptan (48% vs. 14%) (65). Preliminary data also indicate that the drug indeed acts centrally to modulate the sensory input (66) and may be of value in the early, preemptive treatment of migraine (67), yet it produces no more central side effects than sumatriptan (68). Whether these promising results will translate ultimately into a clear-cut therapeutic advance in migraine treatment awaits the outcome of ongoing clinical trials.

VI. FUTURE DIRECTIONS

As a therapeutic breakthrough the significance of sumatriptan is underscored by the fact that, at the time Humphrey and colleagues began their task, there was no real understanding of migraine pathophysiology other than an implied involvement of the trigeminovascular system (69,70). Indeed, it is a testament to receptor-targeted drug discovery that sumatriptan has enabled more direct questions to be asked about the underlying causes of migraine than perhaps at any time in the history of this illness. The ensuing development of 311C90 was a rational attempt to maximize the therapeutic efficacy of this drug class. At least seven other 5-HT$_{1B/1D}$ agonist drugs are now in development, each broadly similar to both sumatriptan and 311C90 in terms of receptor specificity (Fig. 1). None have so far been shown to exhibit the dual central and peripheral inhibitory actions of 311C90. It is therefore difficult to foresee what advance, if any, these newer agents have to offer.

It bears emphasis that in spite of their clear-cut therapeutic efficacy, one of the perceived short-comings of 5-HT$_{1B/1D}$ agonist drugs is vasoconstriction. A target for preclinical and clinical studies has therefore been an agent that produces trigeminal neuroinhibition without vasoconstriction. Three approaches are presently under evaluation, two of which are receptor-targeted: (1) selective antagonists of the sensory neuropeptides, CGRP and substance P, and (2) selective 5-HT$_{1D}$ receptor agonists. The third approach evolved from the serendipitous observation that a novel analog of sumatriptan, CP-122,288, produces potent, selective inhibition of trigeminal neuroactivation by mechanisms that are presently unknown (71). The pursuit and clinical evaluation of so many diverse approaches to migraine acute therapy is unprecedented and reflects the enthusiasm created when a new and effective, receptor-targeted drug is available as a unique tool to probe the pathophysiology of the disease. Inevitably, this must lead to a better understanding of the merits and shortcomings of preclinical paradigms of migraine, diminishing the circularity that plagues this critical step in the drug discovery process and increasing the probability of identifying the most effective approach to the treatment of this debilitating condition.

GLOSSARY

AH25086 : 3-aminoethyl-N-methyl-1H-indole-5-methane carboxamide

Zolmitriptan (311C90) : (S)-4[[3-[2-(dimethylamino)ethyl]-1H-indol-5yl]methyl]-2-oxazolidinone

Rizatriptan (MK-462) : N,N-dimethyl-5-(1H-1,2,4-triazol-1-ylmethyl)-1H-indole-3yl-ethanamine

Naratriptan (GR85548) : N-methyl-3-(1-methyl-4-piperidinyl)-1H-indole-5-ethanesulphonamide

Avitriptan (BMS-180048) : 3-(2-aminoethyl-5(acetamidyl-3-(4-hydroxyphenyl))-propionamidyl-acetamidyloxy)-indole

Eletriptan (UK116,044) : (R)-3-[(1-methyl-2-pyrrolidinyl)methyl]-5-2(phenylsulphonyl)ethyl-1H-indole

IS 159 : 3-(aminoethyl-5-(acetamidyl-3-(4-hydroxyphenyl))-propionamidyl-acetamidyloxy)-1H-indole

CP-122,288 : 5-methylaminosulphonylmethyl-3{N-methylpyrrolidin-2R-yl-methyl}-1H-indole

ACKNOWLEDGMENTS

The author is indebted to numerous colleagues at the former Wellcome Research Laboratories, UK, whose enthusiasm and exceptional skills led to the discovery and subsequent development of 311C90 (zolmitriptan). Thanks are due in particular to A. D. Robertson, S. J. MacLennan, D. Prentice, A. P. Hill, and R. Glen, who contributed much of the preclinical data described in this chapter. Thanks also extend to Dr. Peter Goadsby (Institute of Neurology, London) and Dr. Lars Edvinsson (University Hospital, Lund) for providing a wider perspective through enjoyable and highly productive collaborative research.

REFERENCES

1. 1995, Decision Resources, Inc..
2. Hoyer D, Clarke DE, Fozard JR, et al. International Union of Pharmacology classification of receptors for 5-hydroxytryptamine (serotonin). Pharmacol Rev 1994; 46:157–203.
3. Hartig PR, Hoyer D, Humphrey PPA, Martin GR. Alignment of receptor nomenclature with the human genome: classification of 5-HT$_{1B}$ and 5-HT$_{1D}$ receptor subtypes. Trends Pharmacol Sci 1996; 17:103–105.
4. Hartig PR, Branchek TA, Weinshank RL. A subfamily of 5-HT$_{1D}$ receptor genes. Trends Pharmacol Sci 1992; 13:152–159.
5. Oksenberg D, Masters SA, O'Dowd BF et al. A single amino-acid difference confers major pharmacological variation between human and rodent 5-HT$_{1B}$ receptors. Nature 1992; 360:161–163.
6. Parker EM, Grisel DA, Iben LG, Shapiro RA. A single amino-acid difference accounts for the pharmacological distinctions between the rat and human 5-hydroxytryptamine$_{1B}$ receptors. J Neurochem 1993; 60:380–383.
7. Peroutka SJ. Serotonin receptor subtypes. Their evolution and clinical relevance. CNS Drugs 1995; 4(Suppl 1):18–28.
8. Humphrey PPA, Feniuk W, Perren MJ. 5-HT in migraine: evidence from 5-HT$_1$-like receptor agonists for a vascular aetiology. In: Sandler M, Collins GM, eds. Migraine: A Spectrum of Ideas. New York: Oxford University Press, 1990:147–172.
9. Saxena PR. The effects of antimigraine drugs on the vascular responses by 5-hydroxytryptamine and related biogenic substances on the external carotid bed of dogs: possible pharmacological implications to their antimigraine action. Headache 1972; 12:44–54.
10. Saxena PR. Selective vasoconstriction in carotid vascular bed by methysergide: possible relevance to its antimigraine effect. Eur J Pharmacol 1974; 27:99–105.
11. Apperley E, Humphrey PPA, Levy GP Evidence for two types of excitatory receptor for 5-hydroxytryptamine in dog isolated vasculature. Br J Pharmacol 1980; 68:215–224.
12. Feniuk W, Humphrey PPA, Perren MJ, Watts AD. A comparison of 5-hydroxytryptamine receptors mediating contraction in rabbit aorta and dog saphenous vein. Evidence for different receptor types obtained by use of selective agonists and antagonists. Br J Pharmacol 1985; 86:697–704.

13. Feniuk W, Humphrey PPA, Perren MJ. The selective carotid arterial vasocon-
    strictor action of GR43175 in anaesthetised dogs. Br J Pharmacol 1985;
    96:83–90.
14. Bradley PB, Engel G, Feniuk W et al. Proposals for the classification and
    nomenclature of functional receptors for 5-hydroxytryptamine. Neuropharma-
    cology 1986; 25:563–575.
15. Feniuk W, Humphrey PPA, Watts AD. 5-Hydroxytryptamine-induced relax-
    ation of isolated mammalian smooth muscle. Eur J Pharmacol 1983; 96:71–78.
16. Feniuk W, Humphrey PPA, Watts AD. 5-Carboxamidotryptamine—a potent
    agonist at 5-hydroxytryptamine receptors mediating relaxation. Br J Pharmacol
    1984; 86:697–704.
17. Doenicke A, Siegel E, Hadoke M, Perrin VL. Initial clinical study of
    AH25086B (5-HT$_1$-like agonist) in the acute treatment of migraine. Cephalal-
    gia 7:438–439.
18. Brand J, Hadoke M, Perrin VL. Placebo-controlled study of a selective 5-HT$_1$-
    like agonist, AH25086B, in relief of acute migraine. Cephalalgia 7:402–403.
19. Kauman AJ, Parsons AA, Brown AM. Human arterial constrictor serotonin
    receptors. Cardiovasc Res 1993; 27:2094–2103.
20. Connor HE, Beattie DT, Feniuk W et al. Use of GR55562, a selective 5-
    HT$_{1D}$ antagonist, to investigate 5-HT$_{1D}$ receptor subtypes mediating cerebral
    vasoconstriction. Cephalalgia 1995; 15(Suppl 14):99.
21. Martin GR. Vascular receptors for 5-hydroxytryptamine: distribution, function
    and classification. Pharmacol Ther 1994; 62:283–324.
22. Martin GR, Prentice DJ, MacLennan SJ. The 5-HT receptor in rabbit sapheno-
    pus vein. Pharmacological identity with the 5-HT$_{1D}$ recognition site? Fund
    Clin Pharmacol 1991; 5(5):417.
23. Connor HE, GlaxoWellcome Medicines Research Centre, Herts., UK. Per-
    sonal communication.
24. Martin GR. Serotonin receptor involvement in the pathogenesis and treatment
    of migraine. In: Goadsby PJ, Silberstein SD, eds. Blue Books of Practical
    Neurology: Headache. London: Butterworth-Heineman, 1996 (in press).
25. Humphrey PPA, Feniuk W, Perren MJ, Connor HE, Oxford AW, Coates,
    IH, Butina D. GR43175, a selective agonist for the 5-HT$_1$-like receptor in dog
    isolated saphenous vein. Br J Pharmacol 1988; 94:1123–1132.
26. Parsons AA, Whalley ET. GR43175 is an agonist at the 5-HT$_1$-like receptor
    mediating contraction of human isolated basilar artery. Cephalalgia 1989;
    9(Suppl 9):47–51.
27. Connor HE, Feniuk W, Humphrey PPA. Characterization of 5-HT receptors
    mediating contraction of canine and primate basilar artery by the use of
    GR43175, a selective 5-HT$_1$-like receptor agonist. Br J Pharmacol 1989;
    96:379–387.
28. Humphrey PPA, Feniuk W, Marriott AS, Tanner RJN, Jackson MR, Tucker
    ML. Pre-clinical studies on the anti-migraine drug, sumatriptan. Eur Neurol
    1991; 31:282–290.
29. Humphrey PPA, Feniuk W. Mode of action of the anti-migraine drug sumatrip-
    tan. Trends Pharmacol Sci 1991; 12:444–446.

30. Humphrey PPA, Apperley E, Feniuk W, Perren MJ. A rational approach to identifying a fundamentally new drug for the treatment of migraine. In: Saxena PR, Wallis DI, Wouters W, Bevan P, eds. Cardiovascular Pharmacology of 5-Hydroxytryptamine. Dordrecht: Kluwer, 1990:417–431.

31. Friberg L, Olesen J, Iversen HK, Sperling B. Migraine pain associated with middle cerebral artery dilatation: reversal by sumatriptan. Lancet 388; 13–17.

32. Caekebeke JFV, Zwetsloot CP, Jansen JC, Saxena PR, Ferrari MD. Sumatriptan increases the cranial blood flow velocity during migraine attacks: a transcranial doppler study. In: Olesen J, ed. Migraine and Other Headaches: The Vascular Mechanisms. New York: Raven Press, 1991:331–334.

33. Caekebeke JFV, Ferrari MD, Zwetsloot CP, Jansen J, Saxena PR. Antimigraine drug sumatriptan increases blood flow velocity in large cerebral arteries during migraine attacks. Neurology 1992; 42:1522–1526.

34. Ferrari MD, Haan J, Blokland JA, Arndt JW, Minnee P, Zwinderman AH, Pauwels EK, Saxena PR. Cerebral blood flow during migraine attacks without aura and effect of sumatriptan. Arch Neurol 1995; 52:135–139.

35. Moskowitz MA. Neurogenic versus vascular mechanisms of sumatriptan and ergot alkaloids in migraine. Trends Pharmacol Sci 1992; 13:307–311.

36. Humphrey PPA, Goadsby PJ. Controversies in headache. The mode of action of sumatriptan is vascular? A debate. Cephalalgia 1994; 14:401–410.

37. Goadsby PJ, Edvinsson L. The trigeminovascular system and migraine: studies characterising cerebrovascular and neuropeptide changes seen in humans and cats. Ann Neurol 1993; 33:48–56.

38. Edvinsson L, Goadsby PJ. Neuropeptides in migraine and cluster headache. Cephalalgia 1994; 14:320–327.

39. Bouchelet I, Cohen Z, Seguela P, Hamel E. Differential expression of sumatriptan-sensitive 5-HT$_1$ receptors in human neuronal and vascular tissues. In: Sandler M, Ferrari M, eds. Migraine Research—Towards the Third Millennium. 1996 (in press).

40. Fowler PA, Lacey LF, Thomas M, Keene ON, Tanner RJN, Baber NS. The clinical pharmacology, pharmacokinetics and metabolism of sumatriptan. Eur Neurol 1991; 31:291–294.

41. MacLennan SJ, Martin GR. Effect of the thromboxane A$_2$-mimetic U44619 on 5-HT$_1$-like and 5-HT$_2$ receptor-mediated contraction of the rabbit isolated femoral artery. Br J Pharmacol 1992; 107:418–421.

42. Craig DA, Martin GR. 5-HT$_{1B}$ receptors mediate potent contractile responses to 5-HT in rat caudal artery. Br J Pharmacol 1993; 109:609–611.

43. Martin GR, MacLennan SJ, Maxwell M, Smith RR. A general model of pharmacological synergism. Br J Pharmacol 1996 (in press).

44. Martin GR, MacLennan SJ. Analysis of the 5-HT receptor in rabbit saphenous vein exemplifies the problems of using exclusion criteria for receptor classification. Naunyn-Schmiedeberg's Arch Pharmacol 1990; 342:111–119.

45. MacLennan SJ, Martin GR. Actions of non-peptide ergot alkaloids at 5-HT$_1$-like and 5-HT$_2$ receptors mediating vascular smooth muscle contraction. Naunyn-Schmiedeberg's Arch Pharmacol 1990; 342:120–129.

46. Sicuteri F. Prophylactic and therapeutic properties of 1-methyl-lysergic acid butanolamide in migraine. Int Arch Allergy 1959; 15:300–307.

47. Sicuteri F, Florence University, Italy: Personal communication.

48. Bredberg U, Eyjolfsdottir GS, Paalzow L, Tfelt-Hansen P, Tfelt-Hansen V. Pharmacokinetics of methysergide and its metabolite methylergometrine in man. Eur J Clin Pharmacol 1986; 30:75–77.

49. Marshall GR, Barry CD, Bosshard HE, Dammkoehler RA, Dunn DA. The conformational parameter in drug design: the active analogue approach. Computer-Assisted Drug Design, ACS Symposium Series 112, American Chemical Society, Washington, DC, 1979.

50. Glen RC, Martin GR, Hill AP, Hyde RM, Woollard PM, Salmon JA, Buckingham J, Robertson AD. Computer-aided design and synthesis of 5-substituted tryptamines and their pharmacology at the 5-HT$_{1D}$ receptor: discovery of compounds with potential anti-migraine properties. J Med Chem 1995; 38:3566–3580.

51. Martin GR, Leff P, MacLennan SJ. Tryptamine fingerprints in the classification of 5-hydroxytryptamine receptors. In: Saxena PR, Wallis DI, Wouters W, Bevan P, eds. Cardiovascular Pharmacology of 5-Hydroxytryptamine. Dordrecht, Holland: Kluwer Academic Publishers, 1990:157–162.

52. Martin GR, Robertson AD, MacLennan SJ, Prentice DJ, Barrett VJ, Buckingham J, Honey A, Giles H, Moncada S. Receptor specificity and trigeminovascular actions of a novel 5-HT$_{1D}$ receptor partial agonist, 311C90 (zolmitriptan). Br J Pharmacol 1996 (in press).

53. Martin GR. Pre-clinical profile of the novel 5-HT$_{1D}$ receptor agonist 311C90. In: Clifford-Rose, ed. New Advances in Headache Research, Vol 4. Nishimura: Smith-Gordon, 1994:3–4.

54. Martin GR. Inhibition of the trigemino-vascular system with 5-HT$_{1D}$ agonist drugs: selectively targeting additional sites of action. Eur Neurol 1996; 36(Suppl 2):13–18.

55. Buzzi MG, Moskowitz MA. The antimigraine drug sumatriptan (GR43175) selectively blocks neurogenic plasma protein extravasation from blood vessels in dura mater. Br J Pharmacol 1990; 99:202–206.

56. Goadsby PJ, Edvinsson L. Peripheral and central trigeminovascular activation in cat is inhibited by the serotonin (5-HT)-1D receptor agonist 311C90. Headache 1994; 34:394–399.

57. Goadsby PJ, Hoskin KL. Trigeminal neuronal activity is inhibited by intravenous administration of 311C90 in the cat. Cephalalgia 1995; 15(Suppl 14):106.

58. Kaube H, Hoskin KL, Goadsby PJ. Inhibition by sumatriptan of central trigeminal neurones only after blood-brain barrier disruption. Br J Pharmacol 1993; 109:788–792.

59. Mills A, Rhodes P, Martin GR. [^{3}H]311C90 binding sites in cat brain stem: implications for migraine treatment. Cephalalgia 1995; 15(Suppl 14):116.

60. Knight YE, Goadsby PJ. Central nervous system distribution of [^{3}H]311C90 in cat. Cephalalgia 1995; 15(Suppl 14):214.

61. Mills A, Rhodes P, Martin GR. In vitro autoradiographic studies with [^{3}H]311C90 in cat brain stem: relevance to antimigraine actions of 311C90 (zolmitriptan). Eur J Pharmacol 1996 (in press).

62. Knight YE, Goadsby PJ. Central nervous system distribution of [^{3}H]-311C90 in cat. Cephalagia 1996; 15(Suppl 14):214.
63. Bates D, Ashford E, Dawson R, Ensink FB, Gilhus NE, Olesen J, Pilgrim AJ, Shevlin P. Subcutaneous sumatriptan during the migraine aura. Sumatriptan Aura Study Group. Neurology 1994; 44:1587–1592.
64. Monstad I, Krabbe A, Micieli G, Prusinski A, Cole J, Pilgrim A, Shevlin P. Preemptive oral treatment with sumatriptan during a cluster period. Headache 1995; 35:607–613.
65. Dixon R, On N, Posner J. High oral bioavailability of the novel 5-HT1D agonist, 311C90. Cephalalgia 1995; 15(Suppl 14):218.
66. Schoenen J, University of Liège, Belgium. Personal communication.
67. Dowson A, Rampul-Gokulsing S, Klein K, Cox R, Wilkinson M, Gross M. Can oral 311C90, a novel 5-HT$_{1D}$ agonist, prevent migraine headache when taken during an aura? Cephalalgia 1995; 15(Suppl 14):173.
68. Diener HC, Klein KB for the Multinational Oral 311C90 and Sumatriptan Comparative Study Group. (In preparation.)
69. Moskowitz MA, Rheinhard JF, Romero J, Melamed E, Pettibone DJ. Neurotransmitters and the fifth cranial nerve: is there a relation to the headache phase of migraine? Lancet 1979; 2:883–885.
70. Penfield W, McNaughton FL. Dural headache and innervation of the dura mater. Arch Neurol Psych 1940; 44:43–75.
71. Lee WS, Moskowitz MA. Conformationally restricted sumatriptan analogues, CP-122,288 and CP-122,638 exhibit enhanced potency against neurogenic inflammation in dura mater. Brain Res 1993; 626:303–305.
72. Current Contents—Serotonin ID Research Alert, 1996; 1(1):39–50.
73. Hamel E. 5-HT$_{1D}$ receptors: pharmacology and therapeutic potential. Current Drugs—Serotonin ID Research Alert 1996; 1(1):19–29.

8

Histamine H₂-Receptor Antagonists

Michael E. Parsons
University of Hertfordshire, Hatfield, Hertfordshire, England

I. ACID-RELATED DISEASES

Prior to the mid-1970s the available medical therapy for the so-called "acid-related diseases," i.e., peptic ulcer and gastroesophageal reflux disease (GORD), was inadequate.

The choice was between nonselective anticholinergic drugs such as probanthine, with their associated side effects of dry mouth and blurred vision, or antacids, which, to be effective, had to be taken frequently and in large quantities. The latter often led to low patient compliance, and gastrointestinal side effects such as diarrhea or constipation were not uncommon.

Surgery was in some cases an effective alternative to drug therapy, and operations ranging from total or partial gastrectomy to highly selective vagotomy were performed. However, vagotomy is an expensive and fairly complex procedure and surgery is associated with occasional mortality and significant morbidity with, depending on the operation, up to 22% of patients having dumping syndrome (1).

All these therapies are directed at the control of gastric acid, and to a lesser extent, pepsin secretion. This was because, until recently, acid was considered the primary factor in the etiology of peptic ulcer disease and GORD. Clearly the fact that acid-inhibiting or -neutralizing drugs were effective therapy in this disease area confirmed the importance of acid. However, the fact that not all individuals even with hyperacidity developed ulceration clearly indicated that other factors were involved and the concept of the gastrointestinal mucosal barrier and its breakdown evolved (2). More recently the role of the bacterium *Helicobacter pylori* has received much

195

attention, and convincing evidence has accumulated for its etiological importance particularly in duodenal ulcer but also in gastric ulcer and perhaps gastric cancer (3).

II. HISTAMINE H₂-RECEPTOR ANTAGONISTS: THE BACKGROUND

In the early 1960s the focus in the field of upper gastrointestinal diseases was on acid secretion. The control of gastric acid secretion was known to be complex, with three primary chemical stimulants, gastrin, acetylcholine, and histamine, being involved. However, the relationships between these secretagogues and their relative importance in the *physiological* control of acid secretion was poorly understood. Part of the problem was the absence of suitable analytical tools such as receptor antagonists. As noted above, nonselective anticholinergic drugs could block the effect of acetylcholine and the activity of the parasympathetic vagus nerves, which were known to be important controlling factors since the turn of the century. However, no antagonists for the acid secretory stimulant effect of either gastrin or histamine were known.

In the case of histamine this was surprising since antihistamine drugs were first described in the mid 1930s (4). Over the subsequent two decades increasingly potent and specific antihistamines, such as benadryl and mepyramine, were developed, which would inhibit the effects of histamine in, for example, contracting smooth muscle from the gastrointestinal and respiratory tract. However, during the same period a number of actions of histamine, e.g., increase in the rate of beating of the isolated atrium, inhibition of contractions of the rat uterus, and stimulation of gastric acid secretion, were found to be refractory to inhibition by these classical antihistaminic drugs. As early as 1948 it had been suggested that there may be two "receptive substances" for histamine (5).

III. THE HISTAMINE H₂-RECEPTOR RESEARCH PROGRAM

Data such as these stimulated the start of a research program in 1964 at SmithKline and French Research that was directed at establishing the existence of histamine receptor heterogeneity and the discovery of selective antagonists of the receptor not blocked by the classic antihistamines such as mepyramine.

Clearly the program needed a chemical starting point and this was based on analogy with the catecholamine field. The β-adrenoceptor antagonists bear a fairly close structural relationship to the agonists epinephrine and norepinephrine whereas the α-adrenoceptor antagonists are structurally very dissimilar. The classical antihistamines are also chemically remote from histamine, and by analogy, the structure of histamine itself was taken as the chemical starting point.

Five bioassay systems were established to quantify the agonist and antagonist activity of novel compounds. Two assays were set up in which the histamine response was blocked by mepyramine, the histamine-induced contraction of the isolated guinea pig ileum and the histamine-induced contraction of the body of the stomach of the anesthetized rat in vivo. The three mepyramine-insensitive histamine responses were stimulation of gastric acid secretion in the anesthetized rat, the contraction frequency of the guinea pig right atrium in vitro, and inhibition of the electrically evoked contractions of the rat uterus in vitro.

Using these bioassays to measure agonist potency, it was found that methylation of histamine in the 4-position of the imidazole ring to give 4-methyl histamine produced a compound that had significantly greater potency than histamine in the mepyramine-insensitive assays and that the reverse was true for 2-methyl histamine (Table 1). These data provided support for the concept of heterogeniety in the histamine receptor population. Two years after the commencement of the research program Ash and Schild published a paper (6) that defined the pharmacological receptors involved in the mepyramine-sensitive histamine responses as H_1 receptors.

In the first 4 years of the program some 200 close analogs of histamine were synthesized and tested. Refined pharmacological data were provided to the medicinal chemists to support their structure-activity analysis, and a close interaction between the biologists and chemists was established, which provided the basis for subsequent success. Initial chemical effort concentrated on modification of the imidazole ring but did not lead to compounds exhibiting the desired antagonist activity. Attention was then directed toward modification of the side chain, and in 1968 a simple guanidine analog of histamine, N-α-guanylhistamine, was found to be a very weak histamine antagonist at non-H_1 receptors; this proved to be the essential lead compound (Fig. 1). It was then found that potency could be enhanced by lengthening the side chain and a series of related compounds (amides) were synthesized, but they proved to be partial agonists having both inhibitory and stimulant properties.

Clearly it was critical to separate these activities and it appeared that these compounds might mimic histamine because they had too great a chemical resemblance to it. In an attempt to separate these activities the strongly basic amide group was replaced by nonbasic groups which, though polar, would not be charged. Synthesis of a thiourea derivative produced a compound that lacked agonist activity but was only a weak antagonist. Further extension of the side chain increased activity and led to the synthesis

Table 1 The Agonist Activities of 4-Methyl and 2-Methyl Histamine (relative to histamine = 100%)

Compound	Assay				
	Mepyramine sensitive		Mepyramine insensitive		
	Guinea pig ileum in vitro	Rat stomach contractions in vivo	Guinea pig atrium in vitro	Rat uterus in vitro	Rat gastric secretion in in vivo
4-Methyl histamine	0.2	0.3	43.0	25.3	38.9
2-Methyl histamine	16.5	18.6	4.4	2.1	2.0

HISTAMINE

$CH_2.CH_2.NH_2$

THE 'LEAD' COMPOUND
N^a-guanylhistamine
A weak antagonist and partial agonist.

$CH_2.CH_2.NH$ — $\overset{C.NH_2}{\underset{+NH_2}{\|}}$

SK&F 91486
Increased activity by lengthening the
sidechain.

CH_2 $CH_2.CH_2.NH.\overset{C.NH_2}{\underset{+NH_2}{\|}}$

THIOUREA ANALOGUE
A weak antagonist without partial
agonist activity.

$CH_2.CH_2.CH_2.NH.\overset{C.NH_2}{\underset{S}{\|}}$

SK&F 91863
Further sidechain extension adds
potency to antagonist effect.

CH_2 $CH_2.CH_2.CH_2.NH.\overset{C.NH_2}{\underset{S}{\|}}$

BURIMAMIDE
The first H_2-receptor antagonist
to be administered to humans.

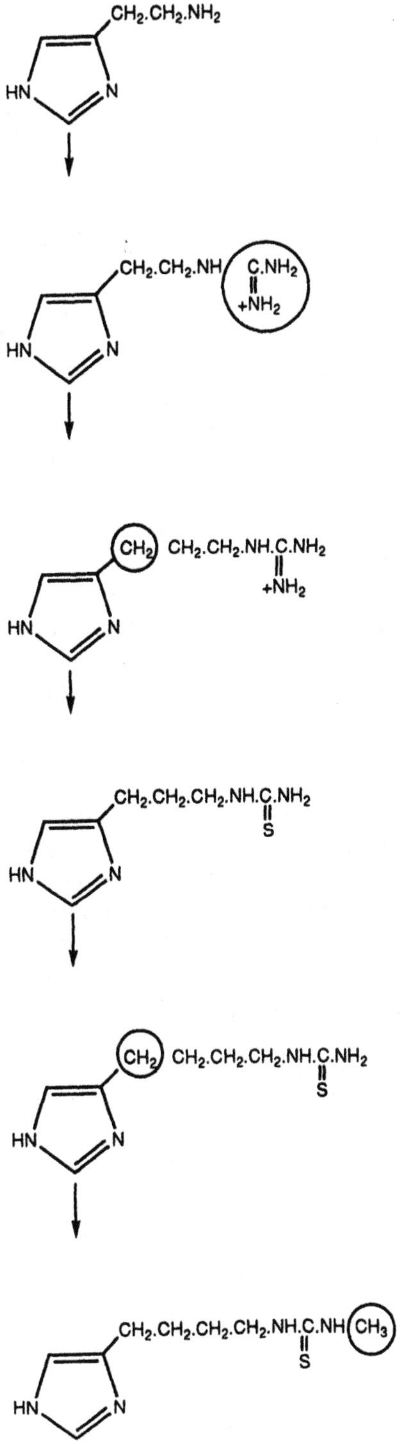

$CH_2.CH_2.CH_2.CH_2.NH.\overset{C.NH}{\underset{S}{\|}}CH_3$

Figure 1 The development of burimamide.

of the first competitive antagonist at non-H_1 histamine receptors, burimamide, and thereby permitted the definition of histamine H_2 receptors (7). Burimamide was the 700th compound tested and showed a reasonable degree of selectivity for the histamine H_2 receptor having no significant interaction with β-adrenoceptors, histamine H_1 receptors, or muscarinic receptors. The compound was found to inhibit histamine-stimulated gastric acid secretion in a range of experimental animals (rat, cat, and dog) and also in healthy human volunteers (8).

However, burimamide lacked sufficient potency after the oral route of administration to allow its clinical development. A study of the chemical properties of burimamide suggested that the electronic influences on the imidazole part of the structure should be altered. Substitution of a methyl group in the 4 position of the imidazole ring and a sulfur atom in the side chain led to the synthesis of metiamide (Fig. 2), a compound some 10 times more potent than burimamide with antisecretory activity that could be detected after oral administration. Double-blind studies in duodenal ulcer patients demonstrated that metiamide could produce a profound inhibition of gastric acid secretion after oral dosing (9). Metiamide was taken into clinical trials in 1972 and was shown to be an effective treatment for duodenal ulcer disease. However, it had to be withdrawn because a small number of patients developed a reversible granulocytopenia. Concern was expressed that this toxic effect would be a common feature of all H_2 antagonists since histamine H_2 receptors had been identified on a range of white cell types including lymphocytes (10). Alternatively H_2 blockade may have affected maturation of the bone marrow. However, the problem turned out to be one of chemistry rather than pharmacology, the toxic effect of metiamide being associated with the thiourea moiety in the side chain of the molecule.

Investigation of nonthiourea alternatives led to the synthesis of a cyanoguanidine analog, cimetidine (11), which did not possess this toxicity problem, and which was taken into clinical trials. The advent of histamine H_2-receptor antagonists in general and cimetidine in particular led to two major breakthroughs. The first was in our understanding of the physiological control of gastric acid secretion and the second (which was a reflection of the first) was a revolution in the treatment of acid-related diseases.

IV. THE ROLE OF HISTAMINE IN THE CONTROL OF ACID SECRETION

As noted above, the exact physiological role of histamine had been a subject of some controversy with some advocating a pivotal position (12,13), while others relegated it to a purely *pharmacological* role unrelated to normal *physiology* (14). It became apparent at an early stage (7) that H_2 antagonists

BURIMAMIDE

$$CH_2.CH_2.CH_2.CH_2.NH.\underset{\underset{S}{\|}}{C}.NH.CH_3$$

THIABURIMAMIDE
Incorporation of an electron-withdrawing
atom in replacement of a methylene group
increased potency.

$$CH_2.S.CH_2.CH_2.NH.\underset{\underset{S}{\|}}{C}.NH.CH_3$$

METIAMIDE
Incorporation of an electron-releasing
methyl group at the vacant 4 (5) position
of the ring stabilized the favored tautomer
and created a molecule ten times more
potent than burimamide *in vitro*.

$$CH_2.S.CH_2.CH_2.NH.\underset{\underset{S}{\|}}{C}.NH.CH_3$$

METIAMIDE ISOSTERES
Replacement of the thiourea sulfur atom
(=S) of metiamide was carried out with
various isoteres (oxygen; imino nitrogen).
The urea and guanidine isosteres lacked
potency, however. For the guanidine this
was attributed to the positive charge.
Reduction of the charge by substitution
of electron-withdrawing elements
reduced the pK$_a$ to approach that of
thiourea (metamide). The cyanoguanidine
was a more potent antagonist and was
selected for development.

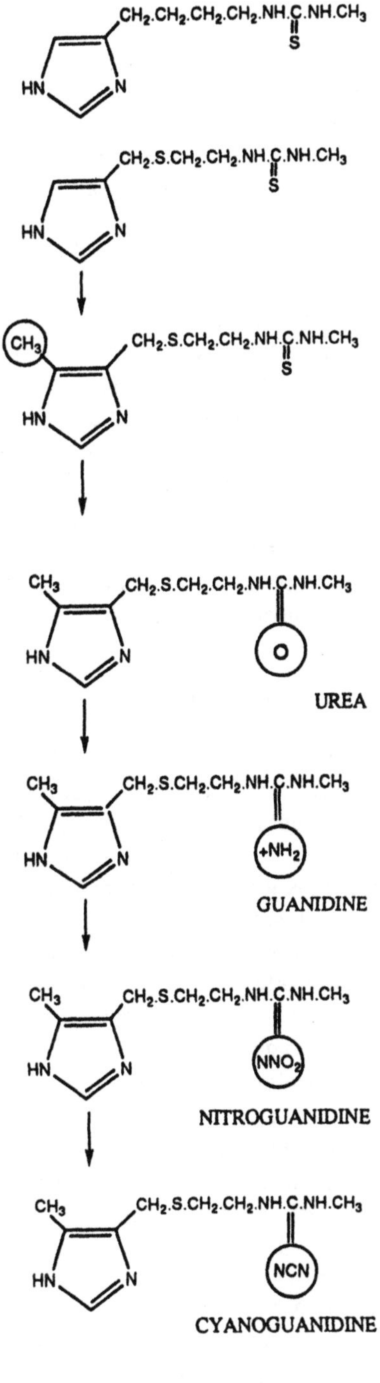

$$CH_3 \quad CH_2.S.CH_2.CH_2.NH.C.NH.CH_3$$
O
UREA

$$CH_3 \quad CH_2.S.CH_2.CH_2.NH.C.NH.CH_3$$
+NH$_2$
GUANIDINE

$$CH_3 \quad CH_2.S.CH_2.CH_2.NH.C.NH.CH_3$$
NNO$_2$
NITROGUANIDINE

$$CH_3 \quad CH_2.S.CH_2.CH_2.NH.C.NH.CH_3$$
NCN
CYANOGUANIDINE

TAGAMET ® (cimetidine, SK&F)

Figure 2 The development of cimetidine.

could block the acid secretory stimulant effect of gastrin (or the synthetic analog pentagastrin) in a manner qualitatively and quantitatively similar to their effects against histamine. Subsequently it was shown that H_2 antagonists could inhibit virtually all forms of basal and stimulated acid secretion in both experimental animals and humans, and this placed histamine firmly in a key physiological controlling position. The exact nature of this key role has been the subject of considerable debate and has led to two hypotheses reviewed by Black and Shankley (15). The "transmission" hypothesis proposes that the other main secretory stimulants gastrin and acetylcholine act via release of histamine from endogenous stores in the gastric mucosa, which then acts as the final common mediator on H_2 receptors on the parietal cell. The "permissive" hypothesis states that all three mediators, histamine, gastrin, and acetylcholine, act directly on their discrete receptors on the parietal cell, but gastrin and acetylcholine may only work in the presence (or with the permission) of histamine. Current experimental evidence suggests that a scheme combining aspects of both hypotheses provides the most accurate model for a complex controlling mechanism.

V. CLINICAL USE OF CIMETIDINE (TAGAMET)

The clinical impact of cimetidine (Tagamet) was enormous. It has been estimated that in the first 6 years after its introduction in 1976 it had been used to treat 20 million patients in 123 countries. In duodenal ulcer patients on 4-week therapy the overall healing rate based on 18 separate double-blind trials was 76% compared to 42% for placebo. This was accompanied by a significant reduction in frequency and severity of pain and in consumption of antacids. In economic terms it has been estimated that the total cost of duodenal ulcer disease in 1 year (1977) in the United States could have been reduced by 29% (US$644 million) if 80% of the patients had received Tagamet therapy (16). There has also been a significant reduction in the number of surgical procedures. For example, information from six centers in the United Kingdom was compared on the basis of the number of operations performed 5 years before and 4 years after the introduction of cimetidine. The comparison showed a mean reduction of 39% in the number of operations since the drug was introduced (17).

Although duodenal ulcer was the primary disease target for cimetidine, it has been shown to be effective in gastric ulcer (18) and to a lesser extent in GORD (19). It has also been used with some success in other acid-related diseases such as stress ulcer, Zollinger-Ellison syndrome, pulmonary acid aspiration syndrome, and drug-associated gastroduodenal lesions.

VI. DEVELOPMENTS SINCE CIMETIDINE

Since cimetidine, nearly 30 histamine H$_2$-receptor antagonists have been taken into development although only four have made it to the marketplace. The first was ranitidine (Zantac) in which a furan ring replaces the imidazole in cimetidine and the cyano-guanidine group is replaced by a 1,1-diamino-nitroethine. Ranitidine is more potent than cimetidine and has a somewhat superior side effect profile. Subsequently famotidine (Pepcid), a guanidinothiazole structure, came on to the market and remains the most potent H$_2$ antagonist available for clinical use. The other two marketed compounds, nizatidine and roxatidine, do not represent a significant advance over the already available compounds. Because of the short half-life of both cimetidine and ranitidine, coupled with the perception that gastric acid secretion needed to be controlled throughout a 24-hr period, both SK&F and Glaxo successfully synthesized and started to develop long-acting H$_2$ antagonists (Lupitidine at SK&F, Loxtidine at Glaxo). Lupitidine was some 16 times more potent than cimetidine in the dog and had an extended duration of antisecretory activity (19). Loxtidine behaved as an unsurmountable H$_2$ antagonist in vitro and had a very prolonged duration of action in the dog (20). In humans a dose of 40 mg twice daily maintained intragastric pH at values above 4.5 throughout 24 hr (21).

Both compounds, however, were withdrawn from development because of toxic effects uncovered in the rat 2-year carcinogenicity studies. Carcinoid tumors were found in the stomachs of the rats, and this was believed to be in part at least a consequence of the prolonged achlorhydria caused by the high doses of these long-acting antisecretory drugs. This achlorhydria leads to sustained hypergastrinemia because the control of the release of the hormone gastrin is partly influenced by intragastric pH, an elevated pH leading to increased secretion. In addition, to be a secretory stimulant, gastrin has trophic effects on tissue and it is this effect that is believed to lead to gastric hyperplasia and carcinoid formation (22).

While the search for long-acting H$_2$ antagonists was going on, clinical pharmacology studies were investigating optimal dosing regimes for the available H$_2$ antagonists. The result of the studies led to a reduction in four times a day to twice-daily dosing (23). Subsequently a single nighttime dose was found to be as effective as twice-daily dosing (24) and nighttime dosing has become the standard treatment for duodenal ulcer. This also indicates the importance of overnight gastric acidity in the pathogenesis and symptomatology of duodenal ulcer disease.

Although H$_2$ antagonists are still widely used in the treatment of acid-related diseases, their dominance is being eroded by two alternative therapeutic approaches. The relative lack of therapeutic success of H_2 antagonists in GORD is primarily because they do not control gastric acid secretion adequately. The H^+/K^+-ATPase (proton pump) inhibitors such as omeprazole (25), which target the final step in acid production, provide more effective control of acid secretion and are a significant therapeutic advance in this field. Recently the recognition that the bacterium *H. pylori*, which occurs in the stomach, is a major etiological factor in duodenal ulcer disease and possibly in gastric ulcer and gastric cancer has led to the use of antibiotics to eradicate the organism (26). Current therapeutic regimens normally involve the use of two antibiotics plus a proton pump inhibitor. This combination of drugs has led to a marked reduction in the relapse rate of duodenal ulcers on cessation of therapy compared to that seen with H_2 antagonists.

VII. LESSONS LEARNED FROM THE DISCOVERY AND DEVELOPMENT OF HISTAMINE H_2-RECEPTOR ANTAGONISTS

It is clear from the above discussion that selective and competitive receptor antagonists can be discovered on the basis of manipulation of the structure of natural substances. For a more detailed discussion of the structure-activity relationships in the histamine field, see Ganellin (27). This is clearly not unique to histamine, and other examples can be drawn from the peptide field, e.g., spantide as a substance P antagonist. Whether this will lead to the development of drugs as therapeutically useful as cimetidine is another matter.

To be successful, receptor-based drug discovery requires close interaction between the pharmacologists and the medicinal chemists with the former setting up appropriate and robust in vitro and in vivo bioassays. This is of course true for other approaches to drug discovery, but the ability of the pharmacologist to provide prompt and accurate biological data to support the chemist's S.A.R. is particularly critical in a receptor-based approach.

The development of drugs discovered from the receptor-based approach suffers the same high attrition rate associated with all drug development programs. As noted earlier, although 30 histamine H_2-receptor antagonists entered development, only four reached the marketplace. The reason for failure ranged from an unacceptable side-effect profile to lack of clinical efficacy compared to drugs of the same class already on the market. Targeting a specific receptor should lead to an inbuilt selectivity since actions at other receptors should be weeded out during the development process. However, toxicity associated with the chemistry of the molecule cannot normally be foreseen although metiamide provides an example of a toxic effect that was (in part) anticipated and an action taken to circumvent it (the introduction of cimetidine).

Would the research to discover H$_2$ antagonists have been accelerated if the benefits of molecular biology had been available at the time? Both H$_1$ and H$_2$ receptors have been subsequently cloned and expressed and are available for ligand-binding studies. However, I think the functional pharmacological assays used to differentiate the histamine receptors turned out to be accurate and provided more than adequate assays to support the medicinal chemistry and its structure-activity analysis. The availability of expressed-receptor assays might have provided a higher throughput of compounds but would have needed the support of the functional assays.

Finally, research in a particular disease area does not stand still even when a successful drug such as cimetidine is discovered. The proton pump inhibitors represent a significant further advance in antisecretory therapy, and if vaccination can eradicate *H. pylori*, then peptic ulcer may be a disease of the past.

REFERENCES

1. Thompson JL, Wiener J. Evaluation of surgical treatment for duodenal ulcer: short and long term effects. Clin Gastroenterol 1984; 2:569–600.
2. Flemstrom G, Turnberg LA. Gastroduodenal defense mechanisms. Clin Gastroenterol 1984; 13:327–341.
3. Lee A. The nature of *Helicobacter pylori*. Scand J Gastroenterol 1996; 31(Suppl 214):5–8.
4. Bovet D, Staub A-M. Action protectrice des ethers phenoliques an cours de l'intoxication histaminique. CR Soc Biol (Paris) 1937; 124:547–549.
5. Folkow B, Haeger K, Kahlson G. Observations on reactive hyperaemia as related to histamine, on drugs antagonising vasodilation induced by histamine and on vasodilator properties of adenosine triphosphate. Acta Physiol Scand 1948; 15:264–278.
6. Ash ASF, Schild HO. Receptors mediating some actions of histamine. Br J Pharmacol 1966; 27:427–439.
7. Black JW, Duncan WAH, Durant GJ, Ganellin CR, Parsons ME. Definition and antagonism of histamine H$_2$-receptors. Nature 1972; 258:385–390.
8. Wyllie JH, Hesselbo T, Black JW. Effects in man of histamine H$_2$-receptor blockade by burimamide. Lancet 1972; 2:1117–1120.
9. Milton-Thompson GJ, Williams JG, Jenkins DJA, Misiewicz JJ. Inhibition of nocturnal acid secretion in duodenal ulcer by one oral dose of metiamide. Lancet 1974; 1:693–694.
10. Plaut M, Lichtenstein LM, Henney CS. Properties of a subpopulation of T cells bearing histamine receptors. J Clin Invest 1975; 55:856–874.
11. Brimblecombe RW, Duncan WAM, Durant GJ, Emmett JC, Ganelin CR, Parsons ME. Cimetidine—a non thiourea H$_2$-receptor antagonist. J Intern Med Res 1975; 3:86–92.

12. Code CF. Histamine and gastric secretion. In: Wolstenholme GEW, O'Connor CM, eds. Symposium on Histamine in Honour of Sir Henry Dale. Boston: Little, Brown, 1956:189–219.

13. Code CF. Histamine and gastric secretion: a later look, 1955–1965. Fed Proc 1965; 24:1311–1321.

14. Johnson LR. Control of gastric secretion: no room for histamine? Gastroenterology 1971; 61:106–118.

15. Black JW, Shankley NP. How does gastrin act to stimulate acid secretion? TIPS 1987; 8:486–490.

16. The Impact of Cimetidine on the National Cost of Duodenal Ulcers. Bryn Bawr, PA: Robinson Associates Inc, 1978.

17. Wyllie JH, Clark CG, Alexander-Williams J, Bell PRF, Kennedy TL, Kirk RM, Mackay C. Effect of cimetidine on surgery for duodenal ulcer. Lancet 1981; 1:1307–1308.

18. Bacler JP, Morin T, Bernier JJ, Betoume C, Gastard J, Lamborg R, Ribert A, Sarles H, Toulet J. Treatment of gastric ulcer by cimetidine. In: Burland WR, Simkins MA, eds. Cimetidine: Proceedings of the Second International Symposium on H_2-Receptor Antagonists. Amsterdam: Excerpta Medica, 1976:287–292.

19. Blackemore RC, Brown TH, Durant GJ, Ganellin CR, Parsons ME, Rasmusson AC, Rawlings DA. SK & F93479 a potent and long acting histamine H_2-receptor antagonist. Br J Pharmacol 1981; 74:200P.

20. Brittain RT, Jack D. Histamine H_2 antagonists—past, present and future. J Clin Gastroenterol 1983; 5(Suppl):71–79.

21. Boyd EJS, Wormsley KG. Effects of loxtidine, a new histamine H_2-receptor antagonist on 24 hour gastric secretion in man. Eur J Clin Pharmacol 1984; 26:443–448.

22. Larsson H, Carlsson E, Mattson H. Plasma gastrin and gastric enterochromaffin-like cell activation and proliferation: studies with omeprazole and ranitidine in intact and antrectomized rats. Gastroenterology 1986; 90:391–399.

23. Langman MJS, Henry DA, Olgivie A. Ranitidine and cimetidine for duodenal ulcer. Scand J Gastroenterol 1981; 16(Suppl 69):115–117.

24. Lee FI, Reed PI, Crowe JP, McIsaac RL, Wood JR. Acute treatment of duodenal ulcer: a multicentre study to compare ranitidine 150mg twice daily with ranitidine 300mg once at night. Gut 1986; 27:1091–1095.

25. Brunner G, Creutzfeldt W, Harke U, Lamberts R. Therapy with omeprazole in patients with peptic ulcerations resistant to extended high dose ranitidine treatment. Digestion 1988; 39:80–90.

26. Heatley RV. Review article: the treatment of *Helicobacter pylori* infection. Aliment Pharmacol 1992; 6:291–303.

27. Ganellin CR. Chemistry and structure activity relationships of drugs acting at histamine receptors. In: Ganellin CR, Parsons ME, eds. Pharmacology of Histamine Receptors. Bristol: PJG Wright, 1982.

9

Angiotensin Antagonists

Mark J. Robertson
Astra Charnwood, Loughborough, Leicestershire, England

I. INTRODUCTION

Any remaining questions that may have challenged the pathophysiological importance of the renin–angiotensin system (RAS) have been dispelled by the established and growing success of the angiotensin converting enzyme inhibitors (ACEIs) in the clinic. For example, captopril, first introduced in 1981, and then enalapril have become established as front-line treatments for hypertension (1,2), and exciting results have emerged from large clinical trials regarding the usefulness of ACEIs in congestive heart failure (CHF) and diabetic nephropathy (3–6). The realization that ACEIs would have tremendous commercial success caused an enthusiastic response from the pharmaceutical industry to identify other opportunities to interfere therapeutically with the RAS. It was from this intense period of research, during the 1980s, that the first nonpeptide angiotensin receptor antagonists were developed. The prototype compound and most clinically advanced of these, losartan (otherwise known as Ex 89, Dup-89, DuP753, or MK-954), is now marketed for the treatment of hypertension and in phase III trials for heart failure. Many other related ligands are also in various stages of clinical and preclinical development.

This overview outlines the developing rationale for the angiotensin receptor antagonists, as well as their evolution from peptide origins to clinically viable therapeutic agents. It does not, however, attempt to provide an extensive catalog of nonpeptide angiotensin receptor antagonists: descriptions of more complete structure-activity relationships can be found elsewhere (e.g., Refs. 7–12).

II. PHYSIOLOGY OF THE RAS: POTENTIAL THERAPEUTIC TARGETS

The octapeptide angiotensin II (AII) is the main effector hormone of the RAS and an important regulator of cardiovascular homeostasis. Simplistically, AII causes vasoconstriction and salt and water reabsorption and, by virtue of these, influences electrolyte balance, blood volume, and arterial blood pressure (13). These responses are mediated directly by an effect of AII on the systemic vasculature, in the kidney, and through release of aldosterone. The effects on cardiovascular homeostasis are mediated both peripherally (14,15) and centrally (16). It is now understood that overactivity of the RAS is a key factor in the pathogenesis of essential hypertension (13,17) and CHF (18). A schematic of the RAS showing the three established sites for pharmacological/therapeutic intervention is shown in Figure 1.

Besides the well-documented "circulatory" RAS (a system comprised of angiotensinogen from the liver, renin from the kidney, and ACE from the lung), it is now accepted that various tissues (e.g., vascular wall, heart, and brain) contain all the biosynthetic machinery to synthesize AII from

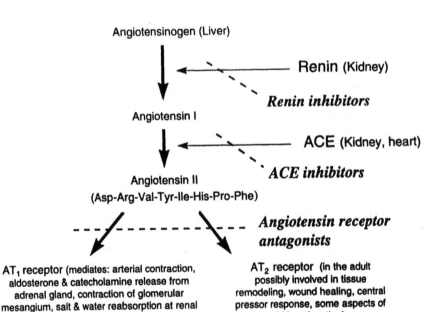

Figure 1 Renin–angiotensin cascade (enzyme/substrate sources for "systemic" RAS only) showing points of potential therapeutic intervention.

angiotensinogen without a requirement for circulating substrates or enzymes (see Ref. 19). The relative importance of these two systems has been a subject of much debate in the literature (e.g., Refs. 20,21), but it is postulated that the therapeutic benefit from ligands that interfere with the RAS manifests from an action at both sites.

In addition to ACE inhibition and angiotensin receptor antagonism, the third target in the cascade to which a large industrial effort has been directed is the identification of ligands that selectively inhibit renin (22). Renin is the enzyme that cleaves the decapeptide angiotensin I (AI) from angiotensinogen and thus provides a substrate for AII production. Many synthetic inhibitors have been described (23); however, clinical trial progress in this area has been hampered owing to a lack of nonpeptide renin inhibitors with good oral bioavailability (24).

III. PHARMACOLOGY OF THE RAS: EVIDENCE FOR ANGIOTENSIN RECEPTOR SUBTYPES AND THEIR FUNCTIONAL ROLES

Development of receptor agonists or antagonists as effective therapeutic agents devoid of side effects depends on a clear understanding of the receptor target in question. In general, it is the ultimate characterization of the receptor subtypes that allows the identification of selective agents that can be applied with confidence to a specified patient population. Data from a broad variety of experiments over the past 25 years have at first suggested and then confirmed the existence of angiotensin receptor subtypes. This provided the information that has allowed rational design of (clinically relevant) selective angiotensin receptor antagonists.

Evidence for more than one angiotensin receptor was originally suggested by anomalous findings using peptide agonists and antagonists in vivo. For example, the peptide antagonist saralasin (Sarcosine[1] Ala[8] -AII) inhibited AII, but to a lesser extent angiotensin III (AIII or des-Asp[1]-AII)-induced increases in plasma aldosterone in the conscious rat (25). Furthermore, although the agonists were equipotent regarding aldosterone release, AII was about 10-fold more potent than AIII as a pressor agent (26). Other work comparing the potencies of AII and AIII in vivo led to similar discrepancies and the suggestion that angiotensin receptors in the kidney and adrenal cortex were different from those in the systemic vasculature (27–30).

Functional (smooth or cardiac muscle contraction) (31,32) and radioligand-binding studies (for review, see Ref. 33) using peptide ligands in vitro supported the hypothesis for organ-dependent, heterogeneous angiotensin receptors. Some of this work even suggested receptor differences within a single compartment of a single organ, e.g., the kidney cortex (34). However, it is now well known that in vivo and even in vitro, metabolism problems can often confound the interpretation of agonist and antagonist rank-order studies [e.g., purine agonist and antagonist rank orders (35)], and to approach the optimal conditions (36) that will make the data from such experiments meaningful, suitable enzyme inhibitors should be included to reduce the chance of metabolism. Such problems are even more apparent with labile ligands such as some peptides [e.g., AIII and Ile[7]-AIII, but not AII in vitro (37)]. However, few groups working in this field at the time appeared to have either considered this complication or had the appropriate enzyme inhibitors available to them. Thus, the first convincing pharmacological evidence for the existence of angiotensin receptor subtypes was only obtained when more stable, nonpeptide, antagonists became available. One of the first reports of this type (38) showed that the nonpeptide ligand Dup-89 (losartan, see Fig. 2) was a potent antagonist of AII-induced contraction of rat pulmonary artery (pA_2 value, 8.4), but did not displace ^{125}I-AII binding in rat brain membranes at concentrations 1000-fold higher. In complete contrast, the reverse specificity was seen for WL-19 (PD 121981; see Fig. 3), which potently displaced ^{125}I-AII in rat brain (IC_{50} value, 19 nM), but did not inhibit AII-induced vascular contraction at a concentration 160-fold higher. It was then shown that both types of AII binding sites were present in rat adrenal in equal amounts (39), but interestingly, the entire functional response to AII in the rat adrenal was blocked by losartan. Similar findings were reported by others around the same time (40), comparing the relative binding displacements of ^{125}I AII by the synthetic pentapeptide CGP 42112A with that of losartan, in human uterus and cultured vascular smooth muscle. Taken together, the results were strongly indicative of different receptors, and as more evidence accrued in support of this concept, various groups laid claim to descriptors for the receptors. The confusion was resolved when it was agreed (41) that the losartan-sensitive site (previously termed AII-1, AII-B, or AII-α) and the PD 121981–sensitive site (previously termed AII-2, AII-A, or AII-β) should be known as AT_1 and AT_2, respectively. Other ligands (PD 123177, PD 123319) subsequently became available from the Parke-Davis laboratories (42), which substantiated this chemical series of tetrahydroimidazopyridines as being selective for the AT_2 receptor.

Notwithstanding the comments made above regarding the potential lability of peptide ligands, CGP 42112A (based on amino acids 4–8 of AII) has proven to be a fairly robust tool for the differentiation of angiotensin receptor subtypes in vitro and remains among the most potent ligands at the AT_2 receptor (IC_{50} value, 0.5 nM in human uterus). Many other nonpeptide AT_2 ligands have now been described (43), and recently, there are

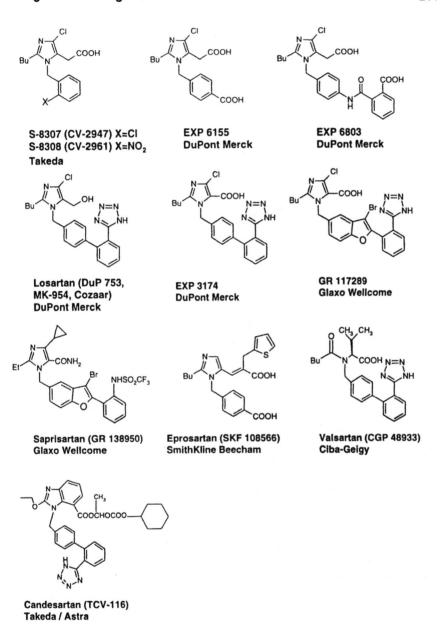

S-8307 (CV-2947) X=Cl
S-8308 (CV-2961) X=NO₂
Takeda

EXP 6155
DuPont Merck

EXP 6803
DuPont Merck

Losartan (DuP 753,
MK-954, Cozaar)
DuPont Merck

EXP 3174
DuPont Merck

GR 117289
Glaxo Wellcome

Saprisartan (GR 138950)
Glaxo Wellcome

Eprosartan (SKF 108566)
SmithKline Beecham

Valsartan (CGP 48933)
Ciba-Geigy

Candesartan (TCV-116)
Takeda / Astra

Figure 2 Nonpeptide AT₁ receptor antagonists.

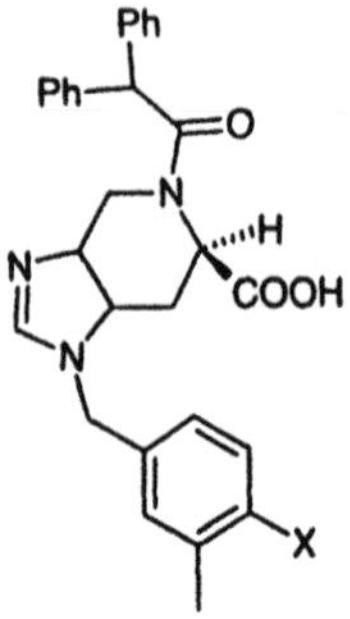

PD 121981 (WL-19) X= OMe
PD 123177 (EXP 655) X=NH₂
PD 123319 X=NMe₂
Parke-Davis

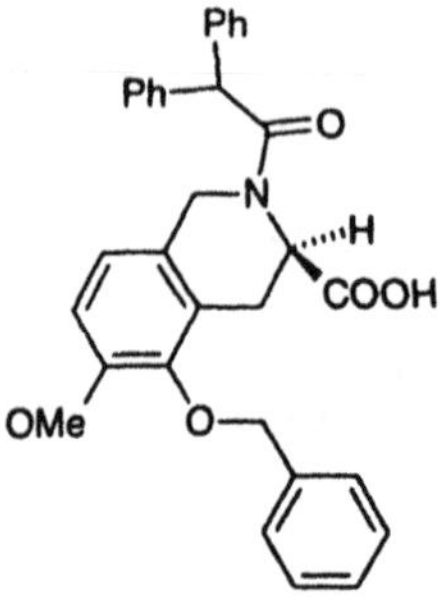

PD 126055
Parke-Davis

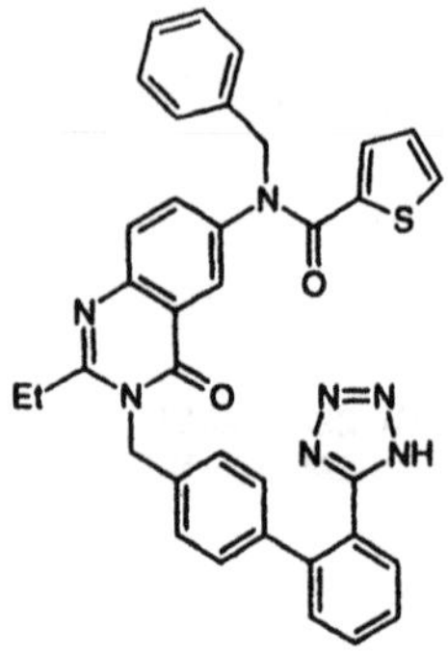

L-161,638
Merck

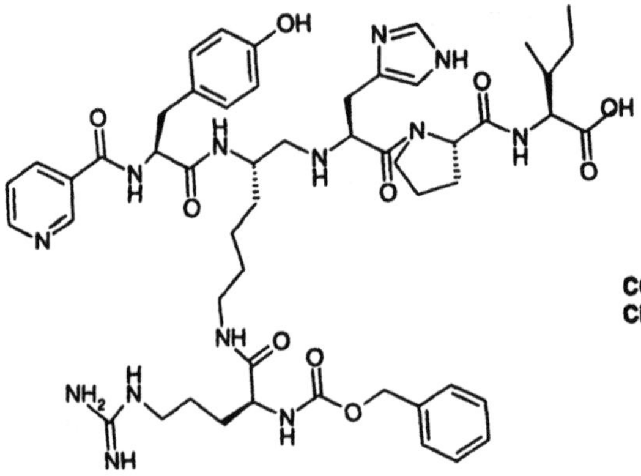

CGP 42112A
Ciba-Geigy

Figure 3 Peptide and nonpeptide AT_2 receptor antagonists.

examples with even greater selectivity [e.g., PD 126055 >100,000-fold selective for AT_2 over AT_1 (44)] and affinity [e.g., L-161,638; IC_{50} value, 60 pM (45)] for AT_2 receptors. Not surprisingly, a large body of work has been directed toward identifying the coupling mechanism and functional consequences of AT_2 receptor activation. A wide variety of signal transduction pathways have been claimed to be associated with the AT_2 receptor (46,47). The AT_2 receptor number is much greater in the fetus than in the adult (48), and it has been suggested that the receptor mediates fetal growth and development. Perhaps in support of this general concept, in adult or young animal models, AT_2 receptor number is increased during tissue repair [e.g., endothelial cells in vitro have suggested that AT_2 stimulation is antiproliferative (51) and that in pathological situations, stimulation of the AT_2 receptor may limit excessive growth. In the adult, AT_2 receptors are located predominantly in the brain, uterus, kidney, and adrenals. Various reports have claimed a role for the renal AT_2 receptor in the adult animal in controlling vascular resistance (52) and the pressure-natriuresis relationship (53).

The functional role of the central AT_2 receptor in the adult has been even more difficult to establish, although there is accumulating evidence for its involvement in the control of cerebral blood flow (54). But despite the availability of a variety of potent, selective, bioavailable AT_2 selective ligands, relatively little progress has been made toward identifying a compelling pathophysiological role for the AT_2 receptor. Even the physiological importance of the AT_2 receptor is still considered debatable (55). In contrast, it is well established that the majority of actions of AII (e.g., on peripheral vasculature, renal tubule, adrenal cortex, CNS) are mediated via the AT_1 receptor (56–59) and until very recently, most Industrial effort has been directed toward the identification of ligands that selectively block the AT_1 subtype.

IV. THE ROLE OF MOLECULAR BIOLOGY IN ANGIOTENSIN RECEPTOR CLASSIFICATION

As described above, during the late 1980s, the classic pharmacological approach gave results providing compelling evidence for angiotensin receptor subtypes. For years, this approach has provided a practically useful starting point to base further work aimed at identifying therapeutically useful receptor agonists and antagonists (e.g., see Chapter 8). Of course, increasingly during the last decade, this method has been complemented or, in some centers, superseded by powerful molecular biology technologies. However, in the angiotensin area, it was purely the pharmacological evidence that led the way and prompted the molecular biology that eventually confirmed the existence of subtypes through cloning and expression of the receptor cDNAs. An early claim (60) that the *mas* oncogene product was a functional angiotensin receptor has not been substantiated. However, expression cloning methods (61,62) allowed expression of the AT_1 receptor in mammalian COS-7 cells. It was found that this receptor belonged to the seven-transmembrane (7-TM), G-protein-coupled superfamily and showed high affinity toward losartan. Homology cloning in various rat tissues revealed two subtypes of the AT_1 receptor (AT_{1A} and AT_{1B}), which are encoded by different genes and are 94% identical (63). A further subtype (AT_3) has been described in rat epithelial cells that has lower affinity for losartan and PD 123177, but high affinity for saralasin (64). The physiological significance of the AT_1 subdivision in the rat and the AT_3 subtype is currently unknown. Early work suggested only a single AT_1 receptor in humans (65,66) (359 aa's with ~95% homology to rodent AT_1), but there is now evidence for two splice variants of human AT_1. Again, the functional and pharmacological properties of the human isoforms are presently indistinguishable (67).

The rat AT_1 clone was described in 1991, but despite much activity, the AT_2 receptor clone was not obtained until 1993. The reason for the delay became partly apparent when it was determined that the AT_2 receptor (363 aa's, also from the 7-TM superfamily) shared only 34% homology with the AT_1 receptor (68). The decision to start (and the resolve to continue) the search for the AT_2 receptor during this period was largely driven by the preexisting pharmacological knowledge. However, the availability of the AT_2 clone, expressed in a suitable vector (i.e., free of endogenous angiotensin receptors), has provided clarity regarding the potential efficacy of the AT_2 receptor ligands. For example, in COS-7 cells in which the AT_2 receptor is transiently expressed, CGP 42112A has been shown to behave as an agonist (69).

V. INDUSTRIAL DEVELOPMENT OF AT_1 RECEPTOR ANTAGONISTS FOR CLINICAL USE

Until the late 1980s, the only available angiotensin receptor antagonists were peptides, but this did not deter important experimental studies in patients, where it was demonstrated that intravenously administered saralasin reduced blood pressure in renovascular hypertensives (70). For the first time, this causatively linked endogenous AII in the etiology of hypertension, beyond the previous association with raised plasma renin activity. The susceptibility of saralasin to degradation when given by the oral route and the short half-life if given intravenously meant that it could not be entertained as a serious drug entity. Furthermore, the unsuitability of saralasin was confirmed when it was shown to act as a partial agonist in human volunteers (71).

The first reports of nonpeptide antagonists of AII were contained in two patents from Takeda in 1982 (72,73) and were apparently discovered by chance in a screening program that was not designed to identify angiotensin receptor antagonists. The best compounds described (S-8307 and S-8308, see Fig. 2) were reevaluated by workers at DuPont (now DuPont Merck) (74,75). These were shown to be weak antagonists of exogenous AII in vitro (pA_2 values, 5.5 and 5.7, respectively, against AII-induced contraction of the rabbit aortic strip, a now established AT_1 receptor effect), and on blood pressure in the pithed rat in vivo. No partial agonist activity was seen; moreover, it was considered surprising that an octapeptide/receptor interaction could be selectively antagonized by such low-molecular-weight compounds with no gross structural similarity to the endogenous agonist. At a high dose and only intravenously, S-8308 (30 mg/kg) lowered blood pressure in a rat renin-dependent model of hypertension. The effects lasted only 30 min, but confirmed the antihypertensive significance of AT_1 receptor blockade. This work prompted the start of an intense period of industrial activity to identify more potent, bioavailable ligands, but it was a series of now classic publications from DuPont Merck that broke new ground toward identifying candidate drugs, apparently based on a molecular model that overlay the Takeda compounds on AII (76). It was anticipated that introduction of a second acidic group to the chemical leads might mimic the charge of either Tyr-4 or Asp-1 of AII. Accordingly, with the diacids EXP6155 and EXP6803 (Fig. 2), greater potencies were achieved (pA_2 values, 6.5 and 7.2, respectively) (77,78). Although intravenous EXP6803 lowered blood pressure for more than 2 hr in the hypertensive rat, it was still not orally active. Before long, however, this was addressed by replacing the amide linker in EXP6803 with a single bond.

The key compound, EXP7711, was slightly weaker than EXP6803 (7), but lowered blood pressure in the rat model after oral administration (ED_{30} value, 11 mg/kg). Finally, replacement of the carboxylic acid in EXP7711 with an alternative acidic group (a tetrazole) resulted in further enhanced potency (pA_2 value, 8.5) but with apparently outstanding oral absorption (ED_{30} values, 0.8 mg/kg and 0.6 mg/kg i.v. and p.o., respectively), duration (3 mg/kg p.o. reduced blood pressure for ~24 hr in the renin-dependent hypertensive rat), selectivity (AT_1 over AT_2), and specificity over a variety of other receptor types (79). This compound, losartan, was now deemed suitable for clinical development, although it was found that losartan was itself metabolized to a more potent compound, EXP3174 (pA_2 value, 10.1, Fig. 2), in vivo (80). The bioavailability of losartan was calculated at only 30%, and one might postulate that metabolism to EXP3174 was more complete after oral than intravenous administration, explaining why losartan is, if anything, more potent by the oral route. Interestingly, losartan was also shown (81) to reduce blood pressure, albeit at 3–10-fold higher doses, in the spontaneously hypertensive rat (SHR), which has a genetic predisposition toward hypertension, but has low plasma renin activity. In these experiments, the high specificity of losartan meant it could be used predictively and allowed the authors to conclude that if losartan lowered blood pressure in this setting, then this defined the model as renin-dependent (probably tissue-renin-dependent). Recently, the hypotensive effect of AT_1 receptor antagonists in SHR has been attributed to an effect predominating in the vasculature of kidney and mesentery (82).

Other groups, which had entered the arena in the late 1980s, also based their AT_1 antagonist chemical programs on derivatives of the Takeda leads, but mostly on the DuPont series. Some of these retained the DuPont biphenyltetrazole moiety, but made dramatic modifications to the rest of the molecule. For example, Ciba-Giegy described valsartan (83) (CGP-48933, recently launched for hypertension and in phase II trials for heart failure, see Fig. 2), in which the imidazole was replaced with N-acyl valine in an attempt to mimic the binding of AII to its receptor (84). Valsartan has nanomolar potency and 24-hr duration in vivo (85). Others workers modified the biphenyl side chain. For example, Glaxo (now GlaxoWellcome) reported that the bromobenzofuran, GR117289 [zolarsartan (86), Fig. 2], although extremely potent (apparent pA_2 value, 9.8), had low bioavailability [20% in humans (10)]. This was replaced as a development candidate by GR138950 (saprisartan, see Ref. 10, now discontinued) in which the diacid was strategically replaced with a monoacid (tetrazole group was replaced by trifluoromethanesulfonamide, Fig. 2), to improve absorption. This was achieved (bioavailability improved to 80–100%) and good potency was maintained (pK_b value ~ 9.5).

SK&F (now SKB) used a different AII scaffold-based model. Independent of the DuPont leads, they used the Takeda patents to develop a series of monophenyl carboxylic acid derivatives (the "imidazoleacrylic acid series") in which the imidazole is aligned with the aromatic component of Pro-7 and 5-substituted, supposedly to mimic the Phe-8 position of AII. This work culminated in eprosartan (8) (SKF 108566; now in phase III, Fig. 2), which has subnanomolar potency, but the magnitude of oral doses used in preclinical studies (87) may suggest a lower bioavailability than some other compounds in the later stages of development.

Takeda, apparently inactive in this area since the original patents, released detailed information on their own extensive program of work in the early 1990s (9). Their most advanced compound is candesartan cilexetil (TCV-116; see Fig. 2), which is being codeveloped with Astra for hypertension (monotherapy and in combination with hydrochlorothiazide, both phase III) and CHF (phase II/III). Candesartan cilexetil is an inactive prodrug (an ethyl ester carbonate) that has good oral bioavailability and is completely converted to the active compound, candesartan (the benzimidazole-7-carboxylic acid, CV-11974), by ester hydrolysis during gastrointestinal absorption. Candesartan was identified from a program in which the 4 and 5 positions of the imidazole in losartan were replaced with a fused ring. This yielded a series of extremely potent and long-lasting AT_1 antagonists: candesartan has a reported K_d value of 0.6 nM (89), and even in SHR, blood pressure was reduced for by >50 mmHg for >24 hr by

1 mg/kg i.v. Orally, candesartan is reported to be 40-fold more potent than losartan in conscious rats (88).

VI. UNUSUAL ANTAGONIST PROFILE IN VITRO

An intriguing pharmacological aspect of the AT_1 receptor antagonist mechanism is highlighted by comparing candesartan with losartan in a simple functional assay in vitro. In rabbit aortic strips, whereas losartan displays apparently normal competitive features [parallel rightward displacement of the AII concentration-response curve, with no change in the maximum response (90)], candesartan appears noncompetitive, causing only collapse with little or no rightward displacement of the AII curve (88). The noncompetitive interaction is subtle, because when the 7-carboxylic acid group of candesartan is replaced with 7-H, the resulting compound behaves competitively like losartan (89).

Normally, agonist curve collapse with increasing antagonist concentration might be attributed to irreversible binding of the antagonist to receptor. However, similar experiments with other AT_1 antagonists show a spectrum of profiles lying between that for losartan and candesartan: most peptide and nonpeptide AT_1 receptor antagonists cause varying degrees of suppression of the maximum response together with a parallel rightward displacement of the AII curve (see Ref. 91). Interestingly, there are even examples where the antagonist causes a *rise* in the maximum response to AII, which is sustained (saturates) as higher concentrations of antagonist cause progressive rightward displacement of the AII curve. This effect may be linked to the observation (91) that losartan is capable of "resensitizing" tissue that has been subjected to AII-induced tachyphylaxis. Various models have been proposed to explain this unusual profile, including ligand-modifying, receptor-transducer coupling (92) and two-state receptor coupled to an allosteric site (93). We have reviewed these and other possibilities and suggested all these observations can be explained by a simple two-state (inactive/activable) receptor model without the need to evoke an allosteric site (94).

VII. NONPEPTIDE AT_1/AT_2 ANTAGONISTS; AT_1 RECEPTOR AGONISTS

The chemical program at Merck identified other imidazole biphenyltetrazoles that were potent AT_1 receptor antagonists, such as L-158,809 (Fig. 4), but their chemical series extended the pharmacology into exciting new directions. L-158,809, an imidazopyridine, is a potent (rat aorta; IC_{50} value, 0.3 nM), long-acting, orally active AT_1 receptor antagonist with >10,000-

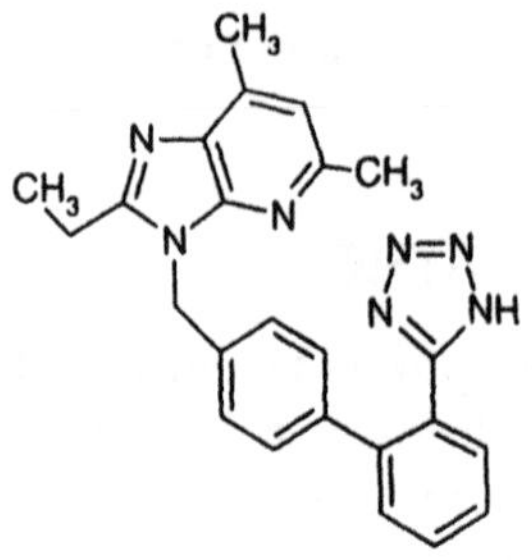

L-158,809
Merck

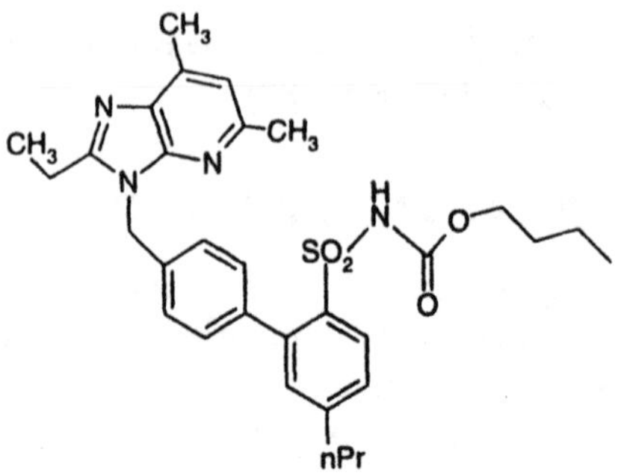

L-162,389
Merck

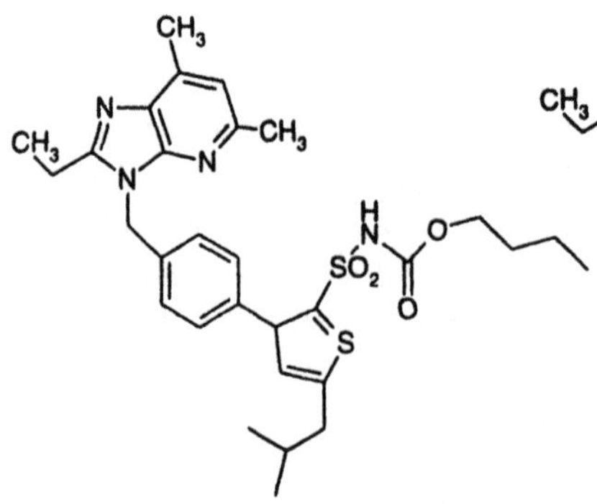

L-162,313
Merck

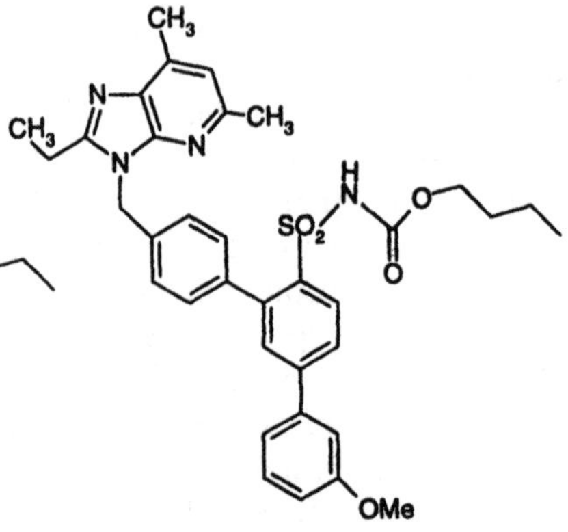

L-163,491
Merck

Figure 4 Merck AT_1, AT_1/AT_2 receptor antagonists and agonists.

fold selective for AT_1 over AT_2 receptors (12). Replacement of the tetrazole with acylsulphonamide moieties generated compounds such as L-162,389, which retains nanomolar potency for AT_1 but also has an IC_{50} value of 6 nM at the AT_2 receptor. The clinical advantage of nonselective over selective AT_1 receptor antagonists is currently unproved (see below). However, an increasing number of nonselective AT_1/AT_2 receptor ligands have now been described by Merck (95,96) and DuPont Merck (97).

A further and even more remarkable disclosure from the Merck chemistry was that L-162,313, a close relative of L-162,389, is a potent *agonist* at the AT_1 and AT_2 receptors (98). Identification of nonpeptide agonists for peptide ligand 7-TM receptors is extremely uncommon. However, other very closely related ligands are not only potent agonists, but also show some selectivity for the AT_1 receptor, e.g., L-163,491: IC_{50} values 1 and 101 nM, at AT_1 and AT_2, respectively (99). A model has been proposed to explain the agonistic requirements of this nonpeptide ligand binding (100). It has now been suggested that biphenylimidazoles antagonists and agonists (e.g., losartan and L-162,313, respectively) and imidazoleacrylic acid antagonists (e.g., eprosartan) share at least some overlap in their binding to the AT_1 receptor (Val-108 in transmembrane loop 3) (101), although it has also been reported that L-162,313 and biphenylimidazole binding sites differ (102). However, it appears to be agreed that the peptide and nonpeptide angiotensin binding sites are separate (101–103). Thus, the surprising conclusion from these findings is that the peptide and nonpeptide agonism can be generated out of different sites on the AT_1 receptor.

VIII. CLINICAL EXPERIENCE WITH AT_1 RECEPTOR ANTAGONISTS

As in animal studies, many AT_1 receptor antagonists have been shown to potently inhibit AII-induced pressor responses in human volunteers [e.g., candesartan cilexetil (104), losartan (105), valsartan (106)]. By far the majority of clinical studies using selective AT_1 receptor antagonists have been conducted in hypertensive patients, although increasing numbers of trials have been designed to investigate the potential of this class of drug in CHF. In hypertension trials, comparisons have been made with other vasodilating agents, but more recently with ACEIs. As suggested by the early animal studies, AT_1 receptor antagonists effectively and predictably lower blood pressure in mild to moderate hypertension (e.g., see Ref. 107). Furthermore, AT_1 receptor antagonists have been shown to be as effective as enalapril in this regard (108,109).

Although ACEIs are fairly well tolerated, aside from symptoms arising from their vasodilating properties (dizziness, syncope, headaches; see Ref. 2), their most prevalent side effect is a disagreeable (but not disabling) dry cough [incidence, 7–15% of patients (110)]. It is thought that the cough is caused by endogenous bradykinin, which is raised in the lung because ACEIs also prevent ACE (also known as kininase II) from carrying out its other main role in the body: that of a major contributor to the metabolism of bradykinin (111). Early theoretical expectations that cough would not occur with AT_1 receptor antagonists (which would not be expected to interfere with bradykinin levels) have been upheld in the clinic (112,113). More generally, AT_1 receptor antagonists appear very well tolerated, and in pooled studies involving more than 2000 hypertensive patients treated with losartan, dizziness (possibly attributable to the fall in blood pressure, and usually reported with antihypertensive drugs) was the only significant side effect with an incidence twofold that of placebo (114). This difference may provide the angiotensin antagonists with a therapeutic advantage over ACEIs in hypertension.

Aside from those compounds mentioned above, other potent, bioavailable, selective AT_1 receptor antagonists in phase III clinical development for the treatment of hypertension include the biphenyl tetrazoles, irbesartan [BMS-186295, SR-47436, Sanofi/Bristol-Myers Squibb (115)] and tasosartan [ANA-756, American Home products (116)], and the benzimidazole, telmisartan [BIBR-277, BIBR-363, Thomae (Boehringer Ingelheim) (117)]. In various ways, these and other emerging development compounds are preclinical improvements on losartan. Some are more potent and some have more duration. With current knowledge, it is more difficult to see how the dozens of other selective AT_1 receptor antagonists currently claimed to be in preclinical to phase II development will impact on what will become a rapidly diminishing commercial arena after the first three or four compounds are launched.

IX. CHALLENGE FOR THE FUTURE

AT_1 receptor antagonists are becoming an increasingly popular choice for the treatment of hypertension, and some believe that this class of compound has the potential to become a first-line treatment option for this indication (114). If the AT_1 receptor antagonist hypertension market grows, it will probably be to the detriment of the ACE inhibitors, especially if the lack of side effects (e.g., cough) becomes established as a tangible and generally accepted advantage. Furthermore, it is believed that AII can be generated by enzymes other than ACE (118,119), so that in theory, AT_1 receptor antagonists should offer a more specific and complete inhibition of the RAS than the ACEIs can. On the other hand, prolonged clinical treatment with a selective AT_1 receptor antagonist could lead to raised plasma AII levels and an action on AT_2 receptors. However, the clinical relevance of this latter scenario is currently unknown and, as detailed above, it is even difficult to predict in the continued absence of a firm (patho)physiological role for the AT_2 receptor. Consequently, the future of nonselective, AT_1/AT_2 receptor antagonists, of the type currently being developed, for example, by Merck or Yamanouchi (e.g., YM-358 in phase II), cannot be clearly ascertained at this time.

Compared to some receptor targets, the identification and development of the AT_1 receptor antagonists for the treatment of hypertension have been relatively straightforward. However, there are still some important milestones ahead for this promising class of drugs. For example, a key issue is whether AT_1 receptor antagonists will have the same impact in the treatment of CHF as ACEIs have shown. Interestingly, a large body of literature suggests that some beneficial consequences of ACEI administration in animal models of hypertension and/or restenosis is, in fact, partly dependent on the potentiation of bradykinin levels and the ensuing release of nitric oxide (see review in Ref. 111). Blood pressure lowering, prevention of vascular intimal thickening after endothelium damage, and cardioprotective effects are attenuated to an extent by the B_2 bradykinin receptor antagonist, icatibant. Thus, the obvious question is whether AT_1 receptor antagonists will be capable of the same efficacy in CHF [as ably demonstrated by ACEI's (4–6)] if they do not raise bradykinin levels. In this context, those workers developing AT_1 receptor antagonists might be encouraged by preclinical reports, which circumstantially suggest that this class of compound is capable of raising NO levels, by an unknown mechanism, but which is probably independent of bradykinin (120,121). In addition, AT_1 receptor antagonists have already shown promising activity in several animal models of cardiac impairment. For example, candesartan has been shown to prevent myocardial hypertrophy (122,123) and reduce intimal thickening in a coronary graft (124) or carotid injury model (125) in rats. In all of these studies, candesartan compared favorably in activity with an ACEI. Finally, the results of several small clinical studies suggest the promising potential of AT_1 receptor antagonists in CHF patients [e.g., candesartan cilexetil (126); losartan (127)], but perhaps eclipsing all of this work are the recently reported data from the ELITE trial (128). In ELITE, losartan was compared to captopril in over 700 elderly, class II–IV heart failure patients. Losartan was better tolerated than captopril and, surprisingly, losartan treatment was associated with an unexpectedly lower mortality than that seen with captopril. One explanation proposed by the authors was that this measured benefit may be a result of the more complete block of AII afforded by the AT_1 receptor antagonists. Future studies will determine whether this trend holds with larger patient groups, with other ACEIs, and with other AT_1 receptor antagonists.

REFERENCES

1. Waeber B, Nussberger J, Brunner HR. Angiotensin converting enzyme inhibitors in hypertension. In: Laragh JH, Brenner BM, eds. Hypertension: Pathophysiology, Diagnosis and Management. New York: Raven Press, 1990:2209–2232.
2. McAreavy D, Robertson JIS. Angiotensin converting enzyme inhibitors and moderate hypertension. Drugs 1990; 40:326–345.
3. Unger T, Gohlke P. Converting enzyme inhibitors in cardiovascular therapy: current status and future potential. Cardiovasc Res 1994; 28:146–158.

4. The CONSENSUS Trial Study Group. Effects of enalapril on mortality in severe congestive heart failure. N Engl J Med 1987; 316:1429–1435.
5. Garg R, Yusuf S. Overview of randomized trials of angiotensin-converting enzyme inhibitors on mortality and morbidity in patients with heart failure. JAMA 1995; 273:1450–1456.
6. The SOLVD Investigators. The effect of enalapril on survival in patients with reduced left ventricular ejection fractions and congestive heart failure. N Engl J Med 1991; 325:293–302.
7. Carini DJ, Duncia JV, Aldrich PE, Chiu AT, Johnson AL, Pierce ME, Price WA, Santella JB, Wells GJ, Wexler RR, Wong PC, Yoo SE, Timmermans PBMWM. Non-peptide angiotensin II receptor antagonists: the discovery of a series of N-(biphenylmethyl) imidazoles as potent orally active antihypertensives. J Med Chem 1991; 34:2525–2547.
8. Keenan RM, Weinstock J, Finkelstein JA, Franz RG, Gaitanopoulous DE, Girard GR, Hill DT, Morgan TM, Samanen JM, Peishoff CE, Tucker LM, Aiyar N, Griffin E, Ohlstein EH, Stack EJ, Weidley EF, Edwards RM. Potent nonpeptide angiotensin II receptor antagonists. 2. 1-(Carboxybenzyl) imidazole-5-acrylic acids. J Med Chem 1993; 36:1880–1892.
9. Kubo K, Kohara Y, Imamiya E, Sigiura Y, Inada Y, Furukawa Y, Nishikawa K, Naka T. Nonpeptide angiotensin II receptor antagonists. Synthesis and biological activity of benzimidazolecarboxylic acids. J Med Chem 1993; 36:2182–2195.
10. Judd DB, Dowle MD, Middlemiss D, Scopes DIC, Ross BC, Jack TI, Pass M, Tranquillini E, Hobson JE, Panchal TA, Stuart PG, Paton JMS, Hubbard T, Hilditch A, Drew GM, Robertson MJ, Clark KL, Travers A, Hunt AAE, Polley J, Eddershaw PJ, Bayliss MK, Manchee GR, Donnelly MD, Walker DG, Richards SA. Bromobenzofuran-based non-peptide antagonists of angiotensin II: GR 138950, a potent antihypertensive agent with high oral bioavailability. J Med Chem 1994; 37:3108–3120.
11. Wexler RR, Greenlee WJ, Irvin JD, Goldberg MR, Prendergast K, Smith RD, Timmermanns PBMWM. Nonpeptide angiotensin II receptor antagonists: the next generation in antihypertensive therapy. J Med Chem 1996; 39(3):625–656.
12. Mantlo NB, Chakravarty PK, Ondeyka DI, Seigl PKS, Chang RS, Lotti VJ, Faust KA, Chen T-B, Schorn TW, Sweet CS, Emmert SE, Patchett AA, Greenlee WJ. Potent, orally active imidazo (4,5-b) pyridine-based angiotensin II receptor antagonists. J Med Chem 1991; 34:2919–2922.
13. Ferrario CM. Importance of the renin-angiotensin-aldosterone system (RAS) in the physiology and pathology of hypertension. Drugs 1990; 39(Suppl 2):1–8.
14. Vallotton MB. The renin-angiotensin system. Trends Pharmacol Sci 1987; 8:69–74.
15. Johnston CI. Biochemistry and pharmacology of the renin-angiotensin system. Drugs 1990; 39(1):21–31.
16. Lippoldt A, Paul M, Fuxe K, Ganten D. The brain renin-angiotensin system: molecular mechanism of cell to cell interactions. Clin Exp Hypertension 1995; 17:251–266.

17. Laragh JH. The renin system and new understanding of the complications of hypertension and their treatment. Arzneim Forsch Drug Res 1993; 43:247–254.

18. Braunwald E. ACE inhibitors: a cornerstone of the treatment of heart failure. N Engl J Med 1991; 325:351–353.

19. Goldfarb DA. The renin-angiotensin system: new concepts in regulation of blood pressure and renal function. Urol Clin North Am 1994; 21:187–194.

20. Unger T, Gohlke P. ACE inhibition in the tissues. Clin Physiol Biochem 1992; 9(3):89–93.

21. Paul M, Bachmann J, Ganten D. The tissue renin-angiotensin systems in cardiovascular disease. Trends Cardiovasc Med 1992; 2(3):94–99.

22. Cody RJ. Renin system inhibition. Beginning the fourth epoch. Circulation 1992; 85(1):362–364.

23. Kleinert HD. Recent developments in renin inhibitors. Exp Opin Invest Drugs 1994; 3(11):1087–1104.

24. Lin C, Frishman WH. Renin inhibition: a novel therapy for cardiovascular disease. Am Heart J 1996; 131(5):1024–1034.

25. Pals DT, Masucci FD, Denning GS, Sipos F, Fessler DC. Role of the pressor action of angiotensin II in experimental hypertension. Circ Res 1971; 29:673–681.

26. Campbell WB, Pettinger WA. Organ specificity of angiotensin II and des-aspartyl angiotensin II in the conscious rat. J Pharmacol Exp Ther 1976; 198:450–456.

27. Blair-West JR, Coghlan JP, Denton DA, Funder JW. The effect of the heptapeptide (2-8) and hexapeptide (3-8) fragments of angiotensin II on aldosterone secretion. J Clin Endocrinol Metab 1971; 32:575–578.

28. Freeman RH, Davis JO, Lohmeier TE. Des-1-Asp-Angiotenisn II: possible intrarenal role in hemostasis in the dog. Circ Res 1975; 37:30–34.

29. Goodfriend TL, Peach MJ. Angiotensin III: (Des-aspartic acid)-angiotensin II. Evidence and speculation for its role as an important agonist in the renin-angiotensin system. Circ Res 1975; 37:138–147.

30. Carey RM, Vaughan ED, Peach MJ, Ayers CR. Activity of [des-Aspartyl[1]]-angiotensin II in man: differences in blood pressure and adrenocortical responses. J Clin Invest 1978; 61:20–31.

31. Moore AF, Hall MM, Khairallah PA. A comparison of the effects of angiotensin II and heptapeptide on smooth muscle (vascular and uterine). Eur J Pharmacol 1976; 39:101–107.

32. Trachte GJ, Peach MJ. A potent noncompetitive angiotensin II antagonist induces only competitive inhibition of angiotensin III receptors. J Cardiovasc Pharmacol 1983; 5:1025–1033.

33. Peach MJ, Dostall DE. The angiotensin II receptor and the actions of angiotensin II. J Cardiovasc Pharmacol 1990; 16:S25–S30.

34. Douglas JG. Angiotensin receptor subtypes of the kidney cortex. Am J Physiol 1987; 253(1, part 2):F1–F7.

35. Crack BE, Pollard CE, Beukers MW, Roberts S, Hunt SF, Ingall AH, McKechnie KCW, IJzerman AP, Leff P. Pharmacological and biochemical

analysis of FPL 67156, a novel, selective inhibitor of ecto-ATPase. Br J Pharmacol 1995; 114:475–481.

36. Furchgott RF. The classification of adrenoceptors (adrenergic receptors). An evaluation from the standpoint of receptor theory. In: Blaschko H, Muscholl E, eds. Catecholamines. Handbook Exp. Pharmacol. Berlin: Springer-Verlag 1972:283–335.

37. Robertson MJ, Cunoosamy MP, Clark KL. Effects of peptidase inhibition on angiotensin receptor agonist and antagonist potency in rabbit isolated thoracic aorta. Br J Pharmacol 1992; 106:166–172.

38. Chang RSL, Lotti VJ. Selective ligands reveal subtypes of angiotensin receptors in rat vasculature and brain. Pharmacologist 1989; 31(3):150 (abstract 183).

39. Chang RSL, Lotti VJ. Two distinct angiotensin II receptor binding sites in rat adrenal revealed by new selective nonpeptide ligands. Mol Pharmacol 1990; 29:347–351.

40. Whitebread S, Mele M, Kamber B, de Gasparo M. Preliminary biochemical characterization of two angiotensin II receptor subtypes. Biochem Biophys Res Commun 1989; 163(1):284–291.

41. Bumpus FM, Catt KJ, Chiu AT, DeGasparo M, Goodfriend T, Husain A, Peach MJ, Taylor DG, Timmermans PBMWM. Nomenclature for angiotensin receptors. Hypertension 1991; 17:720–721.

42. Dudley DT, Panek RT, Major TC, Lu GH, Bruns RF, Klinkefus BA, Hodges JC, Weishaar RE. Subclasses of angiotensin II binding sites and their functional significance. Mol Pharmacol 1990; 38:370–377.

43. Van Atten MK, Ensinger CL, Chiu AT, McCall DE, Nguyen TT, Wexler RR, Timmermans PBMWM. A novel series of selective non-peptide inhibitors of angiotensin II binding to the AT_2 site. J Med Chem 1993; 36(25):3985–3992.

44. Klutchko S, Hamby JM, Hodges JC. Tetrahydroisoquinoline derivatives with AT_2 specific angiotensin II receptor binding inhibitory activity. Bioorg Med Chem Lett 1993; 3:2023–2028.

45. Glinka TW, de Laszlo SE, Tran J, Chang RS, Chen TB, Lotti VJ, Greenlee WJ. L-161,638: a potent AT_2 selective quinazolinone angiotensin II binding inhibitor. Bioorg Med Chem Lett 1994; 4:1479–1484.

46. Brechler V, Levens NR, DeGasparo M, Bottari SP. Angiotensin AT_2 receptor mediated inhibition of particulate guanylate cyclase: a link with protein tyrosine phosphatase stimulation? Receptors Channels 1994; 2(2):79–87.

47. Nahmias C, Strosberg AD. The angiotensin AT_2 receptor: searching for signal-transduction pathways and physiological function. Trends Pharmacol Sci 1995; 16:223–225.

48. Grady EF, Sechi LA, Griffin CA, Schambelan M, Kalinyak JE. Expression of AT_2 receptors in the developing rat fetus. J Clin Invest 1991; 88(3):921–933.

49. Janiak P, Pillon A, Prost JF, Vilaine JP. Role of angiotensin subtype 2 receptor in neointima formation after vascular injury. Hypertension 1992; 20:737–745.

50. Viswanathan M, Saavedra JM. Expression of angiotensin II AT_2 receptors in the rat skin during experimental wound healing. Peptides 1992; 13:783–786.

51. Stoll M, Meffert S, Stroth U, Unger T. Growth or antigrowth: angiotensin and the endothelium. J Hypertension 1995; 13:1529–1534.
52. Clark KL, Robertson MJ, Drew GM. Role of angiotensin AT_1 and AT_2 receptors in mediating the renal effects of angiotensin II in the anaesthetized dog. Br J Pharmacol 1993; 109:148–156.
53. Lo M, Liu K-L, Lantelme P, Sassard J. Subtype 2 of angiotensin II receptors controls pressure-natriuresis in rats. J Clin Invest 1995; 95:1394–1397.
54. Naveri L, Stromberg C, Saaverdra JM. Angiotensin II AT_2 receptor stimulation extends the upper limit of cerebral blood flow autoregulation: agonist effects of CGP 42112 and PD 123319. J Cereb Blood Flow Metab 1994; 14:38–44.
55. Clauser E, Curnow KM, Davies E, Conchon S, Teutsch B, Vianello B, Monnot C, Corvol P. Angiotensin II receptors: proteins and gene structures, expression and potential pathological involvements. Eur J Endocrinol 1996; 134(4):403–411.
56. Wong PC, Scott SD, Zaspel AM, Chiu AT, Ardecky RJ, Smith RD, Timmermans PBMWM. Functional studies of nonpeptide angiotensin II receptor subtype-specific ligands: DuP 753 (AII-1) and PD 123177 (AII-2). J Pharmacol Exp Ther 1990; 255:584–592.
57. Saavedra JM. Brain and pituitary angiotensin. Endocr Rev 1992; 13:329–380.
58. Brown L, Sernia C. Angiotensin receptors in cardiovascular diseases. Clin Exp Pharmacol Physiol 1994; 21:811–818.
59. Dzau VJ, Mukoyama M, Pratt RE. Molecular biology of angiotensin receptors: targets for drug research? J Hypertension 1994; 12(Suppl 2):S1–S5.
60. Jackson TR, Blair LAC, Marshall J, Godert M, Hanley MR. The *mas* oncogene encodes an angiotensin receptor. Nature 1988; 335:437–440.
61. Sasaki K, Yamamo Y, Bardhan S, Iwai N, Murray JJ, Hasegawa M, Matsuda Y, Inagami T. Cloning and expression of a complementary DNA encoding a bovine adrenal angiotensin II type-1 receptor. Nature 1991; 351:230–233.
62. Murphy TJ, Alexander RW, Griendling KK, Runge MS, Bernstein KE. Isolation of a cDNA encoding the vascular type-1 angiotensin II receptor. Nature 1991; 351:233–236.
63. Iwai N, Inagami T. Identification of two subtypes in the rat type 1 angiotensin II receptor. FEBS Lett 1992; 298:257–260.
64. Smith RD. Identification of atypical (non-AT1, non-AT2) angiotensin binding sites with high affinity for angiotensin I on IEC-18 rat intestinal epithelial cells. FEBS Lett 1995; 373:199–202.
65. Takayanagi R, Ohnaka K, Sakai Y, Nakao R, Yanase T, Haji M, Inagami T, Furuta H, Guo D, Nakamuta M, Nawata H. Molecular cloning, sequence analysis and expression of a cDNA encoding human type-1 angiotensin II receptor. Biochem Biophys Res Commun 1992; 183:910–916.
66. Bergsma D, Ellis C, Kumar C, Nuthulaganti P, Kersten H, Elshourbagy N, Griffin E, Stadtel JM, Aiyar N. Cloning and characterization of a human angiotensin II type 1 receptor. Biochem Biophys Res Commun 1992; 183:989–995.

67. Curnow KM, Pascoe L, Davies E, White PC, Corvol P, Clauser E. Alternatively spliced human type 1 angiotensin II receptor mRNAs are translated at different efficiencies and encode two receptor isoforms. Mol Endocrinol 1995; 9(9):1250–1262.

68. Mukoyama M, Nakajima M, Horiuchi M, Sasamura H, Pratt RE, Dzau VJ. Expression cloning of type 2 angiotensin II receptor reveals a unique class of seven-transmembrane receptors. J Biol Chem 1993; 268:24539–24542.

69. Nahmias C, Cazaubon SM, Briend-Sutren MM, Lazard D, Villageois P, Strosberg AD. Angiotensin II AT_2 receptors are functionally coupled to protein tyrosine dephosphorylation in N1E-115 neuroblastoma cells. Biochem J 1995; 306(1):87–92.

70. Brunner HR, Gavras H, Laragh JH, Keenan R. Hypertension in man. Exposure of the renin and angiotensin components using angiotensin II blockade. Circ Res 1974; 35:I-35–I-47.

71. Hollenberg NK, Williams GH, Burger B, Ishikawa I, Adams DF. Blockade and stimulation of renal adrenal and vascular angiotensin II receptors in normal man. J Clin Invest 1976; 57:39–46.

72. Yoshiyasu F, Shoji K, Kohei N. Hypotensive imidazole-5-acetic acid derivatives. Takeda Chemical Industries, Ltd. 1982; US4355040.

73. Yoshiyasu F, Shoji K, Kohei N. Imidazole-5-acetic acid derivatives, their production and use. Takeda Chemical Industries, Ltd. 1982; EU0028834.

74. Wong PC, Chiu AT, Price WA, Thoolen MJMC, Carini DJ, Johnson AL, Taber RI, Timmermans PBMWM. Non-peptide angiotensin II receptor antagonists. I. Pharmacological characterisation of 2-n-butyl-4-chloro-1-(2-chlorobenzyl) imidazole-5-acetic acid, sodium salt (S-8307). J Pharmacol Exp Ther 1988; 247:1–7.

75. Chiu AT, Carini DJ, Johnson AL, McCall DE, Price WA, Thoolen MJMC, Wong PC, Taber RI, Timmermans PBMWM. Non-peptide angiotensin II receptor antagonists. II. Pharmacology of S-8308. Eur J Pharmacol 1988; 157:13–21.

76. Duncia JV, Chiu AT, Carini DJ, Gregory GB, Johnson AL, Price WA, Wells GJ, Wong PC, Calabrese JC, Timmermans PBMWM. The discovery of potent nonpeptide angiotensin II receptor antagonists: a new class of potent antihypertensives. J Med Chem 1990; 33:1312–1329.

77. Chiu AT, Duncia JV, McCall DE, Wong PC, Price WA, Thoolen MJMC, Carini DJ, Johnson AL, Timmermans PBMWM. Non-peptide angiotensin II receptor antagonists. III. Structure-function studies. J Pharmacol Exp Ther 1989; 250:867–874.

78. Wong PC, Price WA, Chiu AT, Thoolen MJMC, Duncia JV, Johnson AL, Timmermans PBMWM. Non-peptide angiotensin II receptor antagonists. IV. Hypertension 1989; 13:489–497.

79. Duncia JV, Carini DJ, Chiu AT, Johnson AL, Price WA, Wong PC, Wexler RR, Timmermans PBMWM. The discovery of DuP 753, a potent, orally active nonpeptide angiotensin II receptor antagonist. Med Res Rev 1992; 12:149–191.

80. Wong PC, Price WA, Chiu AT, Thoolen MJMC, Duncia JV, Carini DJ, Wexler RR, Johnson AL, Timmermans PBMWM. Non-peptide angiotensin II receptor antagonists. IX. Pharmacology of EXP3174: an active metabolite of DuP 753, an orally active antihypertensive agent. J Pharmacol Exp Ther 1990; 255(1):211–217.

81. Wong PC, Price WA, Chiu AT, Thoolen MJMC, Duncia JV, Carini DJ, Wexler RR, Johnson AL, Timmermans PBMWM. Non-peptide angiotensin II receptor antagonists. X. Hypotensive action of DuP 753, an angiotensin II antagonist, in spontaneously hypertensive rats. Hypertension 1990; 15:459–468.

82. Li XC, Widdop RE. Angiotensin type I antagonists CV-11974 and EXP 3174 cause selective renal vasodilatation in conscious spontaneously hypertensive rats. Clin Sci 1996; 91(2):147–154.

83. Muller P, Cohen T, de Gasparo M, Sioufi A, Racine-Poon A, Howald H. Angiotensin II receptor blockade with single doses of valsartan, in healthy normotensive subjects. Eur J Pharmacol 1994; 47(3):231–245.

84. Bulhmayer P, Furet P, Criscione L, de Gasparo M, Whitebread S, Schmidlin T, Lattmann R, Wood J. Valsartan: a potent orally active angiotensin II antagonist developed from the structurally new amino acid series. Bioorg Med Chem Lett 1994; 4:29–34.

85. Criscione L, de Gasparo M, Bulhmayer P, Whitebread S, Ramjoue H, Wood J. Pharmacological profile of valsartan a potent orally active nonpeptide antagonist of the angiotensin II AT_1-receptor subtype. Br J Pharmacol 1993; 110:761–771.

86. Robertson MJ, Barnes JC, Drew GM, Clark KL, Marshall FH, Michel A, Middlemiss D, Ross BC, Scopes D, Dowle MD. Pharmacological profile of GR117289 *in vitro*: a novel, potent and specific non-peptide angiotensin AT_1 receptor antagonist. Br J Pharmacol 1992; 107:1173–1180.

87. Brooks DP, Fredrickson TA, Weinstock J, Ruffolo RR, Edwards RM, Gellai M. Antihypertensive activity of the nonpeptide angiotensin II receptor antagonist, SK&F 108566 in rats and dogs. Naunyn-Schmiederberg's Arch. Pharmacol 1992; 345:673–678.

88. Inada Y, Wada T, Yumiko S, Ojima M, Sanada T, Ohtsuki K, Itoh K, Kubo K, Kohara Y, Naka T, Nishikawa K. Anti-hypertensive effect of a highly potent and long-acting angiotensin II subtype-1 receptor antagonist, (±)-1-(cyclohexyloxycarbonyloxy)ethyl 2-ethoxy-1-[[2'-(1H-tetrazol-5-yl)biphenyl-4-yl]methyl]-1H-benzimidazole-7-carboxylate (TCV-116), in various hypertensive rats. J Pharmacol Exp Ther 1994; 268(3):1540–1547.

89. Noda M, Shibouta Y, Inada Y, Ojima M, Wada T, Sanada T, Kubo K, Kohara Y, Naka T, Nishikawa K. Inhibition of rabbit aortic angiotensin II (AII) receptor by CV-11974, a new nonpeptide AII antagonist. Biochem Pharmacol 1993; 46:311–318.

90. Chiu AT, McCall DE, Price WA, Wong PC, Carini DJ, Duncia JV, Wexler RR, Yoo SE, Johnson AL, Timmermans PBMWM. Non-peptide angiotensin II receptor antagonists. VII. Cellular and biochemical Pharmacology of DuP

753, an orally active antihypertensive agent. J Pharmacol Exp Ther 1990; 252(2):711–718.

91. Robertson MJ, Wragg A, Clark KL. Modulation of tachyphylaxis to angiotensin II in rabbit isolated aorta by the angiotensin AT_1 antagonist, losartan. Regul Peptides 1994; 50:137–145.

92. Wong PC, Timmermans PBMWM. Nonpeptide angiotensin II receptor antagonists: insurmountable angiotensin II antagonism of EXP3892 is reversed by the surmountable antagonist DuP753. J Pharmacol Exp Ther 1991; 258(1):49–57.

93. Purdy RE, Prins BA, Weber MA, Bakhtiarian A, Smith JR, Kim MK, Nguyen T-HT, Weiler EWJ. Possible novel action of ouabain: allosteric modulation of vascular serotonergic (5-HT_2) and angiotensinergic (AT_1) receptors. J Pharmacol Exp Ther 1993; 267:228–237.

94. Robertson MJ, Dougall IG, Harper D, McKechnie KCW, Leff P. Agonist-antagonist interactions at angiotensin receptors: application of a two-state receptor model. Trends Pharmacol Sci 1994; 15:364–369.

95. Glinka TW, de Laszlo SE, Siegl PKS, Chang RSL, Kivlighn SD, Schorn TS, Faust KA, Chen TB, Zingaro GJ, Lotti VJ, Greenlee WJ. A new class of balanced AT_1/AT_2 angiotensin II antagonists: quinazolone AII antagonists with acylsulfonamide and sulfonylcarbamate acidic functionalities. Bioorg Med Chem 1994; 4:81–86.

96. Chang LL, Ashton WT, Flanagan KL, Chen T-B, O'Malley SS, Zingaro GJ, Kivlighn SD, Seigl PKS, Lotti VJ, Chang RSL, Greenlee WJ. Potent orally active angiotensin II receptor antagonists with equal affinity for human AT_1 and AT_2 subtypes. J Med Chem 1995; 4:2337–2342.

97. Quan ML, Chiu AT, Ellis CD, Wong PC, Wexler RR, Timmermans PBMWM. Balanced AT_1/AT_2 receptor antagonists IV. Orally active 5(3-amidopropanoyl) imidazoles possessing equal affinity for the AT_1 and AT_2 receptors. J Med Chem 1995; 38(15):2938–2945.

98. Kivlighn SD, Huckle WR, Zingaro GJ, Rivero RA, Lotti VJ, Chang RSL, Schorn TW, Kevin N, Johnson RG, Greenlee WJ. Discovery of L-162,313: a nonpeptide that mimics the biological actions of angiotensin II. Am J Physiol 1995; 268:R820–R823.

99. Huckle WR, Kivlighn SD, Zingaro GJ, Kevin N, Rivero RA, Chang RSL, Greenlee WJ, Seigl PKS, Johnson RG. Angiotensin II receptor-mediated activation of phosphoinositide hydrolysis and elevation of mean arterial pressure by a nonpeptide, L163,491. Can J Physiol Pharmacol 1994; 72(Suppl 1):543.

100. Underwood DJ, Strader CD, Rivero R, Patchett, AA, Greenlee WJ, Prendergast K. Structural model of antagonist and agonist binding to the angiotensin II, AT_1 subtype, G protein coupled receptor. Chem Biol 1994; 1:211–221.

101. Nirula V, Zheng W, Sothinathan R, Sandberg K. Interaction of biphenylimidazole and imidazoleacrylic acid nonpeptide antagonists with valine 108 in TM III of the AT_1 angiotensin receptor. Br J Pharmacol 1996; 119:1505–1507.

102. Perlman S, Schambye HT, Rivero RA, Greenlee WJ, Hjorth SA, Schwartz TW. Non-peptide angiotensin agonist. Functional and molecular interaction with the AT_1 receptor. J Biol Chem 1995; 270(4):1493–1496.

103. Ji HJ, Leung M, Zhang Y, Catt KJ, Sandberg K. Differential requirements for specific binding of nonpeptide and peptide antagonists to the AT_1 angiotensin receptor. J Biol Chem 1994; 269(24):16533–16536.

104. Delacretaz E, Nussberger J, Biollaz J, Waeber B, Brunner HR. Characterization of the angiotensin II receptor antagonist TCV-116 in healthy volunteers. Hypertension 1995; 25(1):14–21.

105. Burnier M, Hagman M, Nussberger J, Biollaz J, Armagnac C, Brouard R, Waeber B, Brunner HR. Short-term and sustained renal effects of angiotensin II receptor blockade in healthy subjects. Hypertension 1995; 25(4 part 1):602–609.

106. Muller P, Cohen T, deGasparo M, Sioufi A, Raine Poon A, Howard H. Angiotensin II receptor blockade with single doses of valsartan in healthy, normotensive subjects. Eur J Clin Pharmacol 1994; 47(3):231–245.

107. Byyny RL. Losartan potassium lowers blood pressure measured by ambulatory blood pressure monitoring. J Hypertension 1995; 13(Suppl 1):S29–S33.

108. Townsend R, Haggert B, Liss C, Edelman JM. Efficacy and tolerability of losartan versus enalapril alone or in combination with hydrochlorothiazide in patients with essential hypertension. Clin Ther 1995; 17(5):911–923.

109. Holwerda NJ, Fogari R, Angeli P, Porcellati C, Hereng C, Oddou, Stock P, Heath R, Bodin F. Valsartan, a new angiotensin II antagonist for the treatment of essential hypertension: efficacy and safety compared with placebo and enalapril. J Hypertension 1996; 14(9):1147–1151.

110. Lacourciere Y, Lefebvre J, Nahkle G, Faison EP, Snavely DB, Nelson EB. Association between cough and angiotensin converting enzyme inhibitors versus angiotensin II antagonists: the design of a prospective, controlled trial. J Hypertension 1994; 12(Suppl 2):S49–S53.

111. Linz W, Wiener G, Gohlke P, Unger T, Scholkens BA. Contribution of kinins to the cardiovascular actions of angiotensin-converting enzyme inhibitors. Pharmacol Rev 1995; 47(1):25–49.

112. Ramsay LE, Yeo WW. ACE inhibitors, angiotensin II antagonists and cough. The Losartan Cough Study Group. J Hum Hypertension 1995; 9(Suppl 5):S51–54.

113. Waeber B, Brunner HR. Angiotensin II antagonists: a new class of antihypertensive agent. Br J Clin Pract 1996; 50(5):265–268.

114. Goldberg A, Dunlay M, Sweet C. Safety and tolerability of losartan compared with atenolol, felopdipine and ACE-inhibitors. J Hypertension 1995; 13(Suppl 1):S77–S88.

115. Van der Meiracker AH, Admiraal PJJ, Janssen JA, Kroodsma JM, de Ronde WAM, Boomsma F, Sissmann J, Blankestijn PJ, Mulder PGM, Manin't Veld AJ, Schalekamp MADH. Hemodynamic and biochemical effects of the AT_1 receptor antagonist, irbesartan, in hypertension. Hypertension 1995; 25(1):22–29.

116. Klamerus KJ, Vadiei K, Burghart P, Neefe DL, Zimmerman JJ. Ascending, single-dose, safety, tolerance and pharmacokinetics (PK) of 5,8-dihydro-2,4-dimethyl-8-[2'-(1H-tetrazol-5-yl) [1,1'-biphenyl]-4-yl) methyl] pyrido[2,3-

d]pyrimidin-7(6H)-one (ANA-756). Clin Pharmacol Ther 1995; 57(2):203 (PIII-35).

117. Su CAPF, van Heiningen PNM, van Lier JJ, de Bruin H, Kirchgassler KU, Cornelissen PJG, Jonkman JHG. Pharmacodynamics of the AII antagonist BIBR0277SE. Clin Pharmacol Ther 1994; 55:205 (PIII-93).

118. Urata H, Kinoshita A, Misono KS, Bumpus FM, Hussain A. Identification of a highly specific chymase as the major angiotensin II–forming enzyme in the human chymase. J Biol Chem 1990; 265:22348–22357.

119. Miura S, Ideishi M, Sakai T, Motoyama M, Kinoshita A, Sasaguri M, Tanaka H, Shindo M, Arakawa K. Angiotensin II formation by an alternative pathway during exercise in humans. J Hypertension 1994; 12:1177–1181.

120. Anderson IK, Polley JS, Hilditch A, Drew GM. Inhibition of the antihypertensive effect of enalapril and the angiotensin AT_1 receptor antagonist, GR138950, by Nω-nitro-L-arginine methyl ester in conscious renal hypertensive rats. Br J Pharmacol 1995; 114(Proc Suppl):172P.

121. Cachofeiro V, Maeso R, Rodrigo E, Navarro J, Ruilope LM, Lahera V. Nitric oxide and prostaglandins in the prolonged effects of losartan and ramipril in hypertension. Hypertension 1995; 26:236–243.

122. Kagoshima T, Masuda J, Sutani T, Sakaguchi Y, Tsuchihashi M, Tsuruta S, Iwano M, Dohi K, Nakamura Y, Konishi N. Angiotensin II receptor antagonist, TCV-116, prevents myocardial hypertrophy in spontaneously hypertensive rats. Blood Pressure 1994; 5:(Suppl):89–93.

123. Nishikimi T, Yamagishi H, Takeuchi K, Takeda T. An angiotensin II receptor antagonist attenuates left ventricular dilatation after myocardial infarction in hypertensive rat. Cardiovasc Res 1995; 29(6):856–861.

124. Furukawa Y, Matsumori A, Hirozane T, Sasayama S. Angiotensin II receptor antagonist TCV-116 reduces graft coronary artery disease and preserves graft status in a murine model. A comparative study with captopril. Circulation 1996; 93(2):333–339.

125. Kino H, Hama J, Takenaka T, Sugimura K, Kamoi K, Shimada S, Yamamoto Y, Nagata S, Kai T, Horiuchi M, Katori R. Effect of an angiotensin II receptor antagonist, TCV-116, on neointimal formation following balloon injury in the SHR carotid artery. Clin Exp Pharmacol Physiol 1995; 22(Suppl 1):S360–S362.

126. Matsumori A, Kawai T, Sugimoto T. Study of the efficacy and safety of TCV-116, an angiotensin II receptor antagonist for chronic heart failure. Jpn Circ J 1995; 59(Suppl 1):619 (abstract).

127. Crozier I, Ikram H, Awan N, Cleland J, Stephen N, Dickstein K, Frey M, Young J, Klinger G, Makris L, Rucinska E. Losartan in heart failure. Hemodynamic effects and tolerability. Losartan Hemodynamic Study Group. Circulation 1995; 91(3):691–697.

128. Pitt B, Segal R, Martinez FA, Meurers G, Cowley AJ, Thomas I, Deedwania PC, Ney DE, Snavely DB, Chang PI, on behalf of the ELITE Study Investigators. Randomised trial of losartan versus captopril in patients over 65 with heart failure (Evaluation of Losartan in the Elderly Study, ELITE). Lancet 1997; 349:747–752.

10

α-Adrenoceptor Agonists and Antagonists

J. Paul Hieble and Robert R. Ruffolo, Jr.
SmithKline Beecham Pharmaceuticals, King of Prussia, Pennsylvania

I. INTRODUCTION

Compounds interacting with the α-adrenoceptors, such as phenethylamines, imidazolines, and ergot alkaloids, have been used as therapeutic agents for decades. However, many of these drugs also interact with β-adrenoceptors, or with receptors for other biogenic amines. α-Adrenoceptors were subdivided into the α_1- and α_2-adrenoceptor subtypes approximately 25 years ago [see review by Hieble and Ruffolo (1)]. Based upon radioligand-binding and functional studies, and more recently the application of molecular biological techniques, there appear to exist at least seven subtypes of α-adrenoceptors, not including species variants. In concert with the subdivision and subclassification of α-adrenoceptors, there has been a continuing effort to design agonists and antagonists that interact selectively with a particular subtype, in an attempt to apply this subtype selectivity to the design of safer and/or more efficacious drugs. Although agents showing a variety of selectivity profiles have been identified, highly selective agonists or antagonists for most of the adrenoceptor subtypes are not yet available.

Drugs interacting with the α-adrenoceptors come from a variety of different chemical classes, including the quinazoline α_1-adrenoceptor antagonists for hypertension and benign prostatic hyperplasia (BPH), the imidazoline α_2-adrenoceptor agonists for hypertension, glaucoma, and several applications within the central nervous system, and the phenethylamine and imidazoline α_1-adrenoceptor agonists, which are commonly used as nasal decongestants. There is currently great interest in the design of α_1-adrenoceptor antagonists that are selective for the urogenital system (versus the vasculature) for the treatment of BPH, and this form of organ selectivity is based on α_1-adrenoceptor subtype selectivity. The role of α_1-adrenoceptor subtype selectivity is also being investigated as an approach to improving the therapeutic profile (i.e., efficacy and/or safety) of drugs used for the treatment of several of the other indications noted above.

Some of the α_1-adrenoceptor subtypes identified through the techniques of molecular biology have not yet been assigned a definitive functional role. In addition, pharmacological evidence is abundant to suggest the existence of an additional α_1-adrenoceptor that produces smooth muscle contraction, but it has not yet been cloned. Both of these lines of experimentation are continuing, and the ultimate assignment of function to all adrenoceptors is likley to result in additional therapeutic targets for which selective agonists and antagonists are designed. It is probable that several of the α_1-adrenoceptor subtype-selective agonists and antagonists arising from these efforts will become important therapeutic agents, perhaps for clinical indications where currently available agents have been unsuccessful or even for indications where α-adrenoceptor agents are not now considered.

II. USE OF SUBTYPE SELECTIVITY BETWEEN α_1- AND α_2-ADRENOCEPTORS IN THE DESIGN OF IMPROVED THERAPEUTIC AGENTS

A. Antihypertensive Therapy

1. α_1-Adrenoceptor Antagonists as Antihypertensive Drugs

In view of the role that the sympathetic nervous system plays in the control of vascular resistance, and the fact that α-adrenoceptors mediate the pressor effects of neuronally released norepinephrine as well as circulating epinephrine, it is logical that α-adrenoceptor antagonists were among the first pharmacological classes to be evaluated as antihypertensive drugs. However, the initial clinical investigations of the α-adrenoceptor antagonists then available, phenoxybenzamine and phentolamine, yielded disappointing results (2,3). Little reduction of supine blood pressure was produced and tolerance appeared to develop rapidly; the side effects, such as orthostatic hypotension, were often intolerable.

By the time that prazosin, the first member of the quinazoline class of α-adrenoceptor antagonists, was developed, the subclassification of α-adrenoceptors into the α_1- and α_2-adrenoceptor subtypes was well established. Prazosin was shown to be a relatively selective α_1-adrenoceptor antagonist, and soon became the accepted pharmacological tool for producing selective α_1-adrenoceptor blockade. Interestingly, the α_1-adrenoceptor selectivity (versus α_2-adrenoceptors) of prazosin is readily observed in virtually all experimental systems, despite the fact that it is now known that prazonsin has high affinity for several, although not all, of the α_2-adrenoceptor subtypes (see below).

Clinical evaluation of prazosin in hypertensive patients showed the drug to be a highly effective antihypertensive agent, with sustained reductions observed in both standing and supine blood pressure (4). Other quinazolines, such as terazosin and doxazosin, provide a similar degree of efficacy, with the advantage of longer durations of action, allowing for once-daily dosing. The greater degree of efficacy as antihypertensive agents of the quinazolines compared to the earlier α-adrenoceptor antagonists has been widely attributed to the selectivity of the quinazolines for postjunctional vascular α_1-adrenoceptors and the consequent lack of antagonist action at the prejunctional neuroinhibitory α_2-adrenoceptors, such that the feedback loop controlling norepinephrine release from sympathetic nerve terminals is not disrupted (3). Although there is some evidence to refute this (5), it is clear that the nonselective α-adrenoceptor antagonists, which also block the prejunctional neuroinhibitory α_2-adrenoceptor, significantly elevate circulating catecholamine levels in both experimental animals and humans. It appears, therefore, that the elevations in circulating levels of catecholamines, and presumably increases in norepinephrine levels at the vascular neuroeffector junction, resulting from prejunctional α_2-adrenoceptor blockade can counteract the competitive blockade of postjunctional vascular α_1-adrenoceptors in the case of the nonselective α_1-adrenoceptor antagonists, thereby limiting their efficacy.

A newer α-adrenoceptor antagonist, naftopidil, has been shown to have only a small degree of selectivity between α_1- and α_2-adrenoceptors (6), yet the drug appears to be an effective antihypertensive agent. This may relate, in part, to the ability of naftopidil to discriminate between pre- and postjunctional α_2-adrenoceptors in the vasculature.

2. α_2-Adrenoceptor Agonists as Antihypertensive Drugs

The use of α_2-adrenoceptor agonists as centrally acting antihypertensive agents is based on the discovery, 30 years ago, of the hypotensive and bradycardic actions of clonidine (7). Clonidine was originally synthesized as an α_1-adrenoceptor agonist to be used as a nasal decongestant, but clinical evaluation soon showed the drug to possess a different pharmacological profile from other imidazolines that were known at the time, which when administered systemically produced the anticipated increases in blood pressure. The mechanism responsible for the hypotensive action of clonidine was not clarified until a decade later when the subclassification of α-adrenoceptors into the α_1- and α_2-adrenoceptor subtypes was established

(8,9). Indeed, clonidine became one of the primary pharmacological tools used to establish the existence of distinct populations of α-adrenoceptors (9,10).

It is now firmly established that activation of postsynaptic α_2-adrenoceptors in the brainstem by clonidine and clonidine-like drugs inhibits sympathetic outflow from the central nervous system and thereby lowers blood pressure and heart rate (11). Because the activation of specific "imidazoline receptors" that are distinct from α_2-adrenoceptors has also been shown to produce centrally mediated sympatholytic effects (12), the relative role of α_2-adrenoceptors and imidazoline receptors in the antihypertensive actions of systemically administered imidazolines has been debated and remains a matter of current controversy. The action of clonidine itself, however, appears to be mediated primarily by α_2-adrenoceptors in the brainstem rather than by imidazoline receptors inasmuch as the antihypertensive actions of clonidine, when administered systemically, are inhibited by α_2-adrenoceptor antagonists (13–16).

Stimulation of central α_2-adrenoceptors also induces sedation and sleep. Although this effect of clonidine and related α_2-adrenoceptor agonists can be utilized therapeutically, it is most commonly a limiting side effect when these drugs are used in the treatment of hypertension. A great deal of research has been directed toward the separation of the centrally mediated sympatholytic and sedative actions of clonidine-like imidazolines and the discovery of drugs that selectively reduce blood pressure without producing sedation. To date, none of the clonidine-like imidazolines has been demonstrated clinically to produce a convincingly greater degree of separation between the antihypertensive effects and sedation than clonidine. As such, the other clonidine-like, centrally acting antihypertensive agents that have been marketed, such as guanfacine and guanabenz, are superior to clonidine only with respect to duration of action and α_2- versus α_1-adrenoceptor selectivity. All are still associated with a high incidence of sedation.

It has been postulated that some clonidine-like imidazolines having preferential affinity for imidazoline receptors over α_2-adrenoceptors may have an improved therapeutic profile and a lower propensity to produce sedation [see review by Yu and Frishman (17)]. Early clinical results with two such agents, rilmenidine and moxonidine, support this hypothesis, although experience is still limited at this time. Dry mouth, another side effect commonly associated with clonidine therapy, is also prominent with rilmenidine and moxonidine (18,19). In a direct comparison with clonidine, moxonidine was shown to produce a significantly lower incidence of tiredness (13% versus 17%) (20,21); however, the magnitude of this difference is small, particularly if it is assumed that these two drugs act through different receptors within the brainstem. Some experimental evidence in the rabbit using selective antagonists supports the involvement of imidazoline receptors (as opposed to α_2-adrenoceptors) in the hypotensive actions of centrally administered rilmenidine and moxonidine (22). However, this phenomenon could not be demonstrated upon systemic intravenous administration of rilmenidine (23). As such, proof for a role of central imidazoline receptors in the antihypertensive actions of some clonidine-like imidazolines awaits the development of a potent and selective imidazoline receptor agonist that lacks agonist activity at α_2-adrenoceptors.

B. α-Adrenoceptor Agonists as Nasal Decongestants

Vasoconstriction of blood vessels in the nasal mucosa has long been recognized as an important approach to relieving nasal congestion; accordingly, systemic or topical administration of a vasoconstrictor can produce nasal decongestion. Selective α_1-adrenoceptor agonists, such as phenylephrine and naphazoline, nonselective (i.e., mixed α_1-/α_2) α-adrenoceptor agonists, such as tetrahydrozoline, and even relatively selective α_2-adrenoceptor agonists, such as oxymetazoline, are currently employed for this indication.

Recent research in this area has been directed toward highly selective α_1-adrenoceptor agonists (e.g., Abbott-57219) based on the premise that a selective α_1-adrenoceptor agonist will preferentially constrict capacitance vessels in the nasal mucosa to reduce congestion, without compromising nutritional blood flow provided by resistance vessels through α_2-adrenoceptor-mediated vasoconstriction (24). However, potent nonselective α-adrenoceptor agonists (e.g., Abbott-54741) have also been shown to be effective in decreasing resistance to air flow in the nasal cavity (25). As such, it is now postulated that stimulation of both α_1- and α_2-adrenoceptors on nasal vasculature will produce an additive effect by decreasing blood flow to the mucosal capillary bed (mediated by α_2-adrenoceptors) and contraction of postcapillary venules to reduce mucosal volume (mediated by α_1-adrenoceptors) (26).

Because α-adrenoceptor agonists have the capacity to increase systemic arterial blood pressure by producing vasoconstriction in the peripheral systemic vasculature, limiting their absorption from the nasal mucosa is a highly desirable feature. Dose-response studies of the intranasal administration of Abbott-54741 in the dog have shown that near-maximal effects on nasal resistance can be achieved before effects on blood pressure and heart rate are observed (25). This finding appears to result from the relatively high polarity of Abbott-54741, a catecholimidazoline analog, which appears to have only limited absorption through the nasal mucosa. PGE-6201204, a hydrophilic imidazoline (27) that produces selective activation of peripheral α_2-adrenoceptors (28), is also currently being evaluated as a potential nasal decongestant.

Cirazoline, a selective stimulant of the α_1-adrenoceptor, has also been shown to be a potent antagonist of prejunctional α_2-adrenoceptors (29). Because blockade of prejunctional α_2-adrenoceptors at the vascular neuroeffector junction will increase stimulation-evoked neurotransmitter release from sympathetic nerve endings, one could speculate that cirazoline would produce an exaggerated stimulation of postjunctional vascular α_1-adrenoceptors through the activation of postjunctional α_1-adrenoceptors by neuronally released norepinephrine, which would be additive to the direct activation of α_1-adrenoceptors produced by cirazoline itself. Such a pharmacological profile may be highly effective as a topical nasal decongestant.

C. α-Adrenoceptor Agonists and Antagonists for the Treatment of Glaucoma

It is interesting that a variety of agents interacting with adrenoceptors have been shown to reduce intraocular pressure. These include β-adrenoceptor agonists, β-adrenoceptor antagonists, α_2-adrenoceptor agonists, and both α_1- and α_2-adrenoceptor antagonists. This unusual phenomenon likely results from multiple factors contributing to intraocular pressure, and the fact that intraocular pressure may be lowered either by inhibiting the formation of aqueous humor or by decreasing the resistance to its outflow, or both [see review by Potter (30)].

Recent clinical trials have confirmed the efficacy of α_2-adrenoceptor stimulation in decreasing intraocular pressure (31–35). These studies were performed with the selective α_2-adrenoceptor agonists UK 14,304 (brimonidine) and *para*-aminoclonidine. A high density of α_2-adrenoceptors has been observed in the ciliary body (36,37), and their activation is thought to lower intraocular pressure primarily by inhibiting the secretion of aqueous humor from the ciliary epithelium (38–40). α_2-Adrenoceptor activation can have a variety of other effects within the eye, some involving modulation of the sympathetic nervous system (41).

It has also been postulated that imidazoline receptors may be involved in some of the actions of clonidine-like imidazolines in glaucoma (41–43). Radioligand-binding assays provide no evidence for the presence of imidazoline binding sites in the ciliary body of the rabbit (36). However, it is possible that an imidazoline receptor located at another ocular or extraocular site may contribute to the ocular hypotensive actions of those clonidine-like imidazolines that interact with both α_2-adrenoceptors and imidazoline I_1-receptors, such as moxonidine.

Because stimulation of α_1-adrenoceptors can increase intraocular pressure (44,45), a high degree of selectivity for α_2-adrenoceptors is required when an α-adrenoceptor agonist is used for the treatment of glaucoma. Even agents such as UK 14,304, which have a reasonable degree of selectivity for α_2- versus α_1-adrenoceptors, can produce an initial, short-lived increase in intraocular pressure (46), which probably results from α_1-adrenoceptor activation produced by the high local concentrations achieved immediately following topical administration. Hence the design of even more selective α_2-adrenoceptor agonists (or possibly an agent that is both an α_2-adrenoceptor agonist and an α_1-adrenoceptor antagonist) for this indication may be desirable.

α-Adrenoceptor blockade will also lower intraocular pressure. Interestingly, selective α_1-adrenoceptor antagonists, selective α_2-adrenoceptor antagonists, and nonselective α-adrenoceptor antagonists are capable of decreasing intraocular pressure in experimental animals (47–50). A selective α_2-adrenoceptor antagonist, SK&F 86466, was effective in lowering intraocular pressure in the rabbit (48). However, in clinical trials this antagonist did not reduce intraocular pressure in patients with open-angle glaucoma or ocular hypertension (asymptomatic elevation in intraocular pressure) (unpublished observations).

It is likely that α_1- and α_2-adrenoceptor antagonists have different mechanisms of action for decreasing intraocular pressure (49). Because α_1-adrenoceptor blockade appears to be more effective in lowering intraocular pressure when sympathetic innervation is intact, it has been postulated that blockade on an action of norepinephrine is involved, resulting in a decrease in aqueous humor inflow (50). Ocular α_2-adrenoceptors are located both prejunctionally on sympathetic neurons and postjunctionally on cells in the iris/ciliary body (51). The mechanism of action for the α_2-adrenoceptor antagonists has not been clearly established, but may involve blockade of presynaptic α_2-adrenoceptors, which potentiates norepinephrine release (48). In contrast to the β-adrenoceptor antagonists, which represent the primary therapy for glaucoma, α-adrenoceptor antagonists are not currently used clinically for treatment of this disorder.

D. α-Adrenoceptor Antagonists for the Treatment of Benign Prostatic Hyperplasia

The first clinical evidence for the utility of an α-adrenoceptor antagonist in the symptomatic treatment of benign prostatic hyperplasia (BPH) was provided by Caine and colleagues (52) using phenoxybenzamine, an irreversible antagonist that alkylates both α_1- and α_2-adrenoceptor adrenoceptors. The efficacy of phenoxybenzamine has been confirmed in several subsequent studies (53,54), and the intravenous administration of a nonselective α-adrenoceptor antagonist, phentolamine, was shown to relieve the acute urinary retention associated with BPH (55), thereby supporting the notion that an α-adrenoceptor antagonist would be useful in the treatment of BPH. In spite of the fact that nonselective α-adrenoceptor antagonists have been shown to relieve symptoms associated with BPH, because the contractile response of isolated prostatic smooth muscle is mediated by α_1-adrenoceptors (56), the design of drugs for this indication has recently concentrated on selective α_1-adrenoceptor antagonists. Most of the currently marketed drugs are quinazolines, which have a high degree of selectivity for α_1- versus α_2-adrenoceptors (i.e., prazosin, terazosin, doxazosin, alfuzosin, bunazosin). Although these drugs are generally believed to produce fewer side effects than phenoxybenzamine (54), this has not been demonstrated conclusively in side-by-side comparative studies.

More recent data with naftopidil, an antagonist having high affinity for α_2-adrenoceptors in some assays (6,57,58), suggest that the drug is efficacious in the treatment of BPH without producing unacceptable side effects (59). As such, the possibility exists that blockade of α_2-adrenoceptors may be a desirable feature in an α-adrenoceptor antagonist used for the treatment of BPH.

III. DISCOVERY OF NOVEL α-ADRENOCEPTOR SUBTYPES AND THEIR RELATIONSHIP TO THE DESIGN OF NEW THERAPEUTIC AGENTS

A. Benign Prostatic Hyperplasia

The role of α-adrenoceptors in the contraction of prostatic smooth muscle has been known for over 20 years (60,61), and as discussed above, α-adrenoceptor antagonists have been demonstrated to be effective in the treatment of BPH (52). Selective α_1-adrenoceptor blockade is now the most widely employed pharmacotherapy for the management of BPH. The clinical efficacy of the α_1-adrenoceptor antagonists currently employed has been convincingly demonstrated in long-term, double-blind, placebo-controlled trials (62–64). Nevertheless, there is intense interest in the discovery of improved α_1-adrenoceptor antagonists for the treatment of BPH. This effort has focused on the design of antagonists capable of selectively blocking urogenital α_1-adrenoceptors.

Most of the α_1-adrenoceptor antagonists now marketed for BPH, including prazosin, terazosin, doxazosin, bunazosin, and indoramin, were initially developed as antihypertensive drugs. Accordingly, these agents also block postjunctional vascular α_1-adrenoceptors at the same doses required to block α_1-adrenoceptors in the urogenital system, and these drugs are therefore associated with a significant incidence of cardiovascular side effects, such as dizziness, asthenia, orthostatic hypotension, and syncope. These side effects, in particular orthostatic hypotension and syncope, are most severe upon initiation of therapy (first-dose phenomenon). As such, dose titration is required for most of the α_1-adrenoceptor antagonists used in the management of BPH, with therapy typically initiated using a subtherapeutic dose. Because of the effects of the selective α_1-adrenoceptor antagonists on the peripheral vasculature, it is possible that their efficacy in BPH is significantly limited because of unacceptable dose-limiting side effects on the cardiovascular system. In theory, therefore, improved efficacy in BPH could be achieved if higher doses of these drugs could be administered that would produce more complete blockade of prostatic α_1-adrenoceptors without blocking vascular α_1-adrenoceptors.

The initial efforts directed toward the development of uroselective α_1-adrenoceptor antagonists focused on selectivity between the three known α_1-adrenoceptor subtypes. The α_{1A}-adrenoceptor [formerly known as the α_{1C}-adrenoceptor (6)] is the predominant α_1-adrenoceptor subtype present in human prostate based on mRNA levels (65,66). Autoradiographic studies localize the α_{1A}-adrenoceptor to prostatic stromal smooth muscle (67). The ability of α_1-adrenoceptor antagonists to produce antagonism of norepinephrine-induced contraction of isolated strips of human prostate is highly correlated with their affinity for the recombinant α_{1a}-adrenoceptor (68–70). Antagonists having a high degree of selectivity for the α_{1A}-adrenoceptor compared to the two other known α_1-adrenoceptor subtypes (α_{1B} and α_{1D}) have been identified (71,72). Several of these antagonists appear to have a low propensity to induce orthostatic hypotension in the rat (73); however, their urogenital selectivity is not consistently observed in animal models. SNAP 5089, which is approximately 100-fold selective for the α_{1a}-adrenoceptor subtype (70,71), has low potency against norepinephrine-induced contraction of rabbit urethra and is a weak antagonist of norepinephrine-induced increases in urethral perfusion pressure in the anesthetized dog (74). Inconsistent findings have been reported for this compound in isolated human urogenital tissues, with some investigators reporting potent blockade of norepinephrine-induced responses ($K_B =$ 1 nM) (70,75); and others finding low potency ($K_B > 300$ nM) (76).

Another novel α_1-adrenoceptor antagonist, Rec 15/2739 (SB 216469), produced potent blockade of norepinephrine-induced contraction of isolated human prostate ($K_B = 2.7$–4 nM) (77,78) and demonstrated consistent urogenital selectivity in a variety of in vitro and in vivo experimental models (79–85). In contrast to SNAP 5089 and related dihydropyridines, Rec 15/2739 is only moderately selective for the recombinant human α_{1a}-adrenoceptor versus either the α_{1b}- (27-fold) or α_{1d}- (ninefold) adrenoceptor (81). Nevertheless, when SNAP 5089 and Rec 15/2739 are compared for urogenital selectivity in the anesthetized dog, Rec 15/2739 consistently demonstrates greater antagonist potency at the urethral receptor and has greater urogenital selectivity (74), indicating that factors other than selectivity for the α_{1A}-adrenoceptor can contribute to urogenital selectivity. The molecular mechanism responsible for the urogenital selectivity of SB 216469 has not been conclusively established, but may involve blockade of another α_1-adrenoceptor subtype that is less clearly understood, namely the α_{1L}-adrenoceptor.

Studies with another novel α_1-adrenoceptor antagonist, RS 17053, suggest that the α_{1A}-adrenoceptor may not actually mediate contraction of the human prostate. This antagonist has a high affinity for the recombinant bovine and human α_{1a}-adrenoceptor ($K_i < 1$ nM) and produces potent blockade of native α_{1A}-adrenoceptors in the perfused rat kidney ($K_B < 1$ nM) (76) and rat vas deferens ($K_B = 0.3$ nM) (86). RS 17053, however, is a relatively weak antagonist of norepinephrine-induced contraction of human prostate and urethra ($K_B = 30$–100 nM) (76–87). The results with RS 17053 provide support for an earlier suggestion that the contraction of human prostate is mediated by the α_{1L}-adrenoceptor, rather than by the α_{1A}-adrenoceptor (88).

Although the α_{1L}-adrenoceptor has not yet been cloned, functional evidence for its existence has been provided by studies in various blood vessels (89,90) that showing heterogeneity in the antagonist potency of prazosin. The α_{1L}-adrenoceptor is characterized by a low affinity for prazosin; this observation distinguishes this subtype from the other three α_1-adrenoceptor subtypes, all of which show an equivalent high affinity for prazosin.

A functional subclassification of α_1-adrenoceptors into the α_{1L}- and α_{1H}-adrenoceptor subtypes has been proposed, where the α_{1H}-adrenoceptor includes the α_{1A}-, α_{1B}-, and α_{1D}-adrenoceptor subtypes (91). The majority of studies with prazosin in human isolated prostate demonstrate antagonist affinity ($K_B = 2$–10 nM) that is in the range associated with an α_{1L}-adrenoceptor (92). The α_{1L}-adrenoceptors was initially defined by a low affinity for prazosin; however, the difference in the affinity of prazosin between the α_{1L}- and α_{1H}-adrenoceptors is relatively small, only approximately one order of magnitude. RS 17053 may represent an antagonist that can more effectively discriminate between α_{1A}- and α_{1L}-adrenoceptors. Although affinity for the human recombinant α_{1a}-adrenoceptor correlates well with antagonist potency in human prostate for most antagonists, this may merely indicate the most antagonists do not discriminate between α_{1a}- and α_{1L}-adrenoceptors. Compounds with low affinity for the α_{1L}-adrenoceptor, such as RS 17053, clearly deviate from the correlation line; additional examples of compounds having a similar profile have been identified (75).

The relationship between α_1-adrenoceptor subtype selectivity, urogenital selectivity in animal models, and clinical superiority in the treatment of BPH remains to be established. Although clinical results are not yet available for α_1-adrenoceptor antagonists that have a high degree of urogenital selectivity, two drugs showing some degree of α_1-adrenoceptor subtype selectivity, indoramin and tamsulosin, are currently being used for the treatment of BPH. Radioligand binding studies with recombinant human α_1-adrenoceptors have shown indoramin to be selective for the α_{1a}-adrenoceptor subtype, although the affinity ratio between the α_{1a}- and α_{1b}-adrenoceptors ranges from 10-fold (68) to only threefold (81,93).

Placebo-controlled clinical trials have demonstrated the efficacy of indoramin in BPH. Treatment of patients with indoramin does not appear to be associated with a consistent decrease in blood pressure, and, in contrast to prazosin, initial dosing is not commonly associated with symptoms attributable to orthostatic hypotension.

Like indoramin, tamsulosin shows a moderate degree of selectivity for recombinant human α_{1a}- versus α_{1b}-adrenoceptors (13–15-fold) (81,94). However, tamsulosin has essentially equal affinity for the α_{1a}- and α_{1d}-adrenoceptors. In radioligand-binding studies using native human tissues, tamsulosin was found to have 12-fold higher affinity for α_1-adrenoceptors in prostate versus aorta, in contrast to prazosin, which has equivalent affinity for α_1-adrenoceptors in the two tissues (95). However, several reports have shown tamsulosin to have equivalent potency in blocking the response to the α_1-adrenoceptor agonist phenylephrine in the urethra and systemic arterial circulation in the anesthetized dog (84,96). Comparison of the urethral blocking and hypotensive doses of tamsulosin in the anesthetized dog shows this agent to have some degree of urogenital selectivity (10-fold), which is intermediate between that observed for prazosin (1.8-fold) and terazosin (2.9-fold), but substantially less than that observed for Rec 15/2739 (100-fold) (74).

In clinical trials with tamsulosin, dose titration was not required, and no side effects attributable to vascular α_1-adrenoceptor blockade were observed (67,97). Because dose titration is not required with tamsulosin, a significant increase in peak flow rate could be observed following the first dose (67).

B. Hypertension

Selective α_1-adrenoceptor antagonists have been available as antihypertensive drugs for many years. This pharmacological class offers a number of unique and desirable therapeutic activities, particularly with respect to plasma lipoprotein profile, which is clearly favorably influenced by α_1-

adrenoceptor blockade (98). The roles of the individual α_1-adrenoceptor subtypes in the various actions of an α_1-adrenoceptor antagonist in hypertension are still largely unknown. The relative role of the α_1-adrenoceptor subtypes in mediating vasopressor responses in vivo is not yet well characterized.

Administration of chloroethylclonidine, a selective inactivator of α_{1b}- and α_{1d}-adrenoceptors (99), does not attenuate the pressor response to phenylephrine in the rat (100), suggesting a potential role of the α_{1A}-adrenoceptor in the vasoconstrictor response. The α_{1A}-adrenoceptor also mediates vasoconstriction in the perfused rat kidney (71,101,102) and rat caudal artery (103). α_1-Adrenoceptor-mediated effects on renal sodium excretion in both normotensive and hypertensive rats also appear to involve the α_{1A}-adrenoceptor (104,105). Although the α_{1a}-adrenoceptor subtype has not been detected in human aorta (106,107), it is expressed in rat aorta (99), although the α_1-adrenoceptor may not mediate the vasoconstrictor response in the rat aorta (see below). The distribution of α_1-adrenoceptor subtypes in resistance vessels has not been examined in any species. Hence, although selective α_{1A}-adrenoceptor antagonists have been targeted toward the prostate (see above), one cannot assume that such agents will be devoid of cardiovascular or renal effects in the hypertensive patient.

It now appears clear that contraction of the rat aorta is mediated by the α_{1D}-adrenoceptor (108–110). Studies in skeletal muscle arterioles of the rat also show the involvement of α_{1D}-adrenoceptors, while the primary receptor responsible for contraction of venules is the α_{1B}-adrenoceptor (111). The α_{1B}-adrenoceptor has also been reported to mediate contraction of the rat vena cava (112). Both α_{1b}- and α_{1d}-adrenoceptors are expressed in human aorta (106). The functional contractile response in isolated human coronary arteries is mediated by the α_{1B}-adrenoceptor (113).

In view of these confusing and often conflicting findings, it is difficult to predict whether an α_1-adrenoceptor antagonist that is selective for one of the α_1-adrenoceptor subtypes will offer advantages over those agents that are currently available that do not discriminate between α_1-adrenoceptor subtypes. Because the α_1-adrenoceptor subtypes responsible for vascular contraction often vary between different vascular beds (103), it is possible that a subtype-selective antagonist would produce tissue-selective vasodilation. Investigation of this possibility is hindered by the lack of suitable pharmacological tools having sufficient degrees of selectivity. Although highly selective α_{1A}-adrenoceptor antagonists are available (e.g., SNAP 5089, SNAP 5150), selective antagonists for the other two α_1-adrenoceptor subtypes have not yet been well characterized. Cyclazosin (114) can selectively block α_{1B}- versus α_{1A}-adrenoceptors, but does not discriminate between the α_{1B}- and α_{1D}-adrenoceptor subtypes; however, the (+) enantiomer appears to selectively block the α_{1B} (α_{1b}) subtype in both native and recombinant receptor populations (74). AH 11110A shows relatively low affinity for the α_{1B}-adrenoceptor and is only 4–10-fold selective for this receptor versus the other two α_1-adrenoceptor subtypes (74). Although several structural classes of α_1-adrenoceptor antagonists show selectivity for the α_{1D}-adrenoceptor, such as BMY 7378 (110) and SK&F 104856 (6), these molecules also have potent antagonist activity at 5-HT$_{1A}$ receptors and α_2-adrenoceptors, respectively.

The α_{1L}-adrenoceptor also appears to mediate contraction of several blood vessels (113). Many antagonists, including those capable of differentiating between α_{1a}- and α_{1B}-adrenoceptors, do not discriminate between the α_{1a}- and α_{1L}-adrenoceptors (115). The low potency of RS 17053 against phenylephrine-reduced contraction of the rat portal vein (86) suggests that the α_{1L}-adrenoceptor mediates this response, in contrast to the high potency of this antagonist in the perfused rat kidney where α_{1A}-adrenoceptors are presumably involved (76). Again, studies on the potential role of the α_{1L}-adrenoceptor in regulation of vascular tone have been limited by the lack of antagonists that antagonize α_{1L}- but not α_{1H}-adrenoceptors. Some recently identified compounds have this profile (116).

IV. CONCLUSIONS

Drugs interacting with the α-adrenoceptors currently have several important therapeutic applications. Both α_1- and α_2-adrenoceptor agonists are employed topically as nasal decongestants, and α_1-adrenoceptor agonists can be administered intravenously to maintain or elevate blood pressure. α_2-Adrenoceptor agonists are also effective in the treatment of hypertension and glaucoma, with other therapeutic applications resulting from their actions within the central nervous system, such as adjuncts to general anesthesia, analgesia, treatment of opiate withdrawal, and attention deficit hyperactivity disorder. α_1-Adrenoceptor antagonists currently represent first-line therapy for the treatment of BPH and are still commonly used in the treatment of hypertension. Although no highly selective α_2-adrenoceptor antagonists are currently on the market, blockade of central α_2-adrenoceptors may contribute to the antidepressant activity observed clinically with agents such as mianserin and mirtazapine (117).

α_1-Adrenoceptor subtype selectivity is currently being utilized to design novel α_1-adrenoceptor antagonists with selectivity for the urogenital system for the treatment of BPH. Although data in animal models suggest that this represents a valid approach to the treatment of BPH, the clinical superiority of such agents over currently available drugs has not yet been demonstrated. The relationship between α_1-adrenoceptor subtype selectivity and therapeutic activity has not yet been established for α_1-adrenoceptor antagonists when used as antihypertensive agents; however, determination of the subtype(s) responsible for blood pressure reduction and effects on plasma lipoproteins could lead to more effective drugs for the treatment of hypertension.

Subtype selectivity may also offer a novel approach to optimize or improve the therapeutic profile of α_2-adrenoceptor agonists as centrally active antihypertensive agents, or perhaps would result in other therapeutic applications for this class of drugs. While no clear physiological role for the α_{2b}- or α_{2c}-adrenoceptors has yet been identified, it is possible that the development of selective agonists and/or antagonists for these subtypes, or the genetic receptor inactivation studies that are currently being performed on them (118,119), may uncover currently unknown effects mediated by these α_2-adrenoceptor subtypes, which would then serve as additional targets for pharmacotherapy.

Although α-adrenoceptors have been studied for many decades, there are still potentially new therapeutic opportunities for drugs interacting selectively with the various subtypes of the α-adrenoceptors. A key requirement for exploiting these opportunities is the development of subtype-selective agonists and antagonists. Unfortunately, however, the identification of such agents has not kept pace with the discovery of new α-adrenoceptor subtypes through functional and molecular biological studies. Drugs that interact selectively with a single α-adrenoceptor subtype will undoubtedly become important and useful pharmacological tools and could ultimately lead to improvements in therapy and to the identification of new therapeutic targets for α-adrenoceptor activation or blockade.

REFERENCES

1. Hieble JP, Ruffolo RR Jr. α-Adrenoceptors. In: Ruffolo RR, Hollinger MA, eds. G-Protein Coupled Transmembrane Signaling Mechanisms. Boca Raton, FL: CRC Press, 1995:1–34.
2. Miller SI, Ford RV, Moyer JH. Dibenzyline: results of therapy in patients with hypertension and comparisons with hexamethonium, 1-hydrazinophthalazine and semipurified extracts of veratrum. N Engl J Med 1953; 248:576–582.
3. Stokes GS, Marwood JF. Review of the use of alpha-adrenoceptor antagonists in hypertension. Meth Find Exp Clin Pharmacol 1984; 6:197–204.
4. Stanaszek WF, Kellerman D, Brogden RN, Romankiewicz JA. Prazosin update: a review of its pharmacological properties and therapeutic use in hypertension and congestive heart failure. Drugs 1983; 25:339–384.
5. Mulvihill-Wilson J, Gaffney FA, Pettinger WA, Blomqvist CG, Anderson SA, Graham RM. Hemodynamic and neuroendocrine responses to acute and

chronic alpha-adrenergic blockade with prazosin and phenoxybenzamine. Circulation 1983; 67:383–393.

6. Hieble JP, Ruffolo RR Jr, Sulpizio AC, Naselsky DP, Conway TM, Ellis C, Swift AM, Ganguly S, Bergsma DJ. Functional subclassification of α_2-adrenoceptors. Pharmacol Commun 1995; 6:91–97.

7. Graubner W, Wolf M: Kritische Betrachtungen zum Wirkungmechanismus des 2-(2,6-dichlorophenylamino)-2-imidazolinhydrochlorid. Arzneim Forsch 1966; 16:1055–1058.

8. Langer, SZ. Presynaptic regulation of the release of catecholamines. Biochem Pharmacol 1974; 23:1793–1800.

9. Starke K, Endo T, Taube HD. Relative pre- and postsynaptic potencies of α-adrenoceptor agonists in the rabbit pulmonary artery. Naunyn-Schmiedeberg's Arch Pharmacol 1975; 291:55–78.

10. Steinsland OS, Nelson SH. "Alpha-adrenergic" inhibition of the response of the isolated rabbit ear artery to brief intermittent sympathetic nerve stimulation. Blood Vessels 1975; 12:378–379.

11. Hieble JP, Ruffolo RR, Jr. Possible structural and functional relationships between imidazoline receptors and α_2-adrenoceptors. Proc NY Acad Sci 1995; 763:8–21.

12. Ernsberger P, Giuliano R, Willette RN, Reis D. Role of imidazole receptors in the vasodepressor response to clonidine analogs in the rostal ventralaterial medulla. J Pharmacol Exp Ther 1990; 253:408–418.

13. Hieble JP, Kolpak DC. Mediation of the hypotensive action of systemic clonidine in the rat by α_2-adrenoceptors. Br J Pharmacol 1993; 110:1635–1639.

14. Head GA, Godwin SJ, Sannajust F. Differential receptors involved in the cardiovascular effects of clonidine and rilmenidine in conscious rabbits. J Hypertens 1993; 11(5):S322–S323.

15. Vayssettes-Courchay C, Bouysset F, Cordi AA, Laubie M, Verbeuren TJ. A comparative study of the reversal by different α_2-adrenoceptor antagonists of the central sympatho-inhibitory effect of clonidine. Br J Pharmacol 1996; 117:587–593.

16. Ally A, Hand GA, Mitchell JH. Cardiovascular effects elicited by intravenous administration of clonidine in conscious cats. FASEB J 1996; 10:A607.

17. Yu A, Frishman WH. Imidazoline receptor agonist drugs—a new approach to the treatment of systemic hypertension. J Clin Pharmacol 1996; 36(2):98–111.

18. Webster J, Koch HF. Aspects of tolerability of centrally acting antihypertensive drugs. J Cardiovasc Pharmacol 1996; 27(Suppl 3):549–554.

19. Pillion G, Fevier B, Codis P, Schultz D. Long-term control of blood pressure by rilmenidine in high-risk populations. Am J Cardiol 1994; 74:58A-65A.

20. Planitz V. Crossover comparison of moxonidine and clonidine in mild to moderate hypertension. Eur J Clin Pharmacol 1984; 27:147–152.

21. Planitz V. Comparison of moxonidine and clonidine HCl in treating patients with hypertension. J Clin Pharmacol 1987; 27:46–51.

22. Chan CKS, Sannajust F, Head GA. Role of imidazoline receptors in the cardiovascular actions of moxonidine, rilmenidine and clonidine in conscious rabbits. J Pharmacol Exp Ther 1996; 276:411–420.

23. Szabo B, Urban R, Starke K. Sympathoinhibition by rilmenidine in conscious rabbits: involvement of α_2-adrenoceptors. Naunyn-Schmiedeberg's Arch Pharmacol 1993; 348:593–600.

24. DeBernardis JF, Winn M, Kerkman DJ, Kyncl JJ, Buckner S, Horrom B. A new nasal decongestant, A-57219: a comparison with oxymetazoline. J Pharm Pharmacol 1987; 39:760–763.

25. Kyncl JJ, DeBernardis JF, Bush EN, Buckner SA, Brondyk H. Novel adrenergic compounds. I. Receptor interactions of ABBOTT-54741 an alpha-adrenergic agonist. J Cardiovasc Pharmacol 1989; 13:382–389.

26. Johnson DA, Hricik JG. The pharmacology of α-adrenergic decongestants. Pharmacotherapy 1993; 13(6 Pt 2):110S-115S.

27. Rasmusssen, K., Lillibridge, J and Soehner, M.: Pharmacological characterization of PGE-6201204, a novel hydrophilic selective α_2-agonist. FASEB J 1996; 10:A426.

28. Sheldon RJ, Blanton CA, Brown ML, Slusher KL, Henry RT. PGE-6201204: a novel orally active peripherally selective α_2-adrenergic agonist. FASEB J 1996; 10:A426.

29. Ruffolo RR Jr, Waddell JE. Receptor interactions of imidazolines. IX. Cirazoline is an alpha-1 adrenergic agonist and an alpha-2 adrenergic antagonist. J Pharmacol Exp Ther 1982; 222:29–36.

30. Potter DE. Adrenergic pharmacology of aqueous humor dynamics. Pharmacol Rev 1981; 33:133–153.

31. Barnebey H. Long-term efficacy of brimonidine on IOP lowering. Invest Ophthalmol Vis Sci 1995; 37:S1102.

32. Choplin NT and the Brimonidine Study Group. Visual field results from a one-year multi-center randomized double-masked study comparing brimonidine tartrate to timolol maleate in the treatment of ocular hypertension and glaucoma. Invest Ophthalmol Vis Sci 1996; 37(3):S510.

33. Dirks M. Long-term safety and IOP-lowering efficacy of brimonidine tartrate 0.2% in glaucoma and ocular hypertension. Invest Ophthalmol Vis Sci 1996; 37(3):A832.

34. Shelton KF, Ritch R, Robin A. Efficacy of brimonidine when added to maximally tolerated glaucoma medications. Invest Ophthalmol Vis Sci 1996; 37(3):S1101.

35. Pineyro A, Gross RL, Orengonania S. Long-term experience with apraclonidine 0.5% (Iopidine) in clinical practice. Invest Ophthalmol Vis Sci 1996; 37(3):S1100.

36. Jin Y, Verstappen A, Yorio T. Characterization of α_2-adrenoceptor binding sites in rabbit ciliary body membranes. Invest Ophthalmol Vis Sci 1994; 35:2500–2508.

37. Huang Yi, Gil DW, Vanscheeuwijck P, Stamer WD, Regan JW. Localization of α_2-adrenergic receptor subtypes in the anterior segment of the human eye with selective antibodies. Invest Ophthalmol Vis Sci 1995; 36:2729–2739.

38. Chiou GCY. Effects of α_1 and α_2 activation of adrenergic receptors on aqueous humor dynamics. Life Sci 1983; 32:1699–1704.

39. Kintz P, Himber J, deBurlet G, Andermann G. Characterization of α_2-adrenergic receptors, negatively coupled to adenylate cyclase, in rabbit ciliary processes. Curr Eye Res 1988; 7:287–292.

40. Jin Y, Elko EE, Tran T, Yorio T. Inhibition of adenylate cyclase in bovine ciliary process and rabbit iris ciliary body to α_2-adrenergic agonists. J Ocul Pharm 1989; 5:189–197.

41. Ogidigben M, Chu TC, Potter DE. Ocular actions of moxonidine: a possible role for imidazoline receptors. J Pharmacol Exp Ther 1994; 269:897–904.

42. Fuder H, Selbach M. Characterization of sensory neurotransmission and its inhibition via α_{2B}-adrenoceptors and via non- α_2-receptors in rabbit iris. Naunyn-Schmiedeberg's Arch Pharmacol 1993; 347:394–401.

43. Campbell WR, Potter DE. Centrally mediated ocular hypotension: potential role of imidazoline receptors. Ann NY Acad Sci 1995; 763:463–485.

44. Innemee HC, van Zwieten PA. The influence of clonidine on intraocular pressure. Doc Ophthalmol 1979; 46:309–315.

45. Innemee HC, deJonge A, van Meel JCA, Timmermans PBMW, van Zwieten PA. The effect of selective α_1- and α_2-adrenoceptor stimulation on the intraocular pressure in the conscious rabbit. Naunyn-Schmiedeberg's Arch Pharmacol 1981; 316:294–298.

46. Burke JA, Potter DE. Ocular effects of a relatively selective alpha 2 agonist (UK 14,304-18) in cats, rabbits and monkeys. Curr Eye Res 1986; 5:665–676.

47. Sulpizio AC, Deegan J, Horner E, Nelson A, Fowler P, Matthews W. The ocular hypotensive activity of SK&F 86466 in the rabbit following intravenous or oral administration. Pharmacologist 1983; 25:193.

48. Matthews WD, Sulpizio A, Fowler PJ, DeMarinis R, Hieble JP, Bergamini MV. The ocular hypotensive action of SK&F 86466 in the conscious rabbit. Curr Eye Res 1984; 3:737–742.

49. Mittag TW, Tormay A, Severin C, Podos SM. Alpha-adrenergic antagonists: correlation of the effect on intraocular pressure and on α_2-adrenergic receptor binding specificity in the rabbit eye. Exp Eye Res 1985; 40:591–599.

50. Chang FW, Burke JA, Potter DE. Mechanism of the ocular hypotensive action of ketanserin. J Ocul Pharmacol 1985; 1(2):137–147.

51. Crosson CE, Heath AR, DeVries GW, Potter DE. Pharmacological evidence for heterogeneity of ocular α_2-adrenoceptors. Curr Eye Res 1992; 11(10):963–970.

52. Caine M, Pfau A, Perlberg S. The use of alpha-adrenergic blockers in benign prostatic obstruction. Br J Urol 1976; 48:255–263.

53. Caine M, Perlberg S, Meretyk S. A placebo-controlled double blind study of the effect of phenoxybenzamine in benign prostatic obstruction. Br J Urol 1978; 50:551–554.

54. Lepor H. Prostate selectivity of alpha-blockers: from receptor biology to clinical medicine. Eur Urol 1996; 29(Suppl 1):12–16.

55. Caine M, Perlberg S. Dynamics of acute retention in prostatic patient and role of adrenergic receptors. Urology 1977; 9:399–403.

56. Hieble JP, Caine M, Zalaznik E. In vitro characterization of the alpha-adrenoceptors in human prostate. Eur J Pharmacol 1985; 107:111–117.

57. Yamada S, Suzuki M, Kato Y, Kimura R, Mori R, Matsumoto K, Maruyama M, Kawabe K. Binding characteristics of naftopidil and α_1-adrenoceptor antagonists to prostatic α-adrenoceptors in benign prostatic hypertrophy. Life Sci 1991; 50:127–135.

58. Muramatsu I, Yamanaka K, Kigoshi S. Pharmacological profile of the novel α-adrenoceptor antagonist KT-611 (Naftopidil). Jpn J Pharmacol 1991; 55:391–398.

59. Yamanaka N, Yamaguchi O, Kameoka H, Fukaya Y, Yokoto T, Shiraiwa Y. Effect of KT-611 (naftopidil) on the contraction of human prostatic tissue and its use in benign prostatic obstruction. Acta Urol Jpn 1991; 37:1759–1772.

60. Raz S, Ziegler M, Caine M. Pharmacological receptors in the prostate. Br J Urol 1973; 45:663–667.

61. Caine M, Raz S, Ziegler M. Adrenergic and cholinergic receptors in the human prostate, prostatic capsule and bladder neck. Br J Urol 1975; 47:193.

62. Lepor H, Auerbach S, Puras-Baez A, et al. A randomized placebo-controlled multicenter study of the efficacy and safety of terazosin in the treatment of benign prostatic hyperplasia. J Urol 1992; 148:1467–1474.

63. Lepor H, Meretyk S, Knapp-Maloney G. The safety, efficacy and compliance of terazosin therapy for benign prostatic hyperplasia. J Urol 1992; 147:1554–1557.

64. Kirby RS. Profile of doxazosin in the hypertensive man with benign prostatic hyperplasia. Br J Clin Pract 1994; 74(Suppl):23–28.

65. Price DT, Schwinn DA, Lomasney JW, Allen LF, Caron MG, Lefkowitz RJ. Identification, quantification and localization of mRNA for three distinct α_1-adrenergic receptor subtypes in human prostate. J Urol 1993; 150:546–551.

66. Faure C, Pimoule C, Vallancien G, Langer SZ, Graham D. Identification of α_1-adrenoceptor subtypes present in the human prostate. Life Sci 1994; 54:1595–1605.

67. Lepor H. Clinical evaluation of tamsulosin, a prostate selective alpha1C antagonist. J Urol 1995; 153 (Suppl):274A.

68. Forray C, Bard JA, Wetzel JM, Chiu G, Shapiro E, Tang R, Lepor H, Hartig PR, Weinshank RL, Branchek TA, Gluchowski C. The α_1-adrenergic receptor that mediates smooth muscle contraction in human prostate has the pharmacological properties of the cloned human α_{1C}- subtype. Mol Pharmacol 1994; 45:703–708.

69. Forray C, Chiu G, Wetzel JJ, Bard JA, Weinshank RL, Branchek TA, Hartig P, Gluchowski C. Effects of novel alpha-1C adrenergic receptor antagonists on the contraction of human prostate smooth muscle. J Urol 1994; 151 (5):267A.

70. Gluchowski C, Forray C, Chiu G, Branchek TA, Wetzel J, Hartig PR. The use of alpha-1C specific compounds to treat benign prostatic hyperplasia. International Patent Application No. WO94/10989, 1994.

71. Wetzel JM, Miao SW, Forray C, Borden LA, Branchek TA, Gluchowski C. Discovery of α_{1a}-adrenergic receptor antagonists based on the L-type Ca^{2+} channel antagonist niguldipine. J Med Chem 1995; 38(10):1579–1581.

72. Gluchowski C, Wetzel JM, Chiu G, Marzabadi MR, Wong WC, Nagarathnam D. Dihydropyridines and new uses thereof. International Patents Application No. WO 94/22829, October 13, 1994.

73. Gong G, Chiu G, Gluchowski C, Branchek TA, Hartig PR, Pettinger WA, Forray C. α_{1C}-Adrenergic antagonist and orthostatic hypotension in the rat. FASEB J 1994; 8:A353.

74. Leonardi A, Testa R, Motta G, DeBenedetti PG, Hieble P, Giardina D. α_1-Adrenoceptors: Subtype and organ-selectivity of different agents. In: Giardina D, Piergentili A, Pigini M, eds, Perspectives in Receptor Research. Vol. 24. Elsevier, Amsterdam, 1996; pp. 135–152.

75. Gluchowski C, Forray CC, Chiu G, Branchek TA, Wetzel JM, Hartig PA. Use of α_{1C} specific compounds to treat benign prostatic hyperplasia. US Patent No 5,403,847; April 4, 1995.

76. Ford APDW, Arredondo NF, Blue DR, Bonhaus DW, Jasper J, Kava MS, Lesnick J, Pfister JR, Shieh IA, Williams TJ, McNeal JE, Stamey TA, Clarke DE. RS-17053, a selective α_{1A}-adrenoceptor antagonist, displays low affinity for functional α_1-adrenoceptors in prostate of man: implications for adrenoceptor classification. Mol Pharmacol 1996; 49:209–215.

77. Testa R, Guarneri L, Taddei C, Poggesi E, Angelico P, Sartani A, Leonardi A, Gofrit ON, Meretyk S, Caine M. Functional antagonistic activity of REC 15/2739, a novel α_1-antagonist selective for the lower urinary tract, on noradrenaline-induced contraction of human prostate and mesenteric artery. J Pharmacol Exp Ther 1996; 277:1237–1246.

78. Chess-Williams R, Noble AJ, Couldwell C, Rosario DJ, Chapple CR. The α_1-adrenoceptor antagonist SB 216469 (Rec 15/2739) discriminates between α_{1A}-adrenoceptors and the α_1-adrenoceptors of the human prostate. Br J Pharmacol 1996; 117:109P.

79. Testa R, Poggesi E, Taddei C, Guarneri L, Ibba M, Leonardi A. Rec 15/2739, a new α_1-antagonist selective for the lower urinary tract: in vitro studies. Neurol Urodynam 1994; 13:473–474.

80. Testa R, Sironi G, Colombo D, Greto D, Leonardi A. Rec 15/2739, a new α_1-antagonist selective for the lower urinary tract. In vivo studies. Neurol Urodynam 1994; 13:471–473.

81. Testa R, Taddei C, Poggesi E, Destefani C, Cotecchia S, Hieble JP, Sulpizio AC, Naselsky D, Bergsma D, Swift A, Ganguly S, Leonard A. Rec 15/2739 (SB 216469) A novel prostate selective α_1-adrenoceptor antagonist. Pharmacol Commun 1995; 6:79–86.

82. Ruffolo RR Jr, Bondinell W, Ku T, Naselsky DP, Hieble JP. α_1-Adrenoceptors: pharmacological classification and newer therapeutic applications. Proc West Pharmacol Sci 1995; 38:121–126.

83. Auguet M, Delaflotte S, Chabrier PE. Different α_1-adrenoceptor subtypes mediate contraction in rabbit aorta and urethra. Eur J Pharmacol 1995; 287:153–161.

84. Blue D, Zhu OM, Isom P, Young S, Larson M, Clarke D. Effect of α_1-adrenoceptor antagonists in dog prostate/blood pressure models. FASEB J 1996; 10:A425.

85. Katwala SP, Milicic IP, Hancock AA, Kerwin JF, Brune ME. Differential effects of various alpha-1 adrenoceptor antagonists on urogenital and vascular smooth muscle in vivo. FASEB J 1996; 10:A425.

86. Marshall I, Green M, Hussain MB, Burt RP. Differences in affinity for the antagonist RS-17053. Br J Pharmacol 1996; 117:P110.

87. Ford APDW, Arredondo NF, Blue DR, Bonhaus DW, Kawa MS, Williams TJ, Vimont RL, Zhu QM, Pfister JR, Clarke DE. Do $\alpha_{1A}(\alpha_{1C})$ adrenoceptors mediate prostatic smooth muscle contraction in man? Studies with a novel, selective α_{1A}-adrenoceptor antagonist, RS 17053. Br J Pharmacol 1995; 114:24P.

88. Muramatsu I, Oshita M, Ohmura T, Kigoshi S, Akino H, Gobara M, Okada K. Pharmacological characterization of α_1-adrenoceptor subtypes in the human prostate: functional and binding studies. Br J Urol 1994; 74:572–578.

89. Flavahan NA, Vanhoutte PM. α_1-Adrenoceptor subclassification in vascular smooth muscle. Trends Pharmacol Sci 1986; 7:347–349.

90. Muramatsu I, Ohmura T, Kigoshi S, Hashimoto S, Oshita M. Pharmacological subclassification of α_1-adrenoceptors in vascular smooth muscle. Br J Pharmacol 1990; 99:197–201.

91. Oshita M, Takita M, Kigoshi S, Muramatsu L. α_1-Adrenoceptor subtypes in rabbit thoracic aorta. Jpn J Pharmacol 1992; 58(Suppl II):276.

92. Hieble JP, Ruffolo RR Jr. The use of α-adrenoceptor antagonists in the pharmacological management of benign prostatic hypertrophy: an overview. Pharmacol Res 1996; 33:145–160.

93. Tseng-Crank J, Kost T, Goetz A, Hazum S, Roberson KM, Haizlip J, Godinot N, Robertson CN, Saussy D. The α_{1C}-adrenoceptor in human prostate: cloning, functional expression, and localization to specific prostatic cell types. Br J Pharmacol 1995; 115:1475–1485.

94. Shibata K, Foglar R, Horie K, Obika K, Sakamoto A, Ogawa S, Tsujimoto G. KMD-3213, novel, potent, α_{1a}-adrenoceptor-selective antagonist: characterization using recombinant human α_1-adrenoceptors and native tissue. Mol Pharmacol 1995; 48:250–258.

95. Yamada S, Suzuki M, Tanaka C, Mori R, Kimura R, Inagaki O, Honda K, Asano M, Takenaka T, Kawabe K. Comparative study on α_1-adrenoceptor antagonist binding in human prostate and aorta. Clin Exp Pharmacol Physiol 1994; 21:405–411.

96. Kenny BA, Naylor AM, Carter AJ, Read AM, Greengrass PM, Wyllie MG. Effect of α_1-adrenoceptor antagonists on prostatic pressure and blood pressure in the anesthetised dog. Urology 1994; 44:52–57.

97. Abrams PH, Schulman CC, Vaage S. The efficacy and safety of 0.4 mg tamsulosin once daily in symptomatic BPH. J Urol 1995; 153(Suppl):274A.

98. Hieble JP, Ruffolo RR Jr. Effects of α- and β-adrenoceptors on lipids and lipoproteins. In: Witiak DT, Newman HAI, Feller DR, eds. Antilipidemic Drugs: Medicinal, Chemical and Biochemical Aspects. Amsterdam: Elsevier, 1991:301–344.

99. Laz TM, Forray C, Smith KE, Bard JA, Vaysse PJJ, Branchek TA, Weinshank RL. The rat homologue of the bovine α_{1C} adrenergic receptor shows

the pharmacological properties of the classical α_{1A} subtype. Mol Pharmacol 1994; 46:414–422.

100. Piascik MT, Soltis EE, Barron KW. Alpha1-adrenoceptor subypes and the regulation of peripheral hemodynamics in spontaneously hypertensive rats. Pharmacol Commun 1992; 1:131–138.

101. Elhawary AM, Pettinger WA, Wolff, DW. Subtype-selective alpha-1 adrenoceptor alkylation in the rat kidney and its effect on the vascular pressor response. J Pharmacol Exp Ther 1992; 260:709–713.

102. Blue DR, Bonhaus DW, Ford APDW, Pfister JR, Sharif NA, Shieh IA, Vimont RL, Williams TJ, Clarke DE. Functional evidence equating the pharmacologically defined α_{1A} and α_{1C} adrenoceptors: studies in the isolated perfused kidney of the rat. Br J Pharmacol 1995; 115:283–294.

103. Villalobos-Molina R, Ibarra M. α_1-Adrenoceptors mediating contraction in arteries of normotensive and spontaneously hypertensive rats are of the α_{1D} or α_{1A} subtypes. Eur J Pharmacol 1996; 298:257–263.

104. Sattar MA, Johns EJ. The α_1-adrenoceptor subtypes mediating adrenergically induced antinatriuresis and antidiuresis in Wistar and stroke-prone spontaneously hypertensive rats. Eur J Pharmacol 1995; 294:727–736.

105. Sattar MA, Johns EJ. Alpha-1 adrenoceptor subtypes involved in mediating adrenergically induced antinatriuresis and antidiuresis in two kidneys, one-clip Goldblatt and deoxycorticosterone acetate-salt hypertensive rats. J Pharmacol Exp Ther 1996; 277:245–252.

106. Price DT, Lefkowitz RJ, Caron MG, Berkowitz D, Schwinn DA. Localization of mRNA for three distinct α_1-adrenergic receptor subtypes in human tissues: implications for human α_1-adrenergic physiology. Mol Pharmacol 1994; 45:171–175.

107. Hirasawa A, Horie K, Tanaka T, Takagaki K, Murai M, Yano J, Tsujimoto G. Cloning, functional expression and tissue distribution of human cDNA for the α_{1c}-adrenergic receptor. Biochem Biophys Res Commun 1993; 195:902–909.

108. Buckner SA, Oheim KW, Morse PA, Knepper SM, Hancock AA. α_1-Adrenoceptor induced contractility in rat aorta is mediated by the α_{1D} subtype. Eur J Pharmacol 1996; 297:241–248.

109. Kenny BA, Chalmers DH, Philpott PC, Naylor AM. Characterization of an α_{1D} adrenoceptor mediating the contractile response of rat aorta to noradrenaline. Br J Pharmacol 1995; 115:981.

110. Goetz AS, King HK, Ward SD, True TA, Rimele TJ, Saussy DL, BMY 7378 is a selective antagonist of the D subtype of α_1-adrenoceptors. Eur J Pharmacol 1995; 272:R5–R6.

111. Leech CJ, Faber JE. Different α_1-adrenoceptor subtypes mediate constriction of arterioles and venules. Am J Physiol 1996; 270:H710–H722.

112. Sayet I, Neuilly G, Rakotoarisoa L, Mironneau J, Mironneau C. Rat vena cava α_{1B}-adrenoceptors: characterization by [^{3}H]prazosin binding and contraction experiments. Eur J Pharmacol 1993; 264:275–281.

113. Muramatsu I, Ohmura T, Hashimoto S, Oshita M. Functional subclassification of vascular α_1-adrenoceptors. Pharmacol Commun 1995; 6:23–28.

114. Giardina D, Crucianelli M, Melchiorre C, Taddei C, Testa R. Receptor binding profile of cyclazosin, a new α_{1B}-adrenoceptor antagonist. Eur J Pharmacol 287: 13–16, 1995.
115. Muramatsu I, Takita M, Suzuki F, Miyamoto S, Sakamoto S, Ohmura T. Subtype selectivity of a new α_1-adrenoceptor antagonist, JTH-601: comparison with prazosin. Eur J Pharmacol 1996; 300:155–157.
116. Hieble JP, Ruffolo RR Jr. Recent advances in the identification of α_1 and α_2-adrenoceptor subtypes; therapeutic implications. Exp Opinion Invest Drugs 1997; 6:367–387.
117. de Boer T. The effects of mirtazapine on central noradrenergic and serotonergic neurotransmission. Int Clin Psychopharmacol 1995; 10(4):19–23.
118. Link RE, Stevens MS, Kulatunga M, Scheinin M, Barsh GS, Kobilka BK. Targeted inactivation of the gene encoding the mouse α_{2C} adrenoceptor homolog. Mol Pharmacol 1995; 48:48–55.
119. Link RE, Desai K, Hein L, Stevens ME, Chrascinski A, Bernstein D, Barsh GS, Kobilka BK. Cardiovascular regulation in mice lacking alpha 2 adrenergic receptor subtypes b and c. Science 1996; 273:803–805.

P$_{2U}$-Purinoceptor Agonists
Emerging Therapy for Pulmonary Diseases

Eduardo R. Lazarowski, T. Kendall Harden, and Richard C. Boucher
University of North Carolina at Chapel Hill, Chapel Hill, North Carolina

William C. Watt
University of Washington, Seattle, Washington

I. INTRODUCTION

A. Historical Considerations

The term *purinergic* was coined 25 years ago by Burnstock (1) to explain the potential signaling role of ATP in noncholinergic, nonadrenergic transmission. Although the action of ATP on discrete cell surface receptors was recognized during subsequent years, it was not until 1978 (2) that receptors activated by ATP or ADP (P$_2$-purinoceptors, or P2 receptors) were differentiated from those activated by the ATP metabolite adenosine (P$_1$-purinoceptors). P2 purinoceptors were further subdivided into two major groups: P$_{2X}$-purinoceptors, which mediate vasoconstriction and smooth muscle contraction and were most potently activated by α,β-methylene ATP (α,β-meATP), and P$_{2Y}$-purinoceptors, which promoted vasodilatation and smooth muscle relaxation and were potently activated by 2-methyl-thio-ATP (2MeSATP) (3). Additional P$_2$-purinoceptor subtypes have been subsequently proposed. For example, the stimulatory action of ADP on platelets was ascribed to a so-called P$_{2T}$-purinoceptor at which ATP acts as an antagonist (4,5). A pore-forming ATP^{4-} receptor was described on macrophages and mast cells and was termed the P$_{2Z}$-purinoceptor (6) (for a review, see Ref. 7). Intriguingly, the actions of ATP observed in many tissues were mimicked by UTP, and pharmacological selectivity of the involved receptor did not fit any of the receptor categories defined above, i.e., P_{2X}, P_{2Y}, P_{2T}, or P_{2Z}. A receptor equipotently stimulated by ATP and UTP (but not by 2MeSATP or α,β-meATP) was defined as the P_{2U}-purinoceptor, or P2U receptor (1).* The P_{2U}-purinoceptor was subsequently shown to mediate actions of nucleotides in a broad number of tissues and cell types, from astrocytes to blood neutrophils and to airway epithelial cells (reviewed in Ref. 8). Although the P_{2U}-purinoceptor was activated by the pyrimidine UTP, it nevertheless remained under the "purinergic" umbrella, which highlights the responsiveness of this receptor to submicromolar concentrations of ATP. The existence of selective *pyrimidinergic receptors* was brought into consideration after the description of a phospholipase C–linked receptor in C6-2B rat glioma cells, which was potently activated by UDP and to a lesser extent by UTP, but was not activated by ATP or ADP (9).

B.   Biochemical and Molecular Characterization of P2 Receptors

The lack of highly selective and potent agonists and antagonists of P2 receptors precluded a detailed pharmacological classification of these proteins using classic binding assays. Measurements of physiological or biochemical responses to nucleotides were the only approaches available to functionally characterize a putative receptor in a given tissue. Activation of P2 receptors in general is associated with elevation of intracellular calcium by mechanisms that involve either a nucleotide-gated ion channel (P_{2X}- and P_{2Z}-purinoceptors) or a nucleotide-promoted activation of phospholipase C (P_{2Y}- and P_{2U}-purinoceptors). In the last 4 years, several receptors for nucleotides have been cloned and functionally expressed in heterologous systems. Thus, proteins of known amino acid sequence could be assigned to defined functions. According to their signaling mechanisms and deduced structures, P2 receptors have been divided into two main families. The ionotropic P_{2X}- and P_{2Z}-purinoceptors are nonselective cation-gating proteins and are now grouped under the generic name of P2X receptors. At least seven P2X receptors have been cloned, $P2X_1$–$P2X_7$, which are numbered according to their chronological order of publication. The phospholipase C-linked receptors, P_{2Y}- and P_{2U}-purinoceptors, belong to the

---

* The receptor referred to here as the P_{2U}-purinoceptor or P2U receptor is alternatively designated as $P2Y_2$ receptor according to recent recommendations by the International Union of Pharmacology Committee on Drug Classification and Receptor Nomenclature (IUPHAR) (66).

superfamily of G-protein-linked receptors and are now designated under the generic name of P2Y receptors. A receptor (the P2Y$_1$ receptor) has been cloned that exhibits a pharmacological profile like that originaly attributed to the classically described P$_{2Y}$ purinoceptor. The P$_{2U}$ purinoceptor has been cloned and under this scheme is referred to as the P2Y$_2$ receptor. Two novel proteins have been added to this group, P2Y$_4$ and P2Y$_6$ receptors, which are selectively activated by UTP and UDP, respectively (10–12). The P2Y$_6$ receptor was proven recently to represent the uridine nucleotide selective receptor (UDP receptor) previously described in CD6-2B cells (Table 1) (13).

The classification of P2 receptors based on cloned proteins is not complete. For example, the ADP receptor described in platelets (P$_{2T}$-purinergic), has not been cloned. This "receptor" likely represents more than one protein since both Ca^{2+}-channel and G-protein-coupled phospholipase C- and adenylyl cyclase–linked activities have been associated with receptor activation (14). A receptor has been described on C6 rat glioma cells that pharmacologically resembles a classical P$_{2Y}$-purinoceptor (2MeSATP > ATP > α,β meATP). However, this receptor inhibits adenylyl cyclase activity and, unlike the P2Y$_1$ receptor, does not promote activation of phospholipase C or mobilization of intracellular calcium (15). This putative P2Y-like receptor also awaits cloning.

Assay of agonist-promoted responses with heterologously expressed cloned P2 receptors under conditions in which nucleotide metabolism could be controlled revealed that the classically described order of agonist poten-

Table 1 Major Features of Four Cloned G-Protein-Coupled P2 Receptors

Current name (IUPHAR)	Former nomenclature	Agonist potencies	No effect	Effector
P2Y$_1$	P$_{2Y}$-purinoceptor	2MeSADP > ADP > 2MeSATP > ATP $\gg \alpha,\beta$ meATP	UTP	PLC
P2Y$_2$	P$_{2U}$-purinoceptor P2U receptor Nucleotide receptor	UTP $\geq$ ATP = UTPγS > ATPγS $\gg$ 2MeSATP > α,β/meATP	ADP, UDP	PLC
P2Y$_4$	—	UTP $\gg$ ATP	ADP, UDP	PLC
P2Y$_6$	—	UDP $\gg$ ADP	UTP, ATP	PLC

cies for P2X and P2Y receptors may not accurately reflect the true agonist affinities of these receptors. For example, α,β-meATP is a weaker agonist than 2MeSATP at some P2X receptors ($P2X_1$–$P2X_3$) and is inactive at other P2X receptors ($P2X_4$–$P2X_7$). Observation of potent effects of UTP and ATP, which until recently were considered the hallmark for P_{2U}-purinoceptors, could be indicative of $P2Y_1$ and $P2Y_4$ receptors coexpressed in the same tissue. UDP, which appeared to be a low-potency full agonist at the P_{2U}-purinoceptor (16), was recently shown to have no effect on that receptor after enzymatic removal of contaminating UTP from UDP solutions. Conversely, the full-agonist effect of UTP, originally described in native C6-2B rat glioma cells (9) and in C6 glioma cells heterologously expressing $P2Y_6$ receptors (12), could be explained by its conversion to UDP by extracellular nucleotidases.

II. P2U RECEPTOR

The P2U receptor was first described as a receptor that promotes phosphoinositide breakdown and intracellular calcium mobilization in human neutrophils and that was potently activated by ATP, ATPγS, and UTP (17,18; reviewed in Ref. 19). P2U receptors have been associated on the basis of pharmacological data with a variety of cellular responses including superoxide formation in human neutrophils (20–23), contraction of guinea pig trachea (24), changes in membrane potential in myotubes (25), and regulation of ion fluxes in airway epithelial cells (26). However, the identity of this receptor as the $P2Y_2$ receptor has been established unambiguously in only a few cases. For example, physiological, biochemical, and molecular evidence exists for a central role of the $P2Y_2$ receptor in human airway epithelium (26–28).

Rationale for P2Y₂ as Target for Therapy of Lung Diseases

The initial observations that suggested a role for P2 purinoceptors in the regulation of pulmonary defense mechanisms were reported in 1991 (26). These studies focused on identification of agents that activate Cl^- and water secretory pathways that might replace those that are defective in airway epithelium of cystic fibrosis (CF) patients. It was reported that ATP and UTP were equipotent in inducing active Cl^- secretion when applied to the luminal surface of either normal or CF nasal epithelial monolayers. These studies further demonstrated that UTP was a slightly more effective Cl^- secretagog than β-adrenergic receptor agonists in normal subjects. In contrast, UTP was the only effective Cl^- secretagog identified in CF epithelia, and the effects of UTP were approximately twofold greater in CF than in normal preparations. Comparison of the effects of a variety of P2Y and P2X receptor agonists with those of ATP and UTP suggested that the luminal receptor was of the P2U class. A series of experiments also were performed to characterize the action of nucleotides on the basolateral surface of airway epithelial cells. Responses to ATP, UTP, and 2MeSATP were observed that led to the speculation that both a P2U and P2Y-like receptor are expressed on the basolateral surface of human airway epithelial cells.

A series of studies were carried out to define in greater detail the pharmacological specificity, the signal transduction mechanisms involved, and the spectrum of ion transport processes regulated by these receptors. Studies with both primary cultures of human airway epithelial cells and immortalized cell lines revealed that both ATP and UTP promote the formation of inositol phosphates (27). ATP and UTP promote inositol phosphate formation through a common receptor, based on the cross-desensitization observed between these agonists, the nonadditivity of their effects, and their relative potencies. Parallel studies in cultured human airway epithelia and immortalized cell lines revealed that the actions of ATP and UTP on regulation of intracellular Ca^{2+} levels paralleled their effects on inositol phosphate formation.

Double-barreled, ion-selective microelectrodes were used to characterize the physiological actions of ATP and UTP administered to human airway epithelial ion channels involved in regulation of Cl$^-$ secretion. Intriguingly, the actions ATP or UTP on the apical membrane reflected activation of the apical membrane Ca^{2+}-regulated Cl$^-$ conductance, whereas the actions of ATP and UTP on the basolateral membrane were mediated by activation of basolateral K$^+$ channels and, by a hyperpolarizing cellular mechanism, "secondary" activation of Cl$^-$ secretion (29). Further studies also revealed that ATP/UTP applied to the lumenal membrane initiated a small K$^+$ secretory response and appeared to slow the rate of transepithelial Na$^+$ absorption (30). Finally, the apparent differences in the ion transport effects of ATP/UTP on the apical versus basolateral membranes, i.e., sidedness of response, led to a series of studies to determine whether activation of P2 receptors on apical or basolateral membranes was confined to the surface ipsilateral to receptor activation. Using polarized epithelial monolayers in a Ca^{2+} imaging system, Paradiso and colleagues illustrated that Ca^{2+} mobilization and Ca^{2+} entry triggered by P2Y$_2$ receptor activation were confined to the membrane ipsilateral to receptor activation (31). These results led to the concept that human airway epithelial cells could respond selectively to nucleotides presented at the apical versus basolateral membranes, a feature consistent with the need for these cells to confine cellular responses to stimuli presented from the "outside" or "inside" world.

The observation that ATP/UTP regulated the ion transport processes responsible for controlling the volume and composition of the liquid that lines airway surfaces led to a search for possible regulation by these agonists of other processes on airway epithelial surfaces that are important for the mucociliary component of airways defense, i.e., ciliary beat frequency and mucus secretion. In a series of studies in dog and human airway epithelial explants and primary cultures, Davis and colleagues demonstrated that ATP and UTP were effective in stimulating mucin granule discharge from the mucin-containing goblet cell of the superficial airway epithelium (32,33). The equipotent effects of ATP and UTP and the relative ineffectiveness of other nucleotide analogs suggested that this response was mediated by a $P2Y_2$ receptor. These nucleotides were much more effective when applied to the luminal surface as compared to the basolateral surface. Subsequent studies have implicated roles for both Ca^{2+} mobilization and protein kinase C activation in this mucin secretory response. Utilizing explants freshly derived from human airway epithelium, it was demonstrated that ATP and UTP are equipotent also in promoting an increase in ciliary beat frequency. As in studies in normal tissues with respect to Cl^- secretion, the effects of ATP and UTP were greater than those of other physiological agonists tested, e.g., agents that raise cellular cAMP or cellular cGMP.

Observations that ATP/UTP controlled all the elements involved in regulating mucociliary transport rates, i.e., ion/water transport, ciliary beat frequency, and mucin secretion, led to studies to test whether inhaled triphosphate nucleotides regulated the rate of mucociliary transport in humans. "Safety" studies revealed no adverse effects of inhaled ATP/UTP on pulmonary function or systemic cardiovascular parameters in animals or humans, and rate of clearance of inhaled radiotracers in humans was increased approximately threefold by inhaled UTP (34). These results led to studies to determine whether UTP increased mucociliary clearance in cystic fibrosis patients (35). UTP, in combination with the Na^+ channel blocker amiloride, increased the clearance of secretions from the peripheral lung of CF patients to rates that were near-normal. The therapeutic effects of UTP also have been examined in patients with another genetic defect that slows clearance from the lungs, primary ciliary dyskinesia (PCD). These studies provided a clinical context in which the relative contribution of UTP-promoted increases in beat frequency versus UTP-promoted modulation of the composition/biorheological properties of airway secretions could be assessed. Cough clearance of radiotracers in PCD patients was measured and was increased approximately 2.5- to 3-fold after the inhalation of UTP as compared to vehicle (36). Thus, it appears that the therapeutic actions of UTP likely will not only involve regulation of ciliary beat frequency but also involve "improvement" of hydration/biorheological properties of mucus. In summary, acute studies in both healthy subjects and patients with at least two genetic diseases that reduce mucociliary clearance rates predict that UTP-based agonists may be potentially important therapeutic modalities for infectious lung diseases.

III. CLONING AND PHARMACOLOGICAL SELECTIVITY OF THE HUMAN P2Y₂ RECEPTOR

A. Structure

Although knowledge of P2 receptors has advanced significantly in recent years, our understanding of the physiological and pathophysiological role of extracellular nucleotides acting on the human airway epithelial cell and other P2Y$_2$ receptors has been hampered by lack of stable and selective high-affinity agonists and antagonists. The cloning of the murine P2Y$_2$ receptor by Lustig and colleagues (37) offered the possibility of developing a simple screening system based on a homogeneous receptor population, to identify analogs of ATP that could serve as a base for developing new, highly selective, P2Y$_2$ receptor agonists and antagonists. Availability of the deduced amino acid sequence also has led to beginning insights into identification of key residues potentially important for receptor-ligand interaction. Complementary DNA encoding the human P2Y$_2$ receptor was isolated from a human airway epithelial cell cDNA library using a probe based on the murine P2Y$_2$ receptor sequence (28). The human P2Y$_2$ receptor has a predicted structure of seven transmembrane domains with two putative extracellular glycosylation sites close to the amino-terminus and a consensus PKA phosphorylation site in the third cytosolic loop (Fig. 1). The predicted structure of the P2Y$_2$ receptor is analogous to that of other members of the superfamily of G-protein-coupled receptors. Mutagenesis and modeling studies with G-protein-coupled receptors for small molecules such as β-adrenergic, muscarinic cholinergic, and adenosine receptors have implicated regions within the third and seventh transmembrane domains to be important in formation of the ligand-binding pocket (38–43). The most conserved amino acid sequence in P2Y receptors is the sequence **LFLTCIS** in the third transmembrane domain (TM3) (see Fig. 1). The role of this and other motifs is unknown, and only site-directed mutagenesis in combination with detailed pharmacological studies will establish the importance of these residues in ligand interaction. The overall homology of the P2Y$_2$ receptor to other cloned P2Y receptors, the P2Y$_1$ receptor, the P2Y$_4$ receptor, and the P2Y$_6$ receptor, is only 37%, 40%, and 51%, respectively. Surprisingly, the P2Y$_2$ receptor has higher homology (20–25%) to the thrombin, angiotensin II, neuropeptide Y, and somatostatin receptors

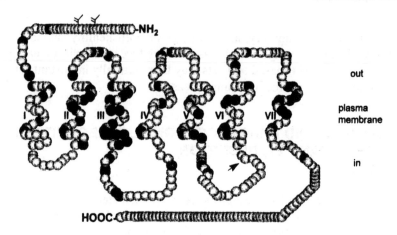

Figure 1 Schematic representation of the human $P2Y_2$ receptor showing the residues conserved among all P2Y receptors (dark circles). Extracellular N-linked glycosylation sites are indicated with branches, and a consensus protein kinase A phosphorylation site is indicated with an arrow. The roman numbers denote the seven transmembrane domains.

than to adenosine receptors (12%). Thus, outside the P2Y receptor family, the $P2Y_2$ receptor has a primary amino acid sequence that is closer to receptors recognizing peptides than, for example, to receptors where adenosine is the natural ligand. Although regions involved in ligand-binding in both P_1- and P_2-purinoceptors may be primarily located along the transmembrane domains, recent studies with A_2-adenosine receptors as well as with peptide-hormone receptors have identified contributions of extracellular receptor domains to ligand-binding (44). The low homology between receptors recognizing ATP ($P2Y_1$ and $P2Y_2$ receptors) and adenosine receptors discourages speculations for drug design based on a putative common nucleoside binding site.

Using a site-directed mutagenesis approach, Erb and colleagues (16) investigated the role of positively charged amino acids within the membrane bilayer on potential interaction with the negatively charged phosphate moieties of nucleotides. Substitution of Lys-289 to Arg resulted in a change of agonist potencies, with ADP and UDP becoming more potent than ATP and UTP for stimulation of calcium mobilization in 1321N1 cells expressing the mutant $P2Y_2$ receptor. The putative role of Lys-289 in establishing receptor selectivity for triphosphates over diphosphates is brought into question by the fact that Lys-289 is also present in the recently cloned $P2Y_6$ receptor, which exhibits high selectivity for diphosphate nucleotides over triphosphate nucleotides. Mutations to Leu of His-262 and Arg-265 in TM6 or Arg-292 in TM7 resulted in a decrease in the potencies of ATP and UTP. These results suggest that these residues may be involved in stabilization of agonist binding and/or receptor activation. However, the validity of this conclusion is uncertain since the level of mutant receptor expression was not quantified in any of these studies. Interestingly, these residues are also conserved in the $P2Y_1$ receptor and have been predicted on the basis of computer modeling studies to be involved in docking of the phosphate chain of ATP (45).

B. Expression and Second-Messenger Signaling

Important insight into agonist-$P2Y_2$ receptor interactions has followed stable expression of the cloned $P2Y_2$ receptor in a cell line, the 1321N1 human astrocytoma cells, that does not natively express a receptor for extracellular nucleotides. 1321N1 cells were infected with retrovirus vector harboring $P2Y_2$ receptor cDNA, selected, and subcloned as a homogenous cell population, HP2U-1321N1 cells (28). The lack of a reliable radioligand-binding assay for $P2Y_2$ receptors, as well as the poor antigenicity of various peptides designed to generate antibodies against the $P2Y_2$ receptor (46), precluded the direct quantification of the receptor after expression in the 1321N1 cells. Therefore, measurement of functional responses was used to follow receptor protein expression. ATP and UTP increased inositol phosphate formation and mobilized calcium from intracellular stores, but no effect on cellular cyclic AMP levels was observed. Nucleotide-promoted responses in HP2U-1321N1 cells were not affected by pertussis toxin, suggesting that the $P2Y_2$ receptor couples to a member of the Gq family of G-proteins. By measuring inositol phosphate accumulation rather than changes in intracellular calcium, responses could be quantified at a step more proximal to the initial agonist-receptor interaction, and therefore, the relative potencies of agonists for promotion of inositol phosphate formation more accurately reflect their relative affinities for the receptor.

C. Pharmacological Properties

Incubation of P2U-1321N1 cells with UTP resulted in a linear formation of inositol phosphates for up to 1 hr (47). This observation contrasted with previous results with $P2Y_2$ receptors natively expressed in airway epithelial cells where desensitization, i.e., cessation of agonist effect, occurred within 5 min of drug challenge (27). The apparent lack of agonist-induced desensitization in HP2U-1321N1 was associated with the presence of a large receptor reserve. The net increase in agonist-stimulated inositol phosphate formation with HP2U-1321N1 cells was 3–5 times larger than stimulation observed with natively expressed $P2Y_2$ receptors. Moreover, consistent with a high level of receptor in 1321N1 cells, the recombinant human P2Y$_2$ receptor was activated by UTP, ATP, and ATPγS with potencies that were 30-fold higher than those observed with natively expressed P2Y$_2$ receptors (Table 2). This feature of the expressed receptor allowed a more detailed examination of the action of nucleotides that were thought to be weak agonists or inactive at natively expressed P2Y$_2$ receptors. For example, 8BrATP, 5BrUTP, 2ClATP, AppNHp, and GTP displayed more potent effects on P2U-1321N1 cells relative to their effects previously observed with P2Y$_2$ receptors endogenously present in CFT43 cells (27,47). ADP and UDP initially were shown to be full agonists in H $\rightarrow$ HP2U-1321N1 cells (16,27). However, we have recently shown that pretreatment of ADP or UDP with hexokinase and glucose to remove contaminating triphosphonucleotides results in the complete loss of effect of nucleoside diphosphates on H $\rightarrow$ HP2U-1321N1 cells (13). Thus, P2Y$_2$ receptors equally recognize adenine and uridine nucleotides but exhibit a high selectivity for nucleoside triphosphates over nucleoside diphosphates. The high sensitivity for nucleotides, as observed with the inositol phosphate assay with HP2U-1321N1 cells, will allow screening of a variety of compounds as candidates not only for potent agonists, but also for partial agonists that may serve as a basis for designing receptor antagonists.

Table 2   Half-Maximal Concentration Values for Nucleotide-Promoted Inositol Phosphate Formation in CF/T43 Human Airway Epithelial Cells Endogenously Expressing the P2U Receptor, and in 1321N1 Human Astrocytoma Cells Heterologously Expressing the Human P2Y$_2$ (HP2U) Receptor

	EC$_{50}$ (nM)	
Nucleotide	CF/T43 cells	1321N1 cells
UTP	3,000	120
ATP	10,000	230
UTPγS	—	240
AppppA	—	720
ATPγS	30,000	1,700
5BrUTP	>100,000	2,000
2ClATP	>100,000	2,300
AppNHp	>100,000	5,600
GTP	>100,000	12,300
GTPγS	>100,000	26,500

Diadenosine polyphosphates (Ap$_n$A) are putative extracellular signaling molecules stored and released from neurosecretory vesicles of neural cells and synaptic terminals and from secretory granules of chromaffin cells and platelets. Extensive studies with both chromaffin and endothelial cells by Miras-Portugal and co-workers (48,49) have indicated that ApppppA and other Ap$_n$A compounds promote responses that are associated with intracellular calcium mobilization. This finding has led to the proposal that diadenosine polyphosphate-selective receptors may exist, i.e., a P$_{2D}$-purinoceptor(s). However, in human airways (50) as well as in other tissues in which actions of Ap$_n$A have been described (51), a similar effect of ATP was also observed. After treatment of ApppppA with apyrase to remove any trace of ATP from ApppppA solutions, ApppppA was shown to display full and potent agonist effects at the human P2Y$_2$ receptor expressed in 1321N1 cells (Table 2). This observation suggests that ApppppA may serve as an endogenous ligand for P2Y$_2$ receptors and establishes ApppppA as an interesting molecule from which novel drugs could be developed in search for stable and potent P2Y$_2$ receptor agonists.

IV. DRUG DESIGN FOR AIRWAY EPITHELIAL P2Y$_2$ RECEPTORS

A. Specificity and Safety

A major consideration in determination of optimal P2Y$_2$ receptor–directed molecules derived from adenine and uridine is selectivity. As described earlier, the P2 receptor family is expanding with at least seven P2X and two P2Y receptors at which ATP is a potent agonist (Table 1). The presence and role in the lung of each receptor are not known. Thus, the administration of ATP to diseased airways potentially could result in unpredictable events via P2X and/or P2Y$_1$ receptors. Rapid conversion of ATP into adenosine on the airway surfaces also should be taken into consideration. Adenosine, via A$_2$-adenosine receptors, may be a physiological regulator of ion transport in normal individuals but is ineffective in CF (52). Moreover, adenosine is known to induce bronchoconstriction in individuals with asthma and, therefore, any protocol designed to treat diseased airways with nucleotides should avoid the potential formation or accumulation of adenosine. Acute adenosine formation should be reduced utilizing either ATPγS or ApppppA as P2Y$_2$ receptor agonists (53). However, the potential action of acid phosphatases and diadenosine polyphosphate hydrolases acting on ATPγS (54) or ApppppA, respectively (55), also could result in accumulation of adenosine. In contrast to the lack of specificity predicted for ATP, only one receptor recognizing UTP has been described in addition to the

P2Y$_2$ receptor, the P2Y$_4$ receptor, and the expression of this receptor appears to be limited, with the largest levels present in placenta (10). Unlike adenosine, uridine, the breakdown product of UTP, is not an agonist on the airway A$_2$-adenosine receptor (52), and no effect of uridine has been described in the lung or other airway tissue. Therefore, UTP likely will be safer than ATP as the base drug from which new analogs are developed for therapeutic use.

B. Stability

Although UTP is the most potent and selective P2U receptor agonist, its therapeutic effects may be compromised by its potentially short duration in vivo. Two factors may contribute to the metabolism of UTP on the airways. First, ectonucleotidases on the mucosal surface of the airway epithelial cells degrade ATP as well as UTP. In vitro studies with polarized human nasal epithelial cells indicate that hydrolysis of UTP (30 μM) occurs at a rate of 4.5 nmol/min/million cells (56). Extrapolating these numbers to a tissue in vivo with an estimated cell density $\sim$2 $\times$ 10^6 cells/cm^2 and assuming that UTP will be evenly distributed along the airways at a concentration of 10–100 μM following a single application of a 10 ml aerosolized UTP (10 mM), the half-life of the drug should be $\sim$1 min in the conducting airways, and only a few seconds in the alveolar surface. This calculation potentially overestimates the lifetime of UTP as it does not take into account the potent action of phosphatases and nucleotidases released into the airway secretions from inflammatory cells. Thus, the benefits of nucleotide-based therapy may be uncertain for patients with infectious lung diseases unless hydrolysis-resistant analogs of UTP are used.

C. Potency

Unlike ATP, hydrolysis-resistant analogs of UTP are not available commercially. Since substitutions in the phosphate chain of adenine nucleotides confer marked stability to the ATP analogs, it is reasonable to speculate that similar substitutions on UTP-based structures will also result in enhanced stability. Although several analogs of ATP have been synthesized as inhibitors of phosphatases and nucleotidases, only a few hydrolysis-resistant ATP analogs have been shown to mimic actions of ATP on P2 receptors. For example, substitution with methylene groups on the pyrophosphate bridges rendered phosphatase-resistant analogs such as α,β-meATP and β,γ-meATP, which are potent agonists at P2X-purinergic receptors but have no effect on either the P2Y$_1$ or the P2Y$_2$ receptor. Although AppNHp, another hydrolysis-resistant ATP analog, is a full agonist at both P2Y$_1$ and P2Y$_2$ receptors, its potency on the latter is almost two orders of magnitude lower than that of ATP. Thus, it is unlikely that α,β-meUTP, β,γ-meUTP, or UppNHp would be potent P2U-receptor agonists. This hypothesis was recently experimentally confirmed by Pendergast and colleagues (57), who reported that chemically synthesized α,β-MeUTP, β,γ-MeUTP, and UppNHp were at least 100 times less potent than UTP at P2U receptors (57).

D. UTPγS Is a Stable and Potent P2Y$_2$ Receptor Agonist

ATPγS exhibits relatively high potency at the P2Y$_2$ receptor, and therefore it was attractive to speculate that its uridine counterpart, i.e., UTPγS, also would be a potent hydrolysis-resistant P2Y$_2$ receptor agonist. Therefore, UTPγS (Fig. 2) was synthesized by nucleoside diphosphate kinase–catalyzed transfer of the gamma-thiophosphate from GTPγS to UDP (56), according to the following reaction:

UDP + GTPγS ↔ UTPγS + GDP

UTPγS was not significantly hydrolyzed following a 10-min incubation with several phosphatases such as alkaline phosphatase or acid phosphatase, or by the ATPase apyrase, all which were shown to rapidly degrade UTP (47). Moreover, in contrast to the lability of UTP in airways, UTPγS was recovered unchanged after a 60-min incubation on the mucosal surface of airway epithelial cells (56) and following a 10-min incubation with a sputum

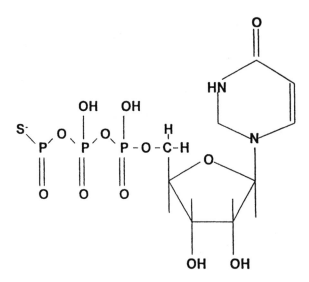

Figure 2 UTPγS.

sample from a cystic fibrosis patient (58). UTPγS was a potent and full agonist for stimulation of inositol phosphate accumulation in HP2U-1321N1 cells, and its EC_{50} value of 220 nM was similar to the EC_{50} values of UTP (120 nM) and ATP (230 nM) (Table 2). Indeed, UTPγS was a more potent $P2Y_2$ receptor agonist than was the adenine nucleotide analog ATPγS. UTPγS also was equipotent with UTP in promoting chloride secretion on cystic fibrosis airway epithelial cells. Thus, UTPγS has proven to be a stable and potent $P2Y_2$ receptor agonist, and therefore, it has potential therapeutic value for treatment of cystic fibrosis. Because standard protocols for enzymatic synthesis of phosphorothioates produce only a few micrograms of UTPγS per reaction, alternative synthetic procedures are needed to scale-up UTPγS production. This will allow tests in vivo of therapeutic potential for lung disease.

V. OTHER NUCLEOTIDE RECEPTORS

A. Adenine Nucleotide Receptors

The expression in the lung of purinoceptors other than the $P2Y_2$ receptor has not been clearly established. Using Northern blot analysis, Valera and colleagues showed that a $P2X_1$ probe hybridized to RNA preparations from different rat tissues, including the lung (59). However, the intensity of the band observed with lung samples was very low relative to other tissues, and the authors concluded that contamination of lung preparations with smooth muscle cells could account for their results. $P2X_4$ (60) and $P2X_5$ (61) receptor transcripts were also reported in lung and bronchi epithelium. No functional studies on P2X receptors in airway tissues have been described.

Although relatively high levels of $P2Y_1$ receptor message have been detected in the lung of several mammalian species, including humans (62), unequivocal evidence for functional $P2Y_1$ receptors in airways is still missing. Mason and colleagues (26) reported a potent stimulatory effect of 2Me-SATP on Cl^- secretion in primary cultures of human nasal epithelial cells. The effect of 2MeSATP was restricted to the basolateral cell surface. Whether the action of 2MeSATP involved the $P2Y_1$ receptor or a ionotropic P2X receptor has not been established.

B. Uridine Nucleotide Receptors

Neither functional nor molecular evidence for the existence of the $P2Y_4$ receptor in airway tissue is available (10,11). Chang and colleagues (12) and Communi and colleagues (63) have reported the presence of $P2Y_6$ receptor transcripts in rat and human lung, respectively. A functional UDP receptor was recently described that regulates phosphoinositide breakdown, calcium mobilization, and chloride secretion in human airway epithelium (64). Interestingly, UDP, a breakdown product of UTP, is more stable than UTP on the nasal epithelial cell surface (59). Although the potential use of UDP for pharmacotherapy has not been explored, the combined targeting of P2Y$_2$ and P2Y$_6$ receptors opens interesting avenues for treatment of cystic fibrosis and other respiratory diseases.

VI. CONCLUSIONS AND FUTURE DIRECTIONS

Despite the impressively rapid expansion of the nucleotide receptor field during the last 5 years, a major issue remains to be solved: the identification of highly selective and potent agonists and antagonists that permit a more direct examination of the role of these receptors in both normal and diseased tissues.

Identification of key residues in the P2Y$_2$ receptor protein that might be involved in ligand-binding will aid in the design of new drugs. Modeling analysis of the three receptors now known to be activated by uridine nucleotides, i.e., P2Y$_2$, P2Y$_4$, and P2Y$_6$ receptors, will augment site-directed mutagenesis studies that not only provide clarification of the role of positively charged amino acids in receptor-phosphate chain interaction, but also lead to identification of residues implicated in binding of both the heterocycle and the ribose moieties. Chimeric constructs between the four cloned P2Y receptors will help expand our knowledge of the importance of residues outside of TM3 and TM7, e.g., within putative extracellular domains, in determining the selectivity of these proteins for nucleotide agonists including adenine versus uridine and diphosphate versus triphosphate selectivity.

Additional insight into the biology of the nucleotide system in airway epithelium is needed. The presence of P2 receptors in the airway lumen predicts the likelihood that a biologically regulated concentration of nucleotide agonist may be maintained on airway surfaces. Preliminary studies have suggested that there is a basal level of ATP on airway surfaces that approximates 3×10^{-7} M (65). This level likely is sustained by poorly defined nucleotide secretory mechanisms that reside in airway epithelial cells in series with cell surface and secreted ATPases that actively metabolize triphosphate nucleotides. These findings have led to speculations that ATP may act as a "lumone" that is secreted in an autocrine/paracrine fashion and that serves normally to coordinate the three major functions, i.e., salt and water transport, ciliary beat frequency, and mucin secretion, that promote effective mucociliary clearance. Future research should be designed to test this hypothesis more definitively, by characterizing nucleotide release mechanisms and their regulation, and by delineating the nucleotide degradation pathways expressed on airway surfaces and in airway secretions. Whether ATP or UTP is the natural ligand on the luminal surface of the airways remains an exciting unanswered question.

Finally, the therapeutic actions of nucleotides in infectious airways diseases will require extension from studies of "surrogate" markers of disease, e.g., measurement of mucociliary clearance rates, to more direct indices of clinical efficacy. Thus, the relative activities of potentially short-acting compounds, e.g., UTP, and potentially longer-acting analogs, e.g., UTPγS, will have to be tested in diseases that are characterized by chronic abnormalities of airway clearance and infection to establish whether one or both of these classes of compounds is active in a particular type of lung disease. Obviously, this is an exciting area of interest owing to the novel nature of this class of compounds for therapy of infectious lung disease and the possibility that they may be effective in nongenetic diseases of mucociliary clearance that are produced as a consequence of inhaled toxins, e.g., chronic bronchitis.

REFERENCES

1. Burnstock G. Purinergic nerves. Pharmacol Rev 1972; 24:509–581.
2. Burnstock G. A basis for distinguishing two types of purinergic receptor. In: Straub LB, ed. Cell Membrane Receptors for Drugs and Hormones: A Multidisciplinary Approach. New York: Raven Press, 1978:107–118.
3. Burnstock G, Kennedy C. Is there a basis for distinguishing two types of P_2-purinoceptor? Gen Pharmacol 1985; 16:433–440.
4. Haslam RJ, Cusack NJ. Blood platelet receptors of ADP and adenosine. In: Burnstock G, ed. Purinergic Receptors: Receptors and Recognition, Series B, Vol 12. London: Chapman and Hall, 1981:221–285.
5. Hourani SMO, Cusack NJ. Pharmacological receptors on blood platelets. Pharmacol Rev 1991; 43:243–298.
6. Cockcroft S, Gomperts BD. The ATP^{4-} receptor of rat mast cells. Biochem J 1980; 188:789–798.
7. Gordon JL. Extracellular ATP: effects, sources and fate. Biochem J 1986; 233:309–319.
8. O'Connor SE, Dainty IA, Leff P. Further subclassification of ATP receptors based on agonist studies. Trends Pharmacol Sci 1991; 12:137–141.
9. Lazarowski ER, Harden TK. Identification of a uridine nucleotide-selective G-protein-linked receptor that activates phospholipase C. J Biol Chem 1994; 269:11830–11836.
10. Communi D, Pirotton S, Parmentier M, Boeynaems J. Cloning and functional expression of a human uridine nucleotide receptor. J Biol Chem 1995; 270:30849–30852.
11. Nguyen T, Erb L, Weisman GA, Marchese A, Heng HHQ, Garrad RC, George SR, Turner JT, O'Dowd BF. Cloning, expression, and chromosomal localiza-

tion of the human uridine nucleotide receptor gene. J Biol Chem 1995; 270:30845–30848.

12. Chang K, Hanaoka K, Kumada M, Takuwa Y. Molecular cloning and functional analysis of a novel P$_2$ nucleotide receptor. J Biol Chem 1995; 270:26152–26158.

13. Nicholas RA, Watt WC, Lazarowski ER, Li Q, Harden TK. Uridine nucleotide selectivity of three phospholipase C-activating P$_2$ receptors: identification of a UDP-selective, a UTP-selective, and an ATP- and UTP-specific receptor. Mol Pharmacol 1996; 50:224–229.

14. Hourani SMO, Hall DA. Receptors for ADP on human blood platelets. Trends Pharmacol Sci 1994; 151:103–107.

15. Boyer JL, Lazarowski ER, Chen X, Harden TK. Identification of a P$_{2Y}$-purinergic receptor that inhibits adenylyl cyclase. J Pharmacol Exp Ther 1993; 267:1140–1146.

16. Erb L, Garrad R, Wang Y, Quinn T, Turner JT, Weisman GA. Site-directed mutagenesis of P$_{2U}$ purinoceptors. Positively charged amino acids in transmembrane helices 6 and 7 affect agonist potency and specificity. J Biol Chem 1995; 270:4185–4188.

17. Stutchfield J, Cockcroft S. Undifferentiated HL60 cells respond to extracellular ATP and UTP by stimulating phospholipase C activation and exocytosis. FEBS Lett 1990; 262:256–258.

18. Dubyak GR, Cowen DS, Meuller LM. Activation of inositol phospholipid breakdown in HL60 cells by P$_2$-purinergic receptors for extracellular ATP. Evidence for mediation by both pertussis toxin-sensitive and pertussis toxin-insensitive mechanisms. J Biol Chem 1988; 263:18108–18117.

19. Dubyak GR, El-Moatassim C. Signal transduction via P$_2$-purinergic receptors for extracellular ATP and other nucleotides. Am J Physiol 1993; 265:C577–C606.

20. Axtell RA, Sandborg RR, Smolen JE, Ward PA, Boxer LA. Exposure of human neutrophils to exogenous nucleotides causes elevation in intracellular calcium, transmembrane calcium fluxes, and an alteration of a cytosolic factor resulting in enhanced superoxide production in response to FMLP and arachidonic acid. Blood 1990; 75:1324–1332.

21. Cowen DS, Lazarus HM, Shurin SB, Stoll SE, Dubyak GR. Extracellular adenosine triphosphate activates calcium mobilization in human phagocytic leukocytes and neutrophil/monocyte progenitor cells. J Clin Invest 1989; 83:1651–1660.

22. Kuhns DB, Wright DG, Nath J, Kaplan SS, Basford RE. ATP induces transient elevations of [Ca^{2+}] in human neutrophils and primes these cells for enhanced O$_2$-generation. Lab Invest 1988; 58:448–453.

23. Walker BAM, Hagenlocker BE, Douglas VK, Tarapchak SJ, Ward PA. Nucleotide responses of human neutrophils. Lab Invest 1991; 64:105–112.

24. Fedan JS, Stem JL, Day B. Contraction of the guinea pig isolated, perfused trachea to purine and pyrimidine agonists. J Pharmacol Exp Ther 1994; 268:1321–1327.

25. Henning RH, Nelemans A, van den Akker J, Den Hertog A. The nucleotide receptors on mouse C2C12 myotubes. Br J Pharmacol 1992; 106:853–858.
26. Mason SJ, Paradiso AM, Boucher RC. Regulation of transepithelial ion transport and intracellular calcium by extracellular adenosine triphosphate in human normal and cystic fibrosis airway epithelium. Br J Pharmacol 1991; 103:1649–1656.
27. Brown HA, Lazarowski ER, Boucher RC, Harden TK. Evidence that UTP and ATP regulate phospholipase C through a common extracellular 5'-nucleotide receptor in human airway epithelial cells. Mol Pharmacol 1991; 40:648–655.
28. Parr CE, Sullivan DM, Paradiso AM, Lazarowski ER, Burch LH, Olsen JC, Erb L, Weisman GA, Boucher RC, Turner JT. Cloning and expression of a human P_{2U} nucleotide receptor, a target for cystic fibrosis pharmacotherapy. Proc Natl Acad Sci USA 1994; 91:3275–3279.
29. Clarke LL, Boucher RC. Chloride secretory response to extracellular ATP in human normal and cystic fibrosis nasal epithelia. Am J Physiol 1992; 263:C348–C356.
30. Clarke LL. Unpublished communication.
31. Paradiso AM, Mason SJ, Lazarowski ER, Boucher RC. Membrane-restricted regulation of Ca^{2+} release and influx in polarized epithelia. Nature 1995; 377:643–646.
32. Davis CW, Dowell ML, Lethem MI, Van Scott M. Goblet cell degranulation in isolated canine tracheal epithelium: response to exogenous ATP, ADP, and adenosine. Am J Physiol 1992; 262:C1313–C1323.
33. Lethem MI, Dowell ML, Van Scott M, Yankaskas JR, Egan T, Boucher RC, Davis CW. Nucleotide regulation of goblet cells in human airway epithelial explants: normal exocytosis in cystic fibrosis. Am J Respir Cell Mol Biol 1993; 9:315–322.
34. Olivier KN, Bennett WD, Hohneker KW, Zemen KL, Edwards LJ, Boucher RC, Knowles MR. Acute safety and effects on mucociliary clearance of aerosolized uridine 5'-triphosphate +/− amiloride in normal human adults. Am J Respir Crit Care Med 1996; 154:217–223.
35. Bennett WD, Olivier KN, Zeman KL, Hohneker KW, Boucher RC, Knowles MR. Effect of uridine 5'-triphosphate plus amiloride on mucociliary clearance in adult cystic fibrosis. Am J Respir Crit Care Med 1996; 153:1796–1801.
36. Noone PG, Bennett WD, Zeman KL, Regnis JA, Carson JL, Boucher RC, Knowles MR. Effects on cough clearance of aerosolized uridine-5'-triphosphate +/− amiloride in patients with primary ciliary dyskinesia. Am J Respir Crit Care Med 1996; 153:A530 (abstract).
37. Lustig KD, Shiau AK, Brake AJ, Julius D. Expression cloning of an ATP receptor from mouse neuroblastoma cells. Proc Natl Acad Sci USA 1993; 90:5113–5117.
38. Strader CD, Candelore MR, Hill WS, Sigal IS, Dixon RAF. Identification of two serine residues involved in agonist activation of the beta-adrenergic receptor. J Biol Chem 1989; 264:13572–13578.
39. Wess J, Blin N, Mutschler E, Blueml K. Muscarinic acetylcholine receptors: structural basis of ligand binding and G protein coupling. Life Sci 1995; 56:915–922.

40. Kim J, Wess J, van Rhee AM, Schoeneberg T, Jacobson KA. Site-directed mutagenesis identifies residues involved in ligand recognition in the human A$_{2a}$ adenosine receptor. J Biol Chem 1995; 270:13987–13997.

41. Olah ME, Jacobson KA, Stiles GL. Identification of an adenosine receptor domain specifically involved in binding of 5′-substituted adenosine agonists. J Biol Chem 1994; 269:18016–18020.

42. Townsend-Nicholson A, Schofield PR. A threonine residue in the seventh transmembrane domain of the human A$_1$ adenosine receptor mediates specific agonist binding. J Biol Chem 1994; 269:2373–2376.

43. Tucker AL, Robeva AS, Taylor HE, Holeton D, Bockner M, Lynch KR, Linden J. A$_1$ adenosine receptors. Two amino acids are responsible for species differences in ligand recognition. J Biol Chem 1994; 269:27900–27906.

44. Kim J, Jiang Q, Glashofer M, Yehle S, Wess J, Jacobson KA. Glutamate residues in the second extracellular loop of the human A$_{2a}$ adenosine receptor are required for ligand recognition. Mol Pharmacol 1996; 49:683–691.

45. van Rhee AM, Fischer B, Van Galen PJM, Jacobson KA. Modelling the P$_{2Y}$ purinoceptor using rhodopsin as template. Drug Des Discov 1995; 13:133–154.

46. Parr CE, Boucher RC. Unpublished.

47. Lazarowski ER, Watt WC, Stutts MJ, Boucher RC, Harden TK. Pharmacological selectivity of the cloned human P2U-receptor: potent activation by diadenosine tetraphosphate. Br J Pharmacol 1995; 116:1619–1627.

48. Pintor J, Miras-Portugal MT. Diadenosine polyphosphates (A$_{Px}$A) as new neurotransmitters. Drug Dev Res 1993; 28:259–262.

49. Miras-Portugal MT, Pintor J, Castro E, Rodriguez-Pascual F, Torres M. Diadenosine polyphosphates from neuro-secretory granules: the search for receptors, signals and function. In: Municio AM, Miras-Portugal MT, eds. Cell Signal Transduction, Second Messengers, and Protein Phosphorylation in Health and Disease. New York: Plenum Press, 1994:169–186.

50. Stutts MJ, Boucher RC. Unpublished.

51. Miras-Portugal MT, Castro E, Mateo J, Pintor J. The diadenosine polyphosphate receptors: P2D purinoceptors. In: P2 Purinoceptors: Localization, Function and Transduction Mechanisms. Chichester: Wiley, 1996:35–52.

52. Lazarowski ER, Mason SJ, Clarke LL, Harden TK, Boucher RC. Adenosine receptors on human airway epithelia and their relationship to chloride secretion. Br J Pharmacol 1992; 106:774–782.

53. Ramos A, Pintor J, Miras-Portugal MT, Rotllan P. Use of fluorogenic substrates for detection and investigation of ectoenzymatic hydrolysis of diadenosine polyphosphates: a fluorometric study on chromaffin cells. Anal Biochem 1995; 228:74–82.

54. Dixon M, Webb EC, Thorne CJR, Tipton KF. Enzyme specificity. In: Enzymes. New York: Academic Press, 1979:231–270.

55. Rodriguez-Pascual F, Torres M, Rotllan P, Miras-Portugal MT. Extracellular hydrolysis of diadenosine polyphosphates, Ap$_n$A, by bovine chromaffin cells in culture. Arch Biochem Biophys 1992; 297:176–183.

56. Lazarowski ER, Watt WC, Stutts MJ, Brown HA, Boucher RC, Harden TK. Enzymatic synthesis of UTPgammaS, a potent hydrolysis resistant agonist of P$_{2U}$-receptors. Br J Pharmacol 1996; 117:203–209.

57. Pendergast W, Siddiqi SM, Rideout JL, James MK, Dougherty RW. Stabilized uridine triphosphate analogs as agonists of the P_{2Y2} purinergic receptor. Drug Dev Res 1996; 37:133 (abstract).

58. Lazarowski ER, Regnis J. Unpublished.

59. Valera S, Hussy N, Evans RJ, Adami N, North RA, Suprenant A, Buell G. A new class of ligand-gated ion channel defined by P_{2X} receptor for extracellular ATP. Nature 1994; 371:516–519.

60. Bo X, Zhang Y, Nassar M, Burnstock G, Schoepfer R. A P2X purinoceptor cDNA conferring a novel pharmacological profile. FEBS Lett 1995; 375:129–133.

61. Collo G, North RA, Kawashima E, Merlo-Pich E, Neidhart S, Surprenant A, Buell G. Cloning of $P2X_5$ and $P2X_6$ receptors and the distribution and properties of an extended family of ATP-gated ion channels. J Neurosci 1996; 16:2495–2507.

62. Janssens R, Communi D, Pirotton S, Samson M, Parmentier M, Boeynaems J. Cloning and tissue distribution of the human $P2Y_1$ receptor. Biochem Biophys Res Commun 1996; 221:588–593.

63. Communi D, Parmentier M, Boeynaems J. Cloning, functional expression and tissue distribution of the human $P2Y_6$ receptor. Biochem Biophys Res Commun 1996; 222:303–308.

64. Lazarowski ER, Paradiso AM, Watt WC, Harden TK, Boucher RC. UDP activates a mucosal-restricted receptor on human nasal epithelial cells that is distinct from the $P2Y_2$ receptor. Proc Natl Acad Sci USA 1997; 94:2599–2603.

65. Donaldson SH, Stutts MJ, Boucher RC, Knowles MR. Adenosine triphosphate levels in nasal surface liquid. Am J Respir Crit Care Med 1996; 153:A854 (abstract).

66. Alexsander SPH, Peters JA. Receptor and ion channel nomenclature supplement. Trends Pharmacol Sci 1997 (Eighth edition) pp. 65–68. (Published by Elsevier Trends Journals.)

12

Muscarinic Receptor Antagonists
Pharmacological and Therapeutic Utility

Richard M. Eglen
Center for Biological Research, Roche Bioscience, Palo Alto, California

I. INTRODUCTION

Muscarinic receptors are pharmacologically identified on the basis of selective antagonism by the antagonist atropine (1–3). The classification of muscarinic receptor *subtypes,* of which there are five, is more problematic. This difficulty arises from a lack of agonists and antagonists selective for one muscarinic receptor subtype over the remaining four. Nonetheless, the therapeutic potential for subtype selective antagonists has ensured that research in the area has continued for at least 25 years, with several compounds in advanced clinical study.

This chapter will describe the various uses of selective muscarinic receptor antagonists, surmountable and unsurmountable, in the operational, i.e., pharmacological, classification of muscarinic receptor subtypes. Irreversible muscarinic antagonists will be addressed since these compounds are finding increasing use in pharmacologically "isolating" a single subtype in tissues that expresses heterogeneous populations. Finally, the potential therapeutic uses of selective antagonists will also be described. The discussion will principally address the use of such compounds in the treatment of smooth muscle disorders, although other uses will be briefly addressed.

This chapter will not describe *all* muscarinic receptor antagonists identified to date with some degree of subtype selectivity. Instead, this review will focus on those compounds that have found extensive use in muscarinic receptor classification, as well as some novel antagonists described recently. Several comprehensive reviews of this area have been recently published (3–5) and the reader is referred to these papers for the extensive bibliographies cited therein.

II. CLASSIFICATION OF MUSCARINIC RECEPTOR SUBTYPES

Reversible Antagonists

Muscarinic receptor genes encode at least five receptor proteins (Table 1; see Ref. 3 for review). Operational criteria have suggested the presence of at least four subtypes, denoted M_1, M_2, M_3, and M_4, a suggestion endorsed by cloning studies, from which four corresponding gene products (m1, m2, m3, m4) have been identified (3). A physiological role for the m5 gene product remains to be identified (4) and, in accordance with IUPHAR nomenclature recommendations (6), will be referred to by the lower-case appellation.

The expression of complementary DNAs encoding muscarinic receptors, including those of human origin, has simplified determination of the pharmacological profile for a single muscarinic receptor subtype (7). Expression has been undertaken with receptors in *Xenopus* oocytes (8), mammalian cells (9), or insect cells (10). In general, the antagonist affinity profiles, determined at cloned receptors, concurs with those generated at endogenously expressed receptors (4). For reasons that are unclear, the correspondence is not exact (11) and it appears that the membrane environment may

Table 1 Characteristics of Muscarinic Receptor Subtypes

Nomenclature	M_1	M_2	M_3	M_4
Receptor gene	M_1	M_2	M_3	M_4
Structure	7TM	7TM	7TM	7TM
Human	460aa	466aa	590aa	479aa
Mouse	460aa	—	—	479aa
Rat	460aa	466aa	589aa	478aa
Porcine	460aa	466aa	590aa	479aa
Intracellular messenger	IP_3/DG	$cAMP/K^+$ channels	IP_3/DG	cAMP

TM, predicted number of transmembrane-spanning domains; aa, amino acid residues; IP_3, inositol (1,4,5) trisphosphate; DG, 1,2 diacylglycerol (mobilization); cAMP, 3′ 5′ cyclic adenosine monophosphate (reduction). A fifth gene, m5, has been cloned but a functional correlate has not been unambiguously demonstrated.
Source: From Ref. 13.

modulate ligand affinity. Consequently, the affinity of antagonists could vary according to the cell type in which the recombinant receptor is expressed (11). Although the extent to which this phenomenon occurs at endogenously muscarinic receptors is unknown, it has clear indications for the screening of novel antagonists.

Nonetheless, until cells expressing recombinant receptors were available, the pharmacological profile of each muscarinic receptor subtype was ambiguous, for two reasons. First, many tissues and cell lines express heterogeneous populations of muscarinic receptor subtypes. Consequently, the binding profile represented an interaction at more than one site and the derived affinities were resultant values. The second problem was the existence of the muscarinic M_4 receptor, pharmacological definition of which was difficult. Currently, implication of this receptor in the mediation of a response remains problematic (3). The physiological importance of this subtype has, therefore, been neglected notably with regard to several prejunctional receptors (3).

These problems not withstanding, muscarinic receptor antagonists can be used in receptor classification, particularly since muscarinic receptor agonists do not discriminate between muscarinic receptor subtypes on an affinity basis (see Refs. 2,12,13 for further discussion). The involvement of a muscarinic receptor in a functional response can be pharmacologically defined by the affinities of several key antagonists (Table 2). However, two important points should be noted in this regard.

First, antagonist affinity values should be measured under equilibrium conditions. Indeed, given the low subtype selectivity of most antagonists between muscarinic receptor subtypes, reducing the experimental errors arising from conditions of disequilibrium is critical. Reassuringly, antagonist

Table 2 Pharmacological Characterization of Muscarinic Receptors

Antagonist	M_1	M_2	M_3	M_4
4-DAMP	8.6 (9.2)	7.8 (8.4)	9.1 (9.3)	n.d. (8.9)
Guanylpirenzepine	7.7	5.6	6.5	6.5
Himbacine	7.2 (7.0)	8.5 (8.0)	7.6 (7.0)	8.8 (8.0)
p-F-HHSiD	7.2 (7.7)	6.0 (6.9)	7.9 (7.8)	n.d. (7.5)
Methoctramine	6.5 (7.3)	7.9 (7.9)	6.0 (6.7)	7.6 (7.5)
Pirenzepine	8.3 (8.2)	6.8 (6.7)	6.9 (6.9)	7.7 (7.4)
PD102807	5.3	5.8	6.3	7.3
Tripitramine	n.d. (8.8)	9.7 (9.6)	6.5 (7.4)	n.d. (7.9)

Source: Modified from Ref. 13.

affinities, when estimated under pharmacologically rigorous conditions, correspond well between laboratories.

Second, some caution must be excercised when comparing antagonist affinities estimated using radioligand binding techniques with those estimated by null methods. Currently, binding is extensively used to generate the selectivity profile of novel ligands. In general, affinity values generated by functional pharmacological methods at recombinant or endogenous receptors correspond to those derived in binding studies. Sometimes, however, the disparity between binding and functional affinities is large, giving rise to difficulties in assessing the selectivity of the ligand. A reason for this discrepancy lies in the use of buffers of different ionic strengths, some of which decrease the binding affinity of antagonists. Ligand affinities at the muscarinic M_2 receptor are particularly susceptible to such changes. Recent data indicate the presence of a cation binding site (an aspartate residue in transmembrane region three) on the muscarinic M_2 receptor at which monovalent ions compete with the charged residues of muscarinic receptor ligands (1). Consequently, the selectivity of novel muscarinic M_3 receptor antagonists, over M_2 receptors, can be compromised by the buffer ionic strength. It is important, therefore, that the subtype selectivity of novel ligands is defined using *both* radioligand-binding and functional approaches.

Experience has shown that the effects of a given agonist can be ascribed to a particular muscarinic receptor subtype by using a relatively small group of selected antagonists (3,4,12). These compounds include atropine (to ensure that the response is, in fact, mediated via a muscarinic receptor), pirenzepine (M_1 selective), methoctramine (M_2/M_4 selective), 4-DAMP, 4-diphenyl-acetoxy-N-methyl piperidine methiodide (M_1/M_3 selective), *para*-fluorohexahydrosiladifenidol (*p*-F-HHSiD; M_3 selective), and himbacine (M_2/M_4 selective). These antagonists have been extensively used to effectively "fingerprint" the muscarinic receptor subtype mediating a particular tissue response. However, novel antagonists, such as tripitramine (M_2 selective), may prove to be equally important discriminative tools.

1. Muscarinic M_1 Receptor Antagonists

Muscarinic M_1 receptors are pharmacologically (3,4) defined by a high affinity toward pirenzepine and 4-DAMP, an intermediate affinity for *p*-F-HHSiD, and a low affinity for methoctramine or himbacine. Pirenzepine is well established as a selective muscarinic M_1 receptor antagonist, although the separation between muscarinic M_1 and M_4 receptors is small (Table 2). Analogs of pirenzepine, such as telenzepine, have a similar selectivity profile (15). Other muscarinic M_1 antagonists, not extensively used, include caramiphen and the iodo or nitro analogs (16). Recently, S-(−)-α-(hydroxymethyl)benzeneacetic acid 1-methyl-4-piperdinyl ester [S-(−)-ET

126] has been reported (17) to act as a selective muscarinic M_1 receptor antagonist. However, affinity data for this antagonist at muscarinic M_4 receptors is unavailable.

Other, less selective, muscarinic M_1 antagonists have also been reported, including some analogs of quinuclidinyl benzilate (QNB) (18). Although not used in receptor classification, these analogs have provided leads for the synthesis of halogenated derivates for use in single-photon emission computed tomography imaging studies, thereby enabling mapping of central muscarinic receptor for distribution in vivo. Other compounds have also been developed for use in positron emission tomography, including [^{11}C]N-methyl-4-piperidyl benzilate [^{11}C]NMBP (19). This ligand, for example, has been shown recently to act as a marker for changes in muscarinic receptor density in the frontal cortex of patients with Parkinson's disease (19).

2. Muscarinic M_2 Receptor Antagonists

The muscarinic M_2 receptor exhibits high affinity toward AF-DX 116, methoctramine, and himbacine but low affinity for pirenzepine, 4-DAMP, and p-F-HHSiD. Minor structural alteration of pirenzepine, a prototypic M_1 receptor antagonist (see above), results in enhanced affinity for the muscarinic M_2 receptor (20). AF-DX 116 is one such compound (21) and has been used extensively in muscarinic receptor classification (3). Analogs of AF-DX 116, including AQ-RA 741 (22), have improved muscarinic M_2 receptor affinity and selectivity. Guanylpirenzepine, in radioligand-binding studies, as a high degree of selectivity for recombinant muscarinic M_2 receptor, over both the muscarinic M_1, M_3, and M_4 receptors (23,24). Other muscarinic M_2 antagonists include several cervane alkaloids, such as imperialine (25) or the chlorinated derivative (26). The latter displays improved muscarinic M_2/M_1 selectivity, lipophilicity, and duration of action in vivo over the parent compound (26).

A problem associated with all of these compounds is the poor discrimination between muscarinic M_2 and M_4 receptors. Recently identified compounds may offer some advantages in this respect. Thus, (S) dimethindene, although possessing histamine H_1 receptor antagonist activity, is selective for muscarinic M_2 receptors over M_1, M_3, or M_4 receptors (27). Tripitramine (28–30) is also more selective for muscarinic M_2 over the M_4 receptor in comparison to other "selective" M_2 antagonists. However, this separation is less at human recombinant muscarinic M_1 and M_2 receptors than at endogenously expressed receptors (29).

Several other selective M_2 receptor antagonists, including gallamine, methoctramine, and to a lesser extent, himbacine allosterically modulate muscarinic receptor function (12). These compounds, with the exception of himbacine, are structurally related to neuromuscular blocking agents, and all appear to act selectively toward the muscarinic M_2 receptor. However, allosterism, most pronounced at this subtype, complicates determination of affinity and thus interpretation of the subtype selectivity (3,31). Some of these compounds will be further described below.

3. Muscarinic M_3 Receptor Antagonists

The muscarinic M_3 receptor exhibits a high affinity for 4-DAMP and p-F-HHSiD but a low affinity for pirenzepine, methoctramine, and himbacine (3). Analogs of 4-DAMP with muscarinic M_3 over M_2 receptor selectivity include pentamethylene bis-4-DAMP (32) and benzyl-4-DAPine (33). However, subsequent studies (34) have not confirmed this selectivity of the benzyl-4-DAPine, possibly due to problems of solubility (35). 3-Halopropylamine derivatives of 4-DAMP are predicted to readily penetrate the blood-brain barrier and form a stable azetidium ion. This process is similar to the formation of aziridinium ion from the alkylating agent 4-DAMP mustard (see below). The 3-chloro derivative of 4-DAMP, N-(3-hydroxypropyl)-4-piperidinyl diphenylacetate (3-CP-4-DAP), forms a stable azetidium ion in aqueous solution. This ion acts as a reversible, high-affinity, ligand for muscarinic M_1, M_3, M_4, and m5 receptors with a 10–14-fold lower affinity for muscarinic M_2 receptors (36). Consequently, 3-CP-4-DAP is a prodrug that enters the CNS and produces long-lasting antagonism of central muscarinic M_2 receptors. Behavior data with this compound, however, are not yet available.

Some muscarinic M_3 receptor antagonists may distinguish between various M_3 receptors, even in tissue from the same species. For example, zamifenacin (37), a novel compound that is structurally related to benzyl-4-DAPine, displays higher affinity for smooth muscle muscarinic M_3 receptors in guinea pig ileum and esophagus than at muscarinic M_3 receptors in trachea and bladder (37,38) (Table 3). Selectivity is also seen in vivo where inhibition of gastrointestinal motility in dogs occurs at doses that do not affect pupil diameter, both of which are mediated by activation of muscarinic M_3 receptors (39). This finding is supported by in vitro data from the same species that show that zamifenacin discriminates, by over two orders of magnitude, between muscarinic M_3 receptors in canine ciliary muscle and ileum (40). Both zamifenacin (41) and a novel, related compound, darifenacin (42), functionally discriminate between muscarinic M_3 receptors in salivary gland and ileum. It is unknown if the structurally related benzyl-4-DAPine exhibits a similar difference in muscarinic M_3 receptor affinities.

The structurally different muscarinic M_3 receptor antagonist p-F-HHSiD (43) also discriminates between smooth muscle M_3 receptors (44,45). This compound separates, by about 10-fold, muscarinic M_3 receptors in guinea

Table 3 Affinities (pA_2) of Zamifenacin at Various Muscarinic M_3 Receptors In Vitro

Species	Preparation	Affinity	Ref.
Canine	Ciliary muscle	<6.0	40
Canine	Ileum	8.6	40
Guinea pig	Ileum	9.3	38
Guinea pig	OMM	8.8	38
Guinea pig	Trachea	8.1	38
Guinea pig	Urinary bladder	7.6	38
Human	Trachea	7.6	47

OMM, esophageal muscularis mucosae.
Source: From Ref. 13

pig (44,45) tracheal and ileal muscarinic M_3 receptors. However, in contrast to zamifenacin (41), p-F-HHSiD does not discriminate between ileum and urinary bladder muscarinic M_3 receptors (44). Interestingly, atypically low-affinity values for this antagonist have also been reported at muscarinic M_3 receptors mediating contraction of human colonic circular muscle (46) and human trachea (47). It is unknown, however, if these differences reflect the existence of different muscarinic M_3 receptor subtypes or if other causes, such as membrane environment, play a role.

One feature of G-protein-coupled receptors is that they are able to exist in an active form; i.e., they may couple and activate the G-protein even in the *absence* of agonist. Recently, it has been shown that this phenomenon holds true for muscarinic receptor subtypes. Most importantly, it has been reported for *both* recombinant and endogenous receptors (48,49). Specifically, for M_1–M_4 receptors expressed in CHO cells, and muscarinic M_2 receptors in rat cardiomyocytes, constitutive activity has been demonstrated (48), with atropine or N-methylscopolamine (NMS) acting as inverse agonists. These data may suggest that some archetypal muscarinic receptor antagonists, such as atropine, have a higher affinity for, and thus stabilize, the inactive conformation of the receptor. In a mutated m5 receptor (denoted as Cam5), atropine, QNB, NMS, 4-DAMP, and pirenzepine also acted as inverse agonists (49). This mutation, i.e., serine 465 for tyrosine and threonine 486 for proline, may alter the spontaneous equilibrium existing between resting and active states of the receptor. These findings indicate that both classic and nonclassic muscarinic antagonists act as inverse agonists. Although it remains to be seen if this phenomenon occurs with endogenous muscarinic M_3 receptors in smooth muscle, it could provide an explanation for the tissue dependence of antagonist affinities at the muscarinic

M_3 receptor. Thus, different degrees of constitutive activity would be expected to vary the affinity of antagonists, such as zamifenacin, between muscarinic M_3 receptors in different tissues.

4. Muscarinic M_4 Receptor Antagonists

The muscarinic M_4 receptor remains difficult to define since both pirenzepine and methoctramine exhibit high affinity for this subtype. The concurrent high affinities of himbacine and p-F-HHSiD, however, serve to distinguish the receptor from muscarinic M_1 and M_2 subtypes, respectively (Table 2). Radioligand-binding studies have also suggested that dicyclomine, DAU 5884, DAU 6202, and AQ-RA 721 may be useful to discriminate muscarinic M_4 from M_2 receptors (24). Tripitramine (29) could also prove useful to distinguish between those responses mediated by M_2 and those by M_4 receptors, the advantage of this compound being the low affinities for other muscarinic receptor subtypes (see above). Finally, a toxin (MTx-3), isolated from venom of *Drendroaspis angusticeps*, has been shown to exhibit high affinities at muscarinic M_4 and M_1 receptors with little binding detectable at muscarinic M_2 and M_3 receptors (50). As such, this ligand is one of the most selective muscarinic receptor antagonists identified to date, although its use is complicated by its apparently pseudoirreversible action (50). In transfected cell lines, clozapine has been shown to act as a full agonist at human muscarinic M_4 receptors, and as an antagonist at muscarinic M_1, and M_2, and M_3 receptors (51). While this selectivity may be due to varying levels of receptor expression between the cells involved, clozapine could provide a novel lead to further define the muscarinic M_4 subtype.

It should also be noted that a novel approach for selective labeling of muscarinic M_4 receptors has been reported (23), in which the distinct kinetics of dexetimide and guanylpirenzepine have been exploited. Thus, incubation with dexetimide, a nonselective antagonist, occupied >90% of all five subtypes (m1–m5). With short incubations with [^{3}H] NMS, the off-rate of dexetimide from m1, m2, and m4 receptors is faster than from m3 and m5 receptors, thus favoring the labeling of the former over the latter. Consequently, guanylpirenzepine and AF-DX 116 can then be used to preferentially label m1 and m2 receptors, thus allowing [^{3}H] NMS to selectively bind to m4 receptors (23). This approach results in selective labeling of this subtype both in homogenate binding and in sections of rat brain (23).

5. Muscarinic M_5 Receptor Antagonists

The pharmacology of the m5 gene product appears to differ from that of other muscarinic receptors (2), although no single ligand is preferential toward the expressed protein. In a baculovirus-infected system in Sf9 insect cells (10) promethazine has shown a 4–7-fold selectivity toward this receptor. It remains to be established if this holds ture in a mammalian expression system. The lack of selective antagonists and limited expression of this receptor accounts for the lack of knowledge regarding the physiological role of the receptor.

III. ALLOSTERIC MUSCARINIC ANTAGONISTS

It has been recognized for many years (see Ref. 12 for review) that some antagonists, such as gallamine (52), act allosterically at muscarinic M_2 receptors. The initial evidence for an allosteric interaction by gallamine was that the Arunlakshana and Schild plots at myocardial muscarinic M_2 receptors exhibited low slopes and curvilinear properties at high antagonist concentrations (50). An allosteric interaction was subsequently shown in radioligand binding studies, where (53) gallamine acted in a negative allosteric fashion or, at least, exhibited a combination of competitive and allosteric antagonism. Similar observations have since been made for stercuronium (54), methoctramikne (55), $C_7/3'$-phthalimidopropyl [heptane-1,7-bis-(dimethyl-3'-phthalimidopropy) ammonium bromide] (56), and himbacine (57).

Since the allosteric properties have been observed on solubilized receptors, and in recombinant systems, it has been suggested (31) that site for these allosteric interactions is near the orthosteric binding site for competitive ligands, although possibly at a more extracellular location. Ligands such as gallamine may create steric hindrance, thus blocking access of the ligand to the orthostatic site. Two other features of allosteric antagonists are notable. First, while compounds such as gallamine *decrease* the binding of [^{3}H] NMS (58), others, such as alcuronium, *increase* the affinity of [^{3}H] NMS (59). Interestingly, the interaction of alcuronium is ligand-specific, in that it is positively allosteric when interacting with [^{3}H] NMS, 4-DAMP and atropine, yet negatively allosteric when interacting with [^{3}H] QNB (59). The structural requirements for determining a positive or negative interaction are, however, unknown (31). A second feature is that the occurrence of allosterism is most pronounced at the muscarinic M_2 receptor, although it can occur, to a lesser degree, at all five subtypes. The reasons for the high sensitivity of the muscarinic M_2 receptor to the phenomenon remains unclear. More importantly, a consequence of allosterism is that the selectivity of the antagonist can be enhanced. This may provide an avenue to novel leads, important for receptor classification and potentially as therapeutics.

IV. IRREVERSIBLE ANTAGONISTS

The lack of subtype-selective muscarinic agonists mandates that alternative approaches be used to study the function of single muscarinic receptor subtypes in cells or tissues expressing more than one receptor. Potentially, selective receptor inactivation allows the use of a nonselective agonist of high intrinsic efficacy to stimulate one muscarinic receptor subtype in pharmacological "isolation."

Several β-haloakylamines, such as dibenamine (N,N-dibenzyl-β-choroethylamine) and phenoxybenzamine, have been used to alkylate numerous neurotransmitter receptor (60,61). Phenoxybenzamine remains a useful compound to estimate muscarinic agonist affinities (62,63) and, under conditions of selective receptor protection, can be used to selectively inactivate muscarinic receptor subtypes (61). One disadvantage of phenoxybenzamine is that it also alkylates other neurotransmitter receptors, including adrenergic and histaminergic receptors (61). N-Ethoxycarbonyl-2-ethoxy-1,2-dihydroquinoline (EEDQ) is also an irreversible antagonist of muscarinic receptors and many monoamine neurotransmitter receptors (61). Radioligand-binding studies using [^{3}H] NMS have shown that EEDQ discriminates between M_1 and M_2 (as defined by those sites exhibiting high and low affinities for pirenzepine, respectively) receptors (64,65). Overall, however, the selectivity between muscarinic receptors subtypes was modest and few investigators have used EEDQ in functional studies to inactivate muscarinic receptors.

Alkylating agents that preferentially interact at muscarinic receptors have also been developed. To this end, a mustard group has been incorporated into several antagonists, including benzilylcholine and propylbenzilylcholine (66,67). With the possible exception of propylbenzilylcholine mustard, none of these compounds have meaningful selectivity between muscarinic receptor subtypes. Propylbenzilylcholine mustard is selective for muscarinic receptors (68). However, propylbenzilylcholine mustard shows little, if any, differential affinity (69) between muscarinic receptor subtypes. This apparent lack of selectivity has, however, been exploited in assays designed to estimate the "receptor reserve" associated with a particular cellular response (e.g., Ref. 70). McLeskey and Wojcik (71), in contrast, have argued that [^{3}H] propylbenzilylcholine mustard *was* selective for the M_2 over M_3 receptors in cultured cerebellar granule cells, stating that previous estimations of reserve associated with these subtypes may be misleading. Takayanagi et al. (72,73) have suggested that propylbenzilylcholine mustard discriminated between two muscarinic receptors in guinea pig smooth muscle. However, these putative "propylbenzilylcholine mustard sensitive and insensitive" receptors were not differentiated by the relatively selective muscarinic M_2 and M_3 antagonists AF-DX 116 or 4-DAMP, respectively. Consequently, the relationship of these sites to muscarinic M_1, M_2, M_3, or M_4 receptors remains obscure.

As described above, the reversible antagonist 4-DAMP has been extensively used in the characterization of muscarinic receptors, as it possesses a 10–20-fold functional selectivity toward M_3 over M_2 receptors (74). However, it does not discriminate between functional M_1 and M_3 receptors (75). A mustard analog of 4-DAMP, 4-diphenylacetoxy-N-(2-chloroethyl)piperidine (4-DAMP mustard), preferentially alkylates guinea pig ileal M_3 receptors in comparison to atrial M_2 receptors (76,77). Radioligand-binding studies in rat cerebral cortex showed that 4-DAMP mustard alkylated M_1, M_2, and M_4 receptors to the same extent (78). Moreover, 4-DAMP mustard also inactivated a population of muscarinic receptors inaccessible to the hydrophilic ligand, [^{3}H] NMS, but accessible to the hydrophobic ligand, [^{3}H] QNB (78).

At muscarinic M_2 and M_3 receptors in guinea pig isolated atria and ileum, respectively, 4-DAMP mustard caused greater dextral shifts in concentration-response curves to carbachol in ileum compared to atria (79). However, as pointed out by Thomas et al. (77), since no washout of 4-DAMP mustard occurred, these shifts were a resultant of both irreversible and reversible antagonism and thus the estimation of the true selectivity of the compound for M_3 over M_2 receptors was ambiguous. Indeed, when 4-DAMP mustard was used under similar conditions in these *with washout,* the dextral shifts were less pronounced (77). However, selective alkylation of muscarinic M_3 receptors was achieved using low concentrations for relatively long periods (77). Overall, these properties suggest that, under appropriate experimental condition, 4-DAMP mustard can be usefully employed to selectively inactivate muscarinic M_3 receptor subtypes.

An analog of 4-DAMP mustard, N-(2-bromoethyl)-4-piperidinyl diphenylacetate (4-DAMP bromo mustard) (80), has been developed to facilitate M_2 and M_3 receptor inactivation studies in vivo. 4-DAMP mustard is of limited use in vivo, since the rate of formation of the aziridinium ion is slow and effects the tissue distribution. In contrast, the formation of the aziridinium ion from 4-DAMP bromo mustard was virtually instantaneous (80) and the CNS penetration of this ion is low. The in vivo used of 4-DAMP bromo mustard may thus enable the selective inactivation of muscarinic M_3 receptors in the periphery.

Attempts to synthesize selective alkylating agents of M_1, M_2, or M_4 receptors have so far been unsuccessful. Baumgold et al. (81) utilized isothiocyanate derivatives of telenzepine and pirenzepine in an attempt to synthesize selective inactivating antagonists of the M_1 receptor. An isothiocyanate analog of the M_1 receptor antagonist, telenzepine, was only fivefold selective for the M_1 (cerebral cortex) over M_2 (myocardium) receptors and no data were given regarding affinity at muscarinic M_3 or M_4 receptors (81).

Various benzilylcholine mustard analogs have been modified to incorporate a photoaffinity label (82). The potential advantage of such compounds is that, until activation by light, antagonism of the receptor is reversible and characterization of the compounds can be undertaken. Moreover, the ability to vary the extent of alkylation, by varying the exposure to light, suggests that they may be useful experimental tools. It is therefore surprising that relatively little work, particularly of a functional nature, has been done with these compounds. [^{3}H] Azido-4-N-methyl-4-piperidyl benzilate ([^{3}H]-azido-4-NMPB, 82) labels muscarinic receptors and could prove useful in characterizing residues essential for ligand-binding (83). Photoaffinity ligands based upon reversible antagonists selective for muscarinic receptor subtypes, such as pirenzepine, methoctramine, tripitramine, or p-FHHSiD, are not available.

V. THERAPEUTIC USES OF SELECTIVE MUSCARINIC RECEPTOR ANTAGONISM

A. Nonselective Muscarinic Antagonists

Classic muscarinic receptor antagonists, such as atropine, do not distinguish between muscarinic receptor subtypes (3). Their therapeutic utility is limited by the occurrence of side effects including mydriasis, xerostomia, CNS disturbances, tachycardia, and constipation. In certain circumstances, however, the nonselective nature of these antagonists may be desirable. For example, atropine and hyoscine are used as premedication for surgical procedures in which blockade of salivary secretion, smooth muscle contraction, along with some degree of cardioprotection, is desirable. However, the nonselective nature of these compounds limits their therapeutic efficacy in chronic treatment of diseases. A summary of several muscarinic antagonists, either in development or commercially launched, is shown in Tables 4 and 5.

B. Muscarinic M$_1$ Antagonists

Pirenzepine, an antagonist with relatively high affinity for the muscarinic M$_1$ and modest affinity for the muscarinic M$_4$ receptor, is approved for clinical use in the treatment of peptic ulcer disease (84). Structurally related compounds in clinical development include telenzepine (13) and nuvenzepine (85). It is arguable that a selective muscarinic M$_3$ receptor antagonist may be useful in the treatment of peptic ulcer disease, given the role of this subtype in regulating parietal cell secretion (86). Although this aspect of muscarinc receptor antagonism has not been extensively explored in clinical studies, nuvenzepine acts as a potent muscarinic M$_3$ antagonist.

Table 4 Some Muscarinic Receptor Antagonists Under Development

Compound	Phase	Receptor selectivity	Indication	Company
Alimentary tract				
Telenzepine Guilden	Pre-registration	M_1/M_4	Antiulcer	Byk
Cimetropium	Launched	NS	Antispasmolytic	Boehringer Ingelheim
Pinaverine	Launched	NS	Antispasmolytic	Solvay
Timepedium	Launched	NS	Antispasmolytic	Tanabe
Tiropramide	Launched	NS	Antispasmolytic	Rotta Research
Trospium	Launched	NS	Antispasmolytic	Madeus
Antiarrythmic				
Otenzepad (AF-DX 116)	Phase II	M2/M4	Bradycardia	Boehringer Ingelheim
Genitourinary				
Cimetropium	Launched	NS	Inhibition of Labour	Boehringer Ingelheim
Darifenacin	Phase III	M_3/M_1	Irritable bowel syndrome	Pfizer
Tolterodine	Phase III	NS	Urinary Incontinence	Pharmacia—Upjohn
Respiratory				
Moguisteine	Launched	NS	Antitussive	Boehringer Mannheim
Tiotropium	Phase II	NS	Antiasthma	Boehringer Ingelheim
Rispenzepine	Phase II	M_3/M_1	Antibrochospastic	Dompe

NS, nonselective.
Source: From Ref. 13.

Table 5　Muscarinic Antagonists Launched

Compound	Phase	Receptor selectivity	Indication	Company
Alimentary tract				
Fenoverine	Launched	NS	Antispasmodic	Synthelabo
Triquizium	Launched	NS	Antispasmodic	Hokuriku
Aclatonium	Launched	NS	Antispasmodic	Toyama
Mebeverine	Launched	NS	Antispasmodic	Reckitt & Coleman
Milverine	Launched	NS	Antispasmodic	Rhone-Poulenc
Otilionium	Launched	NS	Antispasmodic	Menarini
Pramiverine	Launched	NS	Antispasmodic	E. Merck
Rociverine	Launched	NS	Antispasmodic	Guidotti
Tiemonium	Launched	NS	Antispasmodic	AKZO
Trimebutine	Launched	NS	Antispasmodic	Joveinal
Tripibutone	Launched	NS	Antispasmodic	Takeda
Genitourinary				
Rociverine	Launched	NS	Antispasmodic	Guidotti
Flavoxate	Launched	NS	Antispasmodic	Recordati
Respiratory				
Flutropium	Launched	NS	Antitussive	Boehringer Ingelheim
Oxitropium	Launched	NS	Bronchodilator	Boehringer Ingelheim

NS, nonselective.
Source: From Ref. 13.

C.　Muscarinic M_2 Antagonists

Other selective muscarinic receptor antagonists include a lipophilic, centrally acting muscarinic M_2 receptor antagonist, BIBN 99 (87). This compound may be useful in the treatment of Alzheimer's disease, since it could reverse the autoinhibitory control of acetylcholine release (87). Although this hypothesis remains to be clinically proven, studies in aged rats suggest that BIBN 99 augments acetylcholine release (88). Conversely, peripherally acting muscarinic M_2 receptor antagonists, such as AF-DX 116 (Otenzepad), may be useful in the treatment of A-V block (89) or bradycardia occurring after acute myocardial infarction (90).

D.　Muscarinic M_3 Antagonists

An important indication for muscarinic receptor antagonists is to relax smooth muscle, with the degree of relaxation produced depending, to a large extent, upon the level of preexisting parasympathetic nervous tone. In general, muscarinic M_1 receptors are present on parasympathetic ganglia, located in the effector organ, where they serve to modulate cholinergic transmission. At the end-organ terminals, the release of acetylcholine is modulated, usually in an inhibitory fashion, by either a prejunctional M_2, M_3, or M_4 muscarinic autoreceptor.

Contractile responses of smooth muscle are mediated by activation of postjunctional muscarinic receptors, the nature of which varies according to species and anatomical location. Muscarinic M_1 receptors, for example, mediate contraction of canine femoral and saphenous veins (91), while muscarinic M_2 receptors mediate contraction of guinea pig myometrium (92,93). In general, however, muscarinic M_3 receptors mediate contraction of most types of smooth muscle, including guinea pig ileum (94). Surprisingly, these postjunctional muscarinic M_3 receptors are present in low numbers (25% or less), with many smooth muscles processing a preponderant muscarinic M_2 receptor population (94). Selective blockade of muscarinic M_3 receptors may be therapeutically useful in the treatment of respiratory disorders, such as chronic obstructive airway disease (95), gastrointestinal disorders, such as irritable bowel syndrome (40), and urinary tract disorders, such as urge incontinence (96,97). In terms of the former, stimulation of cholinergic nerves provides the major bronchoconstriction control of animal and human airways.

1. Respiratory Tract

Vagal stimulation induces bronchoconstriction and mucus secretion, by activation of muscarinic receptors located on smooth muscle, vascular endothelium, submucosal cells, and neural elements (98,99). Since cholinergic neural mechanisms may contribute to airway narrowing in asthma and chronic obstructive airway disease, muscarinic receptor antagonists are effective in treating acute bronchoconstriction, particularly that occurring in chronic obstructive airway disease (100). Antagonists currently in use for the treatment of this condition are not selective, with the potential shortcomings discussed above. Moreover, a nonselective muscarinic receptor antagonist may also cause a paradoxical bronchoconstriction, due to concurrent antagonism of prejunctional muscarinic autoreceptors, thereby reducing the effectiveness of postjunctional muscarinic M_3 receptor blockade (100).

In lieu of genuinely selective muscarinic M_3 receptor antagonists, some therapeutic approaches to selective blockade exploit differences in receptor kinetics or absorption. Tiotropium bromide (BA 679 BR), for example, is an antagonist with a preferential slow off-rate from muscarinic M_3 receptors with respect to muscarinic M_2 receptors (101,102). Ipratropium, alternatively, is a quaternized derivative of atropine, and poorly absorbed into the systemic circulation when given by inhalation (103). Although it is

nonselective between subtypes, the low systemic absorption of the antagonist facilitates selective antagonism of airway muscarinic receptors.

2. Gastrointestinal and Lower Urinary Tract

In terms of inducing gastrointestinal smooth muscle relaxation, several
relatively old compounds are available including pinaverium, cimetropium,
fendoverine, mebeverine, and milverine. Although lacking selectivity for
muscarinic M_3 receptors, they possess many other properties, such as calcium channel blockade, a property that will contribute to the antispasmodic
actions of the drug. Two other examples of compounds with muscarinic
antagonist properties and nonspecific muscle relaxant properties are terodi-
line (now withdrawn due to the occurrence of Torsades de Pointes) and
imipramine (104). Although these properties may result in preferential
action on smooth muscle, newer compounds, with selectivity for the muscarinic M_1 and M_3 receptor over the M_2 or M_4 receptors, and lacking nonspecific effects, have been clinically evaluated. These include zamifenacin (40),
darifenacin (41), and vamicamide (105).

Currently, muscarinic antagonists are considered front-line therapy for
the treatment of detrusor instability or urge incontinence. Of these, oxybutynin and propantheline are recommended as front- and second-line therapy. Selective muscarinic receptor antagonists may offer therapeutic advantages over these agents in the treatment of urge incontinence. Darifenacin,
for example, is a selective muscarinic M_3 receptor antagonist in phase III
clinical trial for urinary incontinence and irritable bowel syndrome
(106,107). Vamicamide is also a novel compound under development for
the treatment of this condition (105). It is selective toward muscarinic M_1
and M_3 receptors over M_2 receptors in vitro. In vivo, vamicamide dose-
dependently inhibits spontaneous bladder contractions caused by elevations
in the intravesical volume. At 3–10-fold higher doses, no effect was seen
on the contractions of stomach or colon, responses also mediated via activation of muscarinic M_3 receptors (105). A feature of this and other compounds (108,109), including analogs of oxybutynin (110), is that the selectivity for bladder smooth muscle blockade in vivo is more than predicted from
in vitro studies. The reason fo this disparity may be pharmacokinetic, since
compounds such as vamicamide are concentrated in the urinary bladder,
thereby exerting a localized, antispasmodic action (105). Interestingly, a
specific uptake system for quaternary ammonium compounds has also been
identified in mouse urinary bladder, a phenomenon that may explain a
tissue-specific action in this species. However, it is unknown if this system
operates in other smooth muscle types, such as human detrusor muscle
(111), or to what extent it affects antagonist potency in vivo.

VI. CONCLUSION

Muscarinic receptor subtypes, when defined pharmacologically, correspond with those identified on the basis of sequence. Several antagonists (atropine, pirenzepine, methoctramine, p-F-HHSiD, himbacine, and tripitramine) are available that, when used in concert, allow good operational definition of a muscarinic receptor subtype.

Unequivocally selective alkylating compounds for M_1, M_2, M_3, and M_4 receptors are presently unavailable and, lacking subtype selective agonists, it is difficult to examine the function of a single subtype in tissues possessing heterogeneous populations. Nonetheless, 4-DAMP mustard has some use in this respect, although it has only a modest selectivity for M_3 over M_2 receptors. This can be enhanced with concurrent receptor protection with a reversible antagonist such as methoctramine.

Therapeutically, selective M_1 antagonists, such as pirenzepine, are useful in reducing gastric acid secretion. Selective M_2 antagonists may prove useful in the treatment of bradycardia or, if centrally acting, Alzheimer's disease. In at least three areas of smooth muscle pathology (chronic obstructive airway disease, irritable bowel syndrome, and urge incontinence), selective muscarinic M_3 receptor antagonism may be of therapeutic benefit. Although such antagonists remain to be identified, several compounds with muscarinic M_3/M_1 over M_2 receptor selectivity are now in advanced clinical development.

REFERENCES

1. Dale HH. The action of certain esters and ethers of choline and their relation to muscarine. J Pharmacol Exp Ther 1914; 6:147–190.
2. Burgen ASV. The background of the muscarinic system. Life Sci 1995; 56:801–806.
3. Caulfield MD. Muscarinic receptors—characterization, coupling and function. Pharmacol Ther 1993; 58:319–379.
4. Hulme EC, Birdsall NJM, Buckley NJ. Muscarinic receptor subtypes. Annu Rev Pharmacol Toxicol 1990; 30:633–673.
5. Hosey MM. Diversity of structure, signaling and regulation within the family of muscarinic cholinergic receptors. FASEB J 1992; 6:845–852.
6. Kenakin TP, Bond RA, Bonner TI. Definition of pharmacological receptors. Pharmacol Rev 1992; 44:351–362.
7. Brann MR, Ellis J, Jorgensen H, Hill-Eubanks D, Jones SV. Muscarinic acetylcholine receptor subtypes: localization and structure/function. Prog Brain Res 1993; 98:121–127.
8. Fukuda K, Kubo T, Maeda A, Akiba I, Bujo H, Nakai J, Mishina M, Higashida K, Neher E, Marty A, Numa S. Selective effector coupling of muscarinic

acetylcholine receptor subtypes. Trends Pharmacol Sci 1989; 4(Suppl IV):4–10.

9. Dörje F, Wess J, Lambrecht G, Tacke R, Mutschler E, Brann MR. Antagonist binding profiles of five cloned human muscarinic receptor subtypes. J Pharmacol Exp Ther 1991; 256:727–733.

10. Guo ZD, Kameyama K, Rinken A, Haga T. Ligand-binding properties of muscarinic acetylcholine receptor subtypes (m1-m5) expressed in baculovirus-infected insect cells. J Pharmacol Exp Ther 1995; 274:378–384.

11. Richards MH. Pharmacology and second messenger interactions of cloned muscarinic receptors. Biochem Pharmacol 1991; 42:1645–1653.

12. Mitchelson F. Muscarinic receptor differentiation. Pharmacol Ther 1988; 37:357–423.

13. Eglen RM, Watson N. Selective muscarinic receptor agonists and antagonists. Pharm Toxicol 1996; 78:59–68.

14. Hou X, Wehrle J, Menge W, Ciccarelli E, Wess J, Mutschler E, Lambrecht C, Timmerman H. Waelbroeck M. Influence of monovalent cations on the binding of a charged and an uncharged ("carbo"-) muscarinic antagonist to muscarinic receptors. Br J Pharmacol 1996; 117:955–961.

15. Schudt C, Boer R, Eltze M, Reidel R, Grundler G, Birdsall NJ. The affinity, selectivity and biological activity of telenzepine enantiomers. Eur J Pharmacol 1989; 165:87–96.

16. Hudkins RL, Stubbins JF, DeHaven-Hudkins DL. Caramiphen, iodocaramiphen and nitrocaramiphen are potent, competitive muscarinic M1 receptor-selective agents. Eur J Pharmacol 1993; 231:485–488.

17. Ghelardini C, Bartolini A, Galeotti N, Yeodori E, Gualtieri F. S-(−)-ET 126: a potent and selective M1 antagonist in vitro and in vivo. Life Sci 1996; 58:991–1000.

18. Frost JJ. Receptor imaging by positron emission tomography and single-photon emission computed tomography. Invest Radiol 1992; 27(Suppl 2):S54–S58.

19. Asahini M, Shinotoh H, Hirayama K, Suhara T, Shishido F, Inoue O, Tateno Y. Hypersensitivity of cortical muscarinic receptors in Parkinson's disease demonstrated by PET. Acta Neurol Scand 1995; 91:437–443.

20. Jean JC, Davis RE. Recent advances in the design and characterization of muscarinic agonists and antagonists. Annu Rev Med Chem 1994; 29:23–32.

21. Hammer R, Giraldo E, Schiavi GB, Monferini E, Ladinsky H. Binding profile of a novel cardioselective muscarinicreceptor antagonist, AF-DX 116, to membranes of peripheral tissues and brain in the rat. Life Sci 1986; 38:1653–1662.

22. Doods H, Entzeroth M, Mayer N. Cardioselectivity of AQ-RA 741, a novel tricyclic antimuscarinic drug. Eur J Pharmacol 1991; 192:147–152.

23. Ferrari-Dileo G, Waelbroeck M, Mash DC, Flynn DD. Selective labeling and localization of the M4 (m4) muscarinic receptor subtype. Mol Pharmacol 1994; 46:1028–1035.

24. Doods HN, Willim KD, Boddeke HWGM, Enzteroth M. Characterization of muscarinic receptors in guinea-pig uterus. Eur J Pharmacol 1993; 250:223–230.

25. Eglen RM, Harris GC, Cox H, Sullivan AO, Stefanich E, Whiting RL. Characterization of the interaction of the cervane alkaloid imperialine at muscarinic receptors in vitro. Naunyn-Schmiedeberg's Arch Pharmacol 1992; 346:144–149.

26. Baumgold J, Pryzbyc RL, Reba RC. 3-α-Cloroimperialine, an M2 selective muscarinic receptor antagonist that penetrates into the brain. Eur J Pharmacol 1994; 251:315–317.

27. Pfaff O, Hildebrandt C, Waelbroeck M, Hou X, Moser U, Mutschler E, Lambrecht, G. The (S) (+) enantiomer of dimethindene: a novel M2 selective muscarinic receptor antagonist. Eur J Pharmacol 1995; 286:229–240.

28. Melchiorre C, Bolognesi ML, Chiarini A, Minarino A, Spampinato S. Synthesis and biological activity of some methoctramine-related tetraamines bearing a 11-acetyl-5,11-dihydro-6H-pyrido[2,3-b][1,4]-benzodiazepin-6-one moiety as antimuscarinics: a second generation of highly selective M_2 muscarinic receptor antagonists. J Med Chem 1993; 36:3734–3737.

29. Chiarini A, Budresi R, Bolognesi ML, Minarini A, Melchiorre C. In vitro characterization of tripitramine, a polymethylene tetraamine displaying high selectivity and affinity for muscarinic M_2 receptors. Br J Pharmacol 1995; 114:1507–1517.

30. Angeli P, Cantalamessa F, Gulini U, Melchiorre C. Selective blockade of muscarinic M2 receptors in vivo by the new antagonist tripitramine. Naunyn-Schmiedeberg's Arch Pharmacol 1995; 352:304–307.

31. Tucek S, Proska J. Allosteric modulation of muscarinic acetylcholine receptors. Trends Pharmacol Sci 1995; 16:205–212.

32. Barlow RB, Shepherd MK, A search for selective antagonists at M2 muscarinic receptors. Br J Pharmacol 1985; 85:427–435.

33. Barlow RB, Bond S, Holdup DW, McQueen DS, Veale MA, Smith TW, Stephenson GW, Batsanov AS. "FourDAPines," a new class of ileo-selective antimuscarinic drugs. Br J Pharmacol 1992; 106:40P.

34. Caulfied MP, Palazzi E, Lazareno SH, Jones S, Popham A, Birdsall NJM. Lack of significant selectivity of "benzyl-4-DAPine" between four muscarinic receptor subtypes, binding and functional studies. Br J Pharmacol 1993; 108:29P.

35. Barlow RB, Bond SM, Branthwaite AG, Jackson O, McQueen DS, Smith KM, Smith PJ. Selective blockade of M2 and M_3 muscarinic receptors by hexhydrobenzyl-fourdapine and a comparison with zamifenacin. Br J Pharmacol 1995; 116:2897.

36. Ehlert FJ, Oliff HS Griffen MT. The quaternary transformation products of N-(3-chloropropyl)-4-piperidinyl diphenylacetate and N-(2-chloroethyl)-4-piperidinyl diphenylacetate (4-DAMP mustard) have differential affinity for subtypes of the muscarinic receptor. J Pharmacol Exp Ther 1996; 276:405–410.

37. Wallis RM, Alker D, Burgess RA, Cross PE, Newgreen DT, Quinn P. Zamifenacin: a novel gut selective muscarinic receptor antagonist. Br J Pharmacol 1993; 109:36P.

38. Watson N, Reddy H, Stefanich E, Eglen RM. Characterization of the interaction of zamifenacin at muscarinic receptors in vitro. Eur J Pharmacol 1995; 285:135–142.

39. McRitchie B, Merner PA, Dodd MG. In vivo selectivity of the novel muscarinic antagonist, zamifenacin, in the conscious dog. Br. J Pharmacol 1993; 109:38P.

40. McIntyre P, Quinn P. Characterisation and comparison of muscarinic receptors in the dog ciliary muscle with ileum. Br J Pharmacol 1993; 115:139P.

41. Wallis RM. Preclinical and clinical pharmacology of selective muscarinic M_3 receptor antagonists. Life Sci 1995; 56:861–868.

42. Wallis RM, Burges RA, Cross PE, MacKenzie AR, Newgreen DT, Quinn P. Darifenacin, a selective muscarinic M_3 antagonist. Pharmacol Res 1995; 31(Suppl):54.

43. Lambrecht G, Feifel T, Forth B, Strohmann C, Tacke R, Mutschler E. p-Fluoro-hexahydro-sila-difenidol: the first $M_{2\beta}$-selective muscarinic antagonist. Eur J Pharmacol 1988; 152:193–194.

44. Eglen RM, Cornett CM, Whiting RL. Interaction of pFHHSID at tracheal muscarinic receptors in vitro. Naunyn-Schmiedeberg's Arch Pharmacol 1990; 342:394–399.

45. Roffel AF, Hamstra JJ, Elzinga CRS, Zaagsma J. Selectivity profile of some recent muscarinic antagonists in bovine and guinea-pig trachea and heart. Arch Int Pharmacodyn 1994; 328:82–98.

46. Kerr PM, Miller K, Wallis RM, Garland CJ. Characterization of muscarinic receptors mediating contractions of circular and longitudinal muscle of human colon. Br J Pharmacol 1995; 115:1518–1524.

47. Watson N, Magnussen H, Rabe KF. Pharmacological characterization of the muscarinic receptor subtype mediating contraction of human peripheral airways. J Pharmacol Exp Ther 1995; 274:1293–1297.

48. Jakubik J, Bacakova L, El-Fakahany EE, Tucek S. Constitutive activity of the M1-M4 subtypes of muscarinic receptors in transfected CHO cells and of muscarinic receptors in the heart cells revealed by negative antagonists. FEBS Lett 1995; 377:275–279.

49. Spalding TA, Burstein ES, Brauner-Osborne H, Hill-Eubanks D, Brann MR. Pharmacology of a constitutively active muscarinic receptor generated by random mutagenesis. J Pharmacol Exp Ther 1995; 275:1274–1279.

50. Jolkkonen M, van Giersbergen PLM, Hellman U, Wernstadt C, Karlsson E. A toxin from the green mamba *Dendroaspis augusticeps:* amino acid sequence and selectivity for the muscarinic M_4 receptors. FEBS Lett 1994; 352:91–94.

51. Zorn SH, Jones SB, Ward KM, Liston DR. Clozapine is a potent and selective muscarinic M_4 agonist. Eur J Pharmacol 1994; 269:R1–R2.

52. Clark AL, Mitchelson F. The inhibitory effect of gallamine on muscarinic receptors. Br J Pharmacol 1976; 58:323–331.

53. Stockton J, Birdsall NJM, Burgen ASV, Hulme EC. Modification of the binding properties of muscarinic receptors by gallamine. Mol Pharmacol 1983; 23:551–557.

54. Li CK, Mitchelson F. The selective antimuscarinic action of stercuronium. Br J Pharmacol 1980; 70:313–321.

55. Eglen RM, Montgomery WW, Dainty IA, Dubuque LK, Whiting RL. The interaction of methoctramine and himbacine at atrial, smooth muscle and endothelial muscarinic receptors in vitro. Br J Pharmacol 1988; 95:1031–1038.

56. Christopoulos A, Mitchelson F. Assessment of the allosteric interactions of the bisquaternary heptane-1,7-bis(dimethyl-3'-phthalimidopropyl)ammonium bromide at M1 and M2 muscarinic receptors. Mol Pharmacol 1994; 46:105–114.

57. Lee NH, El-Fakahany EE. The allosteric binding profile of himbacine: a comparison with other cardioselective muscarinic antagonists. Eur J Pharmacol 1990; 179:225–229.

58. Dunlap J, Brown JH. Heterogeneity of binding sites on cardiac muscarinic receptors induced by the neuromuscular blocking agents, gallamine and pancuronium. Mol Pharmacol 1983; 24:15–22.

59. Proska J, Tucek S. Competition between positive and negative allosteric effectors on muscarinic receptors. Mol Pharmacol 1995; 48:696–702.

60. Nickerson M. Receptor occupancy and tissue response. Nature 1956; 178:697–698.

61. Furchgott RF, Bursztyn P. Comparison of dissociation constants and of relative efficacies of selected agonists acting at parasympathetic receptors. Ann NY Acad Sci 1967; 144:882–898.

62. Leff P, Dougall IG, Harper D. Estimation of partial agonist affinity by interaction with a full agonist: a direct operational model-fitting approach. Br J Pharmacol 1993; 110:239–244.

63. Eglen RM, Harris GC. Selective inactivation of muscarinic M_2 and M_3 receptors in guinea-pig ileum and atria in vitro. Br J Pharmacol 1993; 109:946–952.

64. Chang KJ, Moran JF, Triggle DJ. Mechanism of cholinergic antagonism by N-ethoxycarbonyl-2-ethoxy-1,2-dihydroquinoline (EEDQ). Pharmacol Res Commun 1970; 2:63–66.

65. Norman, AB, Creese I. Effects of in vivo and in vitro treatments with N-carbonyl-2-ethoxy-1,2-dihydroquinoline on putative muscarinic receptor subtypes in rat brain. Mol Pharmacol 1986; 30:96–103.

66. Gill EW, Rang HP. An alkylating derivative of benzilylcholine with specific and long lasting parasympatholytic activity. Mol Pharmacol 1966; 2:284–297.

67. Young JM, Hiley R, Burgen ASV. Homologues of benzilylcholine mustard. J Pharm Pharmacol 1972;24:950–954.

68. Burgen ASV, Hiley CR, Young JM. The binding of [^{3}H]-propylbenzilylcholine mustard by longitudinal strips from guinea-pig small intestine. Br J Pharmacol 1974; 50:141–151.

69. Norman AB, Eubanks JH, Creese I. Irreversible and quaternary muscarinic antagonists discriminate multiple muscarinic receptor binding sites in rat brian. J Pharmacol Exp Ther 1989; 248:1116–1122.

70. Brown JH, Goldstein D. Differences in muscarinic receptor reserve for inhibition of adenylate cyclase and stimulation of phosphoinositide hydrolysis in chick heart cells. Mol Pharmacol 1986; 30:566–570.

71. McLeskey SW, Wojcik WJ. Propylbenzilylcholine mustard has greater specificity for muscarinic M_2 receptors than for M_3 receptors present in cerebellar granule cell culture from rat. J Pharmacol Exp Ther 1992; 263:703–707.

72. Takayanagi I, Hisayama T, Kiuchi Y, Sudo H. Propylbenziylcholine mustard discriminates between two subtypes of muscarinic cholinoceptors in guinea-pig taenia-caecum. Arch Int Pharmacodyn 1989; 298:210–219.

73. Takayanagi I, Koike K, Saito K. Propylbenziylcholine mustard-sensitive and resisitant muscarinic receptors in cardiac muscle. Gen Pharmacol 1991; 22:691–694.

74. Barlow RB, Berry KJ, Glenton PAM, Nikolaou NN, Soh KS. A comparison of affinity constants for muscarine-sensitive acetylcholine receptors in guinea-pig atrial pace-maker cells at 29°C and in ileium at 29°C and 37°C. Br J Pharmacol 1976; 58:613–620.

75. Brown DA, Forward A, Marsh S. Antagonist discrimination between ganglionic and ileal muscarinic receptors. Br J Pharmacol 1980; 71:362–364.

76. Barlow RB, Shepherd MK, Veale MA. Some differential effects of 4-diphenylacetoxy-N-(2-chloroethyl)-piperidine hydrochloride on guinea-pig atria and ileum. J Pharm Pharmacol 1990; 42:412–418.

77. Thomas EA, Hsin HH, Griffin MT, Hunter AL, Luong T, Ehlert FJ. Conversion of N-(2-chloroethyl)-4-piperidinyl diphenylacetate (4-DAMP mustard) to an aziridinium ion and its interaction with muscarinic receptors in various tissues. Mol Pharmacol 1992; 41:718–726.

78. Waelbroeck M, Renzetti A-R, Tastenoy M, Barlow RB, Christophe J. Inactivation of brain cortex muscarinic receptors by 4-diphenylacetoxy-1-(2-chloroethyl)piperidine mustard. Biochem Pharmacol 1992; 44:285–290.

79. Eglen RM, Harris G. Muscarinic receptor protection studies in isolated functional preparations. Life Sci 1993; 52:571 (abstract 43).

80. Griffin MT, Thomas EA, Ehlert FJ. Kinetics of activation and in vivo muscarinic receptor binding of N-(2-bromoethyl)-4-piperidinyl diphenylacetate: an analog of 4-DAMP mustard. J Pharmacol Exp Ther 1993; 266:301–305.

81. Baumgold J, Karton Y, Malka N, Jacobson KA. High affinity acylating antagonists for muscarinic receptors. Life Sci 1992; 51:345–351.

82. Sokolovsky M. Photoaffinity labeling of muscarinic receptors. Pharmacol Ther 1987; 32:285–292.

83. Brann MR, Ellis J, Jorgensen H, Hill-Eubanks D, Jones SV. Muscarinic acetylcholine receptor subtypes: localization and structure/function. Prog Brain Res 1993; 98:121–127.

84. Carmine AA, Brogden RN. Pirenzepine. A review of its pharmacodynamic and pharmacokinetic properties and therapeutic efficacy in peptic ulcer disease and other allied diseases. Drugs 1985; 30:85–126.

85. Barocelli E, Ballabeni V, Chiavarini M, Molina E, Impiccatore M. Functional comparison between nuvenzepine and pirenzepine on different guinea-pig isolated smooth muscle preparations. Pharmacol Res 1994; 30:161–170.

86. Hirschowitz BI, Keeling D, Lewin M, Okabe S, Parsons M, Sewing K, Wallmark B, Sachs G. Pharmacological aspects of acid secretion. Dig Dis Sci 1995; 40:3S–23S.

87. Doods HN, Entzeroth M, Ziegler H, Schiavi G, Engel W, Mihm G, Rudolf K, Eberlein W. Characterization of BIBN 99, a lipophilic and selective muscarinic M2 receptor antagonist. Eur J Pharmacol 1993; 242:23–30.

88. Doods HN, Quirion R, Mihm G, Engel W, Rudolf K, Entzerof M, Sciavi GB, Ladinsky H, Bechtel WD, Ensinger HA, Mendla KD, Eberlein W.

Therapeutic potential of CNS-active M_2 antagonists: novel structures and pharmacology. Life Sci 1993; 52:497–503.

89. Schulte B, Volz-Zang C, Mutschler E, Horne C, Palm D, Wellstein A, Pitschner HF. AF-DX 116, a cardioselective muscarinic antagonist in humans: pharmacodynamic and pharmacokinetic properties. Clin Pharmacol Ther 1991; 50:372–378.

90. van Zwieten PA, Doods HN. Muscarinic receptors and drugs in cardiovascular medicine. Cardiovasc Drugs Ther 1995; 9:159–167.

91. Eglen RM, Whiting RL. Heterogeneity of vascular muscarinic receptors. J Auton Pharmacol 1990; 19:233–245.

92. Eglen RM, Michel AD, Whitting RL. Characterization of the muscarinic receptor subtype mediating contractions of the guinea-pig uterus. Br J Pharmacol 1989; 96:497–499.

93. Bognar IT, Altes U, Beinhauer C, Kessler I, Fuder H. A muscarinic receptor different from the M1, M2, M3, M4 subtypes mediates contraction of the rabbit iris sphincter. Nauntn-Schmiedeberg's Arch Pharmacol 1992; 345:611–618.

94. Ehlert FJ, Thomas EA. Functional role of M_2 muscarinic receptors in the guinea-pig ileum. Life Sci 1995; 56:965–971.

95. Gross NJ, Skorodin MS. Anticholinergic antimuscarinic bronchodilators. Am Rev Respir Dis 1984; 129:856–870.

96. Taira N. The autonomic pharmacology of the bladder. Annu Rev Pharmacol 1972; 12:197–208.

97. Andersson K-E. Pharmacology of lower urinary tract smooth muscles and penile erectile tissues. Pharmacol Rev 1993; 45:253–308.

98. Barnes PJ. Muscarinic receptor subtypes in airways. Life Sci 1993; 52:521–527.

99. White MV. Muscarinic receptors in human airways. J Allergy Clin Immunol 1995; 95:1065–1068.

100. Doods HN. Selective muscarinic antagonists as bronchodilators. Drug News Perspect 1992; 5:345–352.

101. Maesen FPV, Smeets JJ, Costongs MAL, Cornelissen PJG, Wald FDM. Ba 679 BR, a new long acting antimuscarinic bronchodilator: a pilot dose escalation study in COPD. Eur Respir J 1993; 6:1031–1036.

102. Haddad EB, Mak JC, Barnes PJ. Characterization of [^{3}H]Ba 679 Br, a slowly dissociating muscarinic antagonist, in human lung: radioligand-binding and autoradiographic mapping. Mol Pharmacol 1994; 45:899–907.

103. Lulich KM, Paterson JW, Goldie RG. Ipratropium, sodium chromoglycate and antihistamines. Med J Aust 1995; 162:157–159.

104. Hieble JP, McCafferty GP, Naselsky DP, Bergsma DJ, Ruffolo RR. Recent progress in the pharmacotherapy of diseases of the urinary tract. Eur J Med Chem 1995; 30:269–298.

105. Oyasu H, Yamamoto T, Sato N, Ozaki R, Mukai T, Ozaki T, Nishii T, Sato H, Fujisawa H, Tozuka Z, Koibuchi Y, Honbo T, Esumi K, Ohtsuka M, Shimomura M. Urinary bladder selective action of the new antimuscarinic compound vamicamide. Arzneim Forsch/Drug Res 1994; 44:1242–1249.

106. Swami P, Abrams P, the Darifenacin Study Group. Preliminary dose range study of darifenacin, a novel M3 antagonist in detrusor instability. Proc Int Continence Soc 1995; 48:117 (abstract).
107. Baert L, Leuven G, Dijkman B, the Darifenacin Study Group. Proc Int Continence Soc 1995; 226:214 (abstract).
108. Kaiser C, Spagnuolo CJ, Adams TC, Audia VH, Dupont AC, Hatoum H, Lowe VC, Prosser JC, Sturm BL, Noronha-Blob L. Synthesis and antimuscarinic properties of some substituted 5-(aminomethyl)-3,3-diphenyl-2(3H)-furanones. J Med Chem 1992; 35:4415–4424.
109. Kaiser C, Audia VH, Carter JP, McPherson DW, Waid PP, Lowe VC, Noronha-Blob L. 1-Cycloalkyl-1-hydroxy-1-phenyl-3-(4-substituted piperazinyl)-2-propanones and related compounds. J Med Chem 1993; 36:610–616.
110. Carter JP, Noronha-Blob L, Audia VH, Dupont AC, McPherson DW, Natalie KJ, Rzeszotarski WJ, Spagnuolo CJ, Waid PP, Kaiser C. Analogues of oxybutynin. Synthesis and antimuscarinic and bladder activity of some substituted 7-amino-1-hydroxy-5-heptyn-2-ones and related compounds. J Med Chem 1991; 34:3065–3074.
111. Durant PAC, Shankley NP, Welsh NJ, Black JW. Pharmacological analysis of agonist-antagonist interactions at acetylcholine muscarinic receptors in a new urinary bladder assay. Br J Pharmacol 1991; 104:145–150.

13

NMDA Receptor Antagonists

John A. Kemp and James N. C. Kew
F. Hoffmann-La Roche Ltd., Basel, Switzerland

I. INTRODUCTION

N-Methyl-D-aspartate (NMDA) receptors are a subtype of receptors for the major excitatory transmitter in the mammalian central nervous system (CNS), L-glutamate. Receptors for glutamate are divided into two major classes: 1.) ionotropic receptors, which are ligand-gated cation channels comprising the NMDA, AMPA, and kainate subtypes, and 2.) metabotropic receptors, which are a novel, G-protein-coupled receptor family of which eight subtypes have been cloned to date (1). Interest in the development of specific antagonists for NMDA receptors arose from the findings that these receptors play key roles not only in several important physiological functions, particularly synaptic plasticity (2), but also in neuropathological states such as epilepsy and acute neurodegeneration (3).

The concept of "excitotoxic" neuronal cell damage emerged during the 1970s with the finding that the neurotoxic potency of excitatory amino acids appeared to parallel their excitatory effects on neurons (4). The pattern of selective neuronal vulnerability following injections of excitatory amino acids into the brain and the "axon-sparing" nature of the damage was reminiscent of ischemic neuronal injury. Furthermore, microdialysis studies showed that during and following ischemia there was a massive release of glutamate and aspartate into the extracellular space (5,6). These findings led to the suggestion that the neuronal death caused by periods of cerebral ischemia was due, at least in part, to the overactivation of specific postsynaptic receptors by the excessive release of the endogenous excitatory amino acids, L-glutamate and L-aspartate. The first useful and selective glutamate receptor antagonists available to test this concept were the longer chain, D-amino acid ω-phosphonic acid analogs of glutamate developed in the early 1980s by Watkins and colleagues as NMDA receptor antagonists (7). Thus, D-2-amino-7-phosphonoheptanoic acid (D-AP7) was shown to ameliorate ischemia-induced hippocampal damage (8,9) and hypoglycaemia-induced striatal damage (10). A major step forward in the testing of the hypothesis in vivo came, however, with the discovery that MK-801 (dizocilpine), a potent, orally active anticonvulsant, was a selective, high-affinity, uncompetitive antagonist of NMDA receptors (11). The use of this and similar compounds in a variety of neurodegenerative models in a number of different species confirmed the key role played by NMDA receptors in mediating acute ischemic and traumatic brain injury (see Ref. 12). The marked neurotoxic potential of the NMDA receptor appears to result from its relatively high permeability to calcium (13), a known mediator of cell damage (14), its high affinity for glutamate, and its relative lack of desensitization during prolonged activation (15).

Over the last 15 years considerable effort from both academia and the pharmaceutical industry has been expended on the development of NMDA antagonists, and much progress has been made. A large number of competitive agonist recognition site antagonists and noncompetitive antagonists acting at a variety of sites on the receptor/ion channel complex have been developed (see below). However, blockade of NMDA receptors by these compounds, while potentially beneficial, produced profound CNS side effects and this has limited their therapeutic utility. In humans, these side effects range from light-headedness, dizziness, paresthesia, and agitation at low doses, through nystagmus, hallucinations, somnolence, and blood pressure increases at moderate doses, to catatonia and "dissociative anaesthesia" at high doses (see Ref. 16). Indeed, the work of Lodge and colleagues in the early to mid-1980s (17,18) demonstrated that the dissociative anesthetics phencyclidine and ketamine acted as NMDA antagonists. As a consequence of these side effects, most of the NMDA antagonists currently in clinical development are being considered primarily for the acute treatment of neuronal damage that results from cerebral ischemia following stroke or brain trauma. However, as outlined in the following, the development of subtype-specific and activity-dependent antagonists of NMDA receptors may result in compounds with sufficiently reduced side effects to be considered for chronic treatment.

II. MOLECULAR STRUCTURE OF NMDA RECEPTORS

Receptor cloning studies have identified five NMDA receptor subunits, NMDAR1 and four NR2 subunits, A–D, that are believed to assemble in various combinations to generate heteromeric assemblies predicted to contain five subunits (19). NMDAR1 was first cloned from rat brain (20) and subsequently from mouse brain (21), where it was named $\zeta 1$, and from human brain (22–24). The predicted mature NMDAR1 polypeptide is similar in size to AMPA and kainate receptor subunits and also shares the five hydrophobic domains originally thought to represent an amino-terminal signal peptide and four transmembrane domains (reviewed in Ref. 25). NMDAR1 is expressed widely throughout the brain, and homomers expressed in *Xenopus* oocytes are able to form a low level of functional NMDA receptors that are subject to the well-characterized voltage-dependent blockade by Mg^{2+} (20). Nine NMDAR1 splice variants have been detected including one truncated form that is unable to generate a functional receptor (reviewed in Ref. 25). NMDAR1 splice variants exhibit both distinct expression patterns and functional properties (26–29).

The NR2 subunits, which share around 20% homology with NMDAR1, were subsequently cloned from both rat and mouse (30–34) and were termed NR2A-D and $\varepsilon 1$–$\varepsilon 4$, respectively. Predicted mature NR2-subunit polypeptides also contain the characteristic five hydrophobic domains and differ principally from NMDAR1 in their much larger COOH-termini sequences. Notably, NR2 subunits are unable to form functional NMDA receptors alone; however, coexpression of NMDAR1 with one or more of the NR2 subunits readily generates receptors with distinct functional and pharmacological properties that appear to best resemble native receptors (30,32). To date, splice variants have only been reported for NR2D (35). The expression of the NR2 subunits appears to be regulated in both a regional and developmental manner (reviewed in Ref. 25), which, together with the differential expression of the NMDAR1 isoforms, clearly suggests the existence of a variety of native NMDA receptors.

A number of other proteins have been proposed as prospective NMDA receptor subunits including the 71-kDa glutamate-binding protein (36), GR33 (37), and NMDAR-L or χ-1 (38,39). However, elucidation of the physiological roles of these proteins and their involvement, if any, in NMDA receptor structure and function requires further study.

The NMDA receptor subunits, together with the AMPA and kainate receptor subunits, are all distinguishable from the subunits of the other ligand-gated ion channel superfamily by their relatively high molecular mass, 97–163 kDa, which is approximately twice the size of the subunits for the acetylcholine, GABA, glycine, and 5-HT3 receptors (25). Despite this, and other notable differences, initial predictions of the glutamate receptor subunit transmembrane domain topology were made adopting the four-transmembrane-domain topology model of the nicotinic acetylcholine receptor (reviewed in Ref. 40). Thus, the original models all shared extracellular amino- and carboxy-terminals and a large intracellular loop between transmembrane domains three and four (L3). Subsequently these models were challenged by observations suggesting that the carboxy terminus is in fact intracellular (41–45) and that at least part of L3 is extracellular (46,47). Stern-Bach and colleagues (48) identified two regions of the glutamate receptor subunit that are structurally related to bacterial amino acid–binding proteins that appear to form the glutamate-binding pocket, and notably, one of these regions lies within L3, again suggesting an extracellular position. A series of elegant studies have subsequently established a new transmembrane topology for the glutamate receptor in which the second proposed transmembrane domain does not span the membrane but loops into it with both ends facing into the cytoplasm (49–52). This model accommodates an extracellular amino-terminal, an intracellular carboxy-terminal, and an extracellular L3. Interestingly, the arginine/glutamine editing site in the AMPA and kainate receptors and the corresponding asparagine residue in the NMDA receptor, which lie within the second transmembrane domain and control the ion selectivity and rectification properties of the receptors, are thought to be located within the channel pore (53–56). Thus, the second transmembrane domain is likely to form at least a portion of the channel pore. Although this model satisfies a large body of the experimental evidence, controversy still exists, primarily stemming from the observations of Raymond et al. (57) and Wang et al. (58) whose work on the kainate receptor subunit GluR6 suggests an intracellular position for L3. Although seemingly unlikely, it remains possible that the glutamate receptor subtypes do not share a common transmembrane topology and further investigation of the less well studied receptor subtypes is necessary to address this possibility.

III. TECHNICAL ADVANCES IN THE EVALUATION OF NMDA RECEPTOR AGONISTS AND ANTAGONISTS

Several methodological advances that occurred during the development of NMDA receptor antagonists have contributed significantly to the rapid developments made in this field. One was the discovery that ^{3}H-MK-801, the NMDA receptor open channel blocker, only bound to activated receptors (59). This meant that a high-throughput radioligand binding assay could be used to assess the functional state of the receptor, thus detecting the activity of antagonists acting at any of the sites on the receptor and also distinguishing agonist from antagonist and even partial agonist activity. The discovery of the polyamine site, for example, occurred as a result of the finding that spermine and spermidine potentiated ^{3}H-MK-801 binding to the NMDA receptor in rat brain membranes (60).

Advances in brain slice techniques (61) also allowed quantitative "gut bath" pharmacology to be applied to the brain for the first time. Simple brain slice preparations, such as the hemisected spinal cord, hippocampal slices and cortical wedges, combined with simple population response measurements allowed traditional pharmacological analytical techniques, e.g. Schild analysis, to be applied to the study of NMDA receptor antagonists (62–66). Such preparations, however, have limitations, particularly for the analysis of agonist concentration-response relationships. True equilibrium responses are difficult to achieve as, generally, short application times are necessary because of the toxic effect of prolonged NMDA receptor activation. Furthermore, maximum depolarization of cells, i.e., to the NMDA receptor reversal potential, is often achieved with fairly low levels of receptor occupation. The real breakthrough in the analysis of agonist action came with the development of the patch clamp technique (67) and very rapid drug application devices allowing drug "concentration-jump" experiments to be performed.

Using patch clamp techniques to voltage-clamp cultured or freshly dissociated neurons at a given membrane potential enables the activation of any NMDA receptor (or any other ion channel) to be recorded. There is no issue of "spare receptors"—if a receptor is activated and a current flows, it will be recorded. Furthermore, because, in many cases, the drug equilibration speed around the cell or isolated membrane patch is much faster than the kinetics of the drug receptor interaction, the rate constants of binding and unbinding of both agonists and antagonists can be measured directly. Use of these techniques has led to the conclusion that there are two, equivalent binding sites for both glutamate and glycine per NMDA receptor (15,68,69), that there is a negative allosteric interaction between the glutamate and glycine binding sites (70–73), and that the time course of the NMDA synaptic current is determined by the unbinding rate of glutamate from the receptor (74,75). They have also allowed the quantitative determination of agonist and antagonist affinities for the various sites using both equilibrium concentration-effect and kinetic analysis (76–78) and have enabled the intrinsic activity of partial agonists to be accurately determined (73).

IV. SITES FOR ANTAGONIST ACTION

The NMDA receptor is unique among ligand-gated ion channels in its requirement for two coagonists: glutamate and glycine, for channel activation (79,80). Interestingly, site-directed mutagenesis studies have determined that amino acids in the amino terminal and L3 regions of NMDAR1 homologous to the bacterial amino acid–binding proteins are critical for binding of the coagonist glycine rather than glutamate as occurs in the AMPA and kainate receptors (48,52,81,82). Interestingly, the binding site for glutamate has recently been identified on the homologous regions of the NR2 subunits (151). An allosteric interaction between the glutamate and glycine binding sites of the NMDA receptor has been demonstrated, such that ligand binding at either site can affect the affinity of the other site for its agonist (70–73). The NMDA receptor has a number of regulatory sites, subject to modulation by both endogenous and exogenous compounds (Fig. 1). These include a site within the channel pore where Mg^{2+} binds to confer the well-described voltage dependence of receptor activation (83,84). NMDA receptors are also subject to modulation by Zn^{2+} (85–87), redox state (88), protons (29,89,90), polyamines (29,91–93), and by Mg^{2+} binding at sites distinct from that within the channel pore (94,95) which may be the same as the polyamine binding sites (95). Many of these regulatory sites have attracted attention as possible pharmacological targets to prevent NMDA receptor-mediated neurotoxicity, with the idea that they may have a better side-effect profile than the more conventional NMDA receptor antagonists.

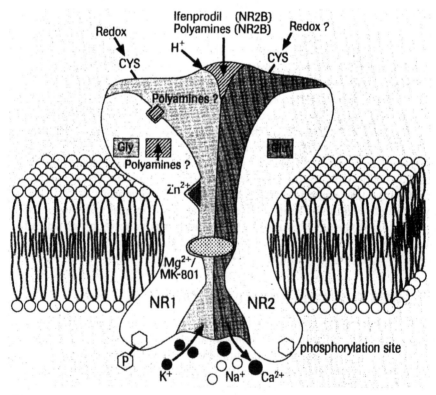

Figure 1 Schematic representation of the NMDA receptor ion channel complex illustrating sites for antagonist action.

The majority of effort in the development of clinically effective NMDA receptor antagonists has, until relatively recently, focused on three pharmacological targets: the glutamate and glycine binding sites and the receptor channel pore. A number of potent compounds selective for each of these sites have been developed, all of which exhibit significant neuroprotection against excitotoxicity in a variety of experimental models.

A. Competitive Glutamate Site Antagonists

A large number of competitive antagonists of the glutamate recognition site of the NMDA receptor have been synthesized based upon the initially developed competitive antagonists, the D-amino acid ω-phosphonic acid analogs of glutamate, D-2-amino-5-phosphonopentanoate (D-AP5) and D-2-amino-7-phosphonoheptanoate (D-AP7), all with essentially similar features. The vast majority are conformationally constrained, α-amino carboxylic acids with an ω-phosphonic acid group an appropriate distance away (reviewed in Ref. 96). A considerable number of structure-activity studies of this site have been performed and pharmacophores for both agonists and antagonists proposed (see Ref. 96). As a whole, competitive antagonists tend to penetrate the blood-brain barrier very slowly because of their highly polar nature. Attempts to overcome this by using less polar substituents has met with limited success as high polarity appears to be a prerequisite for high-affinity binding to the receptor (see Ref. 96). Thus, although they are neuroprotective in vivo, they are most effective when administered prior to the ischemic insult and have a limited therapeutic time window.

Competitive compounds bear an additional potential disadvantage as therapeutic compounds in that, theoretically, they will inhibit NMDA receptors subjected to weaker levels of agonist stimulation more effectively than those subjected to excessive, potentially neurotoxic, levels of glutamate. Thus, such compounds are more likely to target the normal glutamatergic function of the brain than regions of nonphysiological receptor overactivation resulting, for example, from the high levels of extracellular glutamate generated following ischemia. Although competitive antagonists were reported to have a better separation than channel blockers between anticonvulsant and untoward effects in animals, in human clinical trials for the treatment of stroke, selfotel produced dose-related psychomimetic effects and its development was stopped. Similarly, SDZ EAA 494 (D-CPPene) also produced CNS "side effects" in human volunteers and in epilepsy patients was without efficacy at doses that produced adverse events severe enough to stop the trial (see Ref. 98).

B. Channel Blockers

Noncompetitive antagonists possess perhaps a more attractive neuropharmacological profile. The best-characterized class of noncompetitive antagonists are the ion channel blockers, which include the well studied dizocilpine [MK-801] (12) and phencyclidine. These compounds are activity-dependent; i.e., they require channel opening to bind to and block the receptor (11,99,100) and are more accurately described as uncompetitive antagonists. Such activity dependency can, in contrast to the pharmacological profile of competitive antagonists, be seen as a potential desirable feature for a therapeutic receptor blocker. To block channels reversibly during periods of particularly high activity while leaving resting channels relatively unaffected is an attractive strategy for neuroprotection with minimal side effects. However, although the neuroprotective abilities of compounds in this class are now firmly established, a number of associated side effects have become evident, including behavioral (101), cardiovascular (102), and potentially cytotoxic (103,104) effects. Nevertheless, NMDA ion channel blockers, most notably aptiganel (formerly cerestat and CNS 1102), remain in Phase III clinical development. The side effects associated with these compounds seem likely to result, at least in part, from their very effectiveness as receptor blockers. While targeting areas of potentially pathological receptor activation, they also inhibit normal glutamatergic transmission and, due to their high binding affinities, do so in a poorly reversible manner. Several NMDA receptor channel blockers, including ketamine, dextrorphan, and memantine, are better tolerated clinically. However, these compounds exhibit lower affinities for the receptor, and their relatively low potency may limit their effectiveness as neuroprotective compounds (105).

Remacemide, an anticonvulsant in clinical trials for epilepsy, appears to be a NMDA ion channel blocker "prodrug" that has a reduced side-effect profile compared to other channel blockers. The des-glycine metabolite of remacemide is a much more potent blocker of NMDA receptors than remacemide itself, which has only high micromolar affinity (106). Accordingly, pretreatment with remacemide before the onset of ischemia has been shown to be neuroprotective in a cat model of stroke (107). The reasons why a "prodrug" should produce fewer side effects are not entirely clear but may be related to the rate of block of NMDA receptors. Studies with competitive glutamate-site antagonists indicate that rapid i.v. dosing may induce more severe side effects than drug administration by routes with slower rates of adsorption (98).

C. Competitive Glycine Site Antagonists

The next challenge in NMDA receptor antagonist development was to circumvent the problem of "all or nothing" receptor blockade. A number of strategies have been designed to achieve this goal including the development of compounds acting at the glycine site of the NMDA receptor, which has attracted considerable attention as a potentially effective therapeutic target. While both glutamate and glycine are coagonists at the NMDA receptor, glutamate appears to play the neurotransmitter role, being released from presynaptic terminals in an activity-dependent manner. Glycine, on the other hand, plays a modulatory role, apparently continuously present in the extracellular fluid at more constant levels. Thus, although competitive antagonists at the glycine site share some of the same theoretical drawbacks as competitive glutamate site antagonists, a partial blockade of the glycine site should permit a level of physiological NMDA receptor activation while at the same time preventing excessive receptor activation. An attractive way of achieving this could be with partial agonists that would not produce a complete block even when dosed at high levels. In contrast, partial agonists at the glutamate recognition site would produce a constant level of receptor activation, dependent upon their intrinsic activity, and would prevent physiological activation by normal synaptic transmission—a seemingly unattractive concept.

HA-966, one of the first compounds identified to be an NMDA antagonist, was shown to act at the glycine site shortly after its discovery (108,109) and was characterized as a partial agonist (71,109). Structure-activity studies around this lead showed that there was little room for optimization (reviewed in Refs. 110, 111), the best compound being the R-(+)-*cis*-4-methyl analog, L-687,414. However, the advantage of these compounds is that they have relatively good CNS bioavailability and are active in vivo following systemic administration. Thus, L-687,414 was demonstrated to be neuroprotective in a rat model of stroke (112) at doses that were without cardiovascular side effects and did not cause vacuolization in neurons of the cingulate and retrosplenial cortex (113). Furthermore, L-687,414 produced fewer behavioral effects than the open channel blockers (114) and, thus, appears to satisfy the desired criteria of preventing receptor overactivation and being neuroprotective while permitting a "maintenance level" of normal glutamatergic neurotransmission.

Another compound shown to be active at the glycine site was the broad-spectrum excitatory amino acid receptor antagonist kynurenic acid. Although having low affinity and poor selectivity, the scope for optimization of this compound turned out to much greater than that for HA-966. An early finding was that a simple 7-chloro substitution selectively improved affinity for the glycine site by 70-fold (115). A subsequent medicinal chemistry program led to the development of low-nanomolar-affinity compounds with in vivo activity of below 1 mg/kg, exemplified by L-701,324, with an increase in affinity of >10,000-fold over the original lead compound, kynurenic acid. These structure-activity studies, combined with molecular modeling, led to progressively improved pharmacophores of the glycine antagonist binding site on the NMDA receptor, which in turn aided in the subsequent development of more potent compounds (116–121), illustrating how rational drug design can be used to develop potent and selective compounds even from low-affinity, nonselective chemical leads (for reviews see Refs. 110, 111).

D. Redox Site

The redox modulatory site of the NMDA receptor provides another possible therapeutic target. Although redox modulation of the NMDA receptor is complex, generally, reducing agents enhance NMDA-evoked currents, and they are inhibited, importantly not completely, by oxidizing agents (reviewed in Ref. 88). The redox-sensitive sites on the NMDA receptor are thought to be located extracellularly and require the presence of both NMDAR1 and NR2 subunits, presumably for the formation of redox-sensitive disulfide bonds. Two cysteine residues required for redox modulation of recombinant receptors have been identified in the putative L3 domain of NMDAR1 (122). Interestingly, different NR2 subunits appear to confer markedly different redox-sensitive properties on the heteromeric receptor. The native NMDA receptor in a variety of in vitro preparations appears to exist in equilibrium between the fully oxidized and reduced states, presumably maintained as such by endogenous redox modulators (see Ref. 88). Ischemic stroke results in a reducing environment (123), which would be expected to result in an enhancement of NMDA receptor-mediated current, exacerbating the potentially neurotoxic effects of receptor overactivation. Several oxidizing reagents that are able to inhibit NMDA-induced currents have been identified including nitroglycerin and sodium nitroprusside, which are both currently clinically available for cardiovascular indications (reviewed in Ref. 124). The neuroprotective ability of sodium nitroprusside in vivo is unclear. However, in animal models, where its cardiovascular effects have also been accommodated, nitroglycerin has been shown to reduce NMDA-mediated toxicity without behavioural side effects and is undergoing further evaluation for use in stroke (see Ref. 124).

E. Subtype-Selective Antagonists

An alternative approach to the problem of preventing receptor overactivation while permitting enough normal glutamatergic function to avoid unacceptable side effects has been the development of NMDA receptor subunit-selective compounds. In the adult rodent and human brain the predominant NR2 subunits in the forebrain are NR2A and NR2B, with NR2C expressed largely in the cerebellum and various select nuclei, and NR2D expression confined to the diencephalon and midbrain (30,32,34,125). NMDA receptors are thought to be heteromeric complexes (30,32–34). Thus, in the adult forebrain the most abundant heteromeric combinations are likely to be NMDAR1/NR2A and NMDAR1/NR2B although a number of recent studies have suggested that trimeric NMDAR1/NR2A/NR2C (126,127) and NMDAR1/NR2A/NR2B (128) receptors can exist. Clearly, selective blockade of NMDAR1/NR2B receptors while leaving NMDAR1/NR2A receptors relatively unaffected might provide a strategy for neuroprotection with reduced side effects. A number of compounds have been identified that appear to discriminate between NMDA receptors composed of different subunit combinations, the best characterized of which is ifenprodil (129–131), a known neuroprotective agent originally believed to act via its adrenergic receptor activity (132). Subsequently, ifenprodil was identified as an atypical noncompetitive antagonist of NMDA receptors (133) with approximately 400-fold higher affinity for NMDAR1-1a/NR2B than for NMDAR1-1a/NR2A heteromeric receptors (129). Ifenprodil is not a channel blocker and was originally thought to act as an antagonist at the polyamine binding site of the NMDA receptor (134). Accumulating evidence suggests, however, that it binds to a distinct, but closely related site (135,136). Gallagher et al. (136) recently identified an amino acid residue in the NR2B subunit amino-terminal region that is absolutely required for the high-affinity ifenprodil, but not polyamine, interaction with the NMDA receptor. However, both polyamine and ifenprodil binding involve both NMDAR1 and NR2 subunits. Williams and colleagues (137,138) have suggested that three distinct polyamine binding sites exist including one within or near the channel pore. Mutation of amino acid residues in both the amino-terminus and the L3 domain of NMDAR1 that appear important for at least one of the polyamine-mediated effects also reduces the sensitivity of NMDAR1/NR2B receptors to ifenprodil. Thus, it seems likely that the binding sites for ifenprodil and polyamine exhibit an allosteric linkage and perhaps may even overlap. Ifenprodil protects cultured neurons from NMDA-mediated toxicity (139) and is also neuroprotective in in vivo models of cerebral ischemia (140). Notably, ifenprodil lacks the stimulant, amnesic, and discriminative activities exhibited by other NMDA receptor antagonists (141,142), a profile that has been attributed to its selective blockade of NMDA receptors containing the NR2B subunit.

We have recently characterized the mechanism of NMDA receptor antagonism by ifenprodil and have found that it acts by a novel activity-dependent mechanism (143). Using whole-cell voltage clamp recordings from rat cultured cortical neurons in the presence of saturating concentrations of glycine, we have found that ifenprodil antagonizes NMDA receptors in an activity-dependent manner while also increasing the receptor affinity for glutamate recognition–site agonists. Thus, the apparent affinity of ifenprodil for the NMDA receptor is increased in an NMDA concentration-dependent manner. Furthermore, ifenprodil potentiates currents elicited by very low NMDA concentrations due to the increase in affinity for glutamate-site agonists. Thus, with increasing concentrations of NMDA the effect of ifenprodil changes from one of potentiation to one of increasing inhibition. We have formulated a reaction scheme to explain the effects of ifenprodil based on previously described models of NMDA receptor activation and desensitization in which ifenprodil exhibits a 39- and 50-fold higher affinity for the agonist-bound activated and desensitized states of the NMDA receptor relative to the resting, agonist-unbound, state (143). In addition, Ifenprodil binding to the NMDA receptor results in a marked decrease in the channel opening probability and a sixfold higher affinity for glutamate site agonists. It can be seen from Figure 2 that this reaction scheme provides a good description of the experimental results obtained. As previously discussed, activity dependency is a desirable property for therapeutic ion channel blockers. The neuropharmacological profile of ifenprodil is clearly distinct from the activity-dependent ion channel blockers as evidenced by its lack of in vivo side effects (141,142). This novel mechanism of NMDA receptor antagonism, together with the subunit selectivity, probably contributes to this attractive in vivo neuropharmacological profile. Unfortunately, both ifenprodil and its orally active derivative eliprodil have additional pharmacological activity at both adrenoreceptors and voltage-gated calcium channels (132,144,145), which may have compromised their use in the clinic.

Figure 2 Model for the interaction of ifenprodil and NMDA at the NMDA receptor. (A) Reaction scheme for the interaction of NMDA and ifenprodil at the NMDA-receptor channel. Asterisks (*) denote ifenprodil-bound states of the channel. R, the unliganded resting state; A, the double-ligand state; D, the desensitized state of the receptor. States O and O* are open (conductive) states; all other states are shut. Equilibrium constants and receptor open probabilities (p_O) were derived from the best fits of this reaction scheme to the experimental data (see Ref. 143). (B) Three-dimensional plot of the relationship between steady-state current and the concentrations of NMDA and ifenprodil. Circles are mean values of the measured current amplitudes (3–8 cells per data point from a total of 18 cells). The curved surface (mesh) illustrates the best fit of the reaction scheme (A) to the experimental data (see Ref. 143). Note the logarithmic scaling of all three axes. (Adapted from Ref. 143.)

A

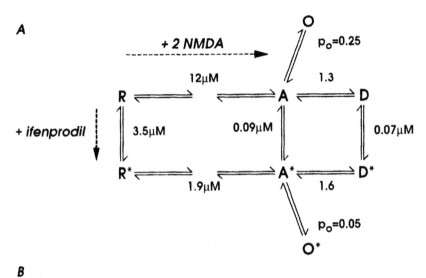

B

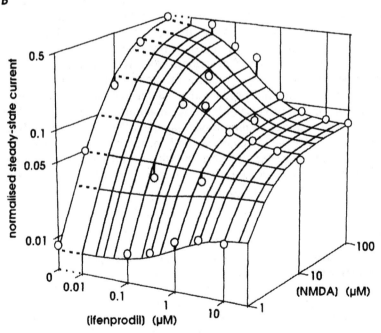

V. CONCLUSION

In conclusion, a number of NMDA receptor antagonists remain in clinical trials of which aptiganel (CNS 1102, cerestat), currently in phase III, is the most advanced. Aptiganel is a classical open channel blocker and, as such, exhibits the expected mechanism-related side effects in humans (146). However, it may be that its long-term therapeutic benefits outweigh any relatively short-term side effects, thus enabling its use clinically. The future of redox compounds such as nitroglycerin awaits confirmation of their neuroprotective ability in animal models as well as further evidence of their applicability in the absence of cardiovascular complications. Two highly selective NR2B selective antagonists have recently been described: Ro 25-6981, which exhibits a >6000-fold increased affinity for heteromeric NMDAR1/NR2B receptors relative to NMDAR1/NR2A receptors expressed in *Xenopus* oocytes (147), and CP-101,606 (148). Importantly, Ro 25-6981 (and probably also CP-101,606) appears to share the activity-dependent mechanism of antagonism exhibited by ifenprodil (147) and is neuroprotective in animal models of cerebral ischemia at doses that are without untoward CNS side effects (149). Together with the further development of such NR2B selective compounds, another future direction in NMDA receptor antagonists might be the development of NR2A selective compounds. To date no compounds significantly selective for NR2A-containing receptors have been described but NR2A-subunit knockout mice do not show severe behavioral abnormalities, suggesting that NR2A selective antagonists may not produce too deleterious side effects (150).

The early preclinical promise of NMDA antagonists has yet to translate to the clinic. The major stumbling block has been the adverse side-effect profile of the nonselective ion channel and competitive glutamate-site antagonists. The next few years will determine whether the more sophisticated approaches outlined above lead to the development of therapeutically useful, well-tolerated NMDA antagonists.

REFERENCES

1. Pin J-P, Duvoisin R. The metabotropic glutamate receptors: structure and functions. Neuropharmacology 1995; 34:1–26.
2. Bliss TV, Collingridge GL. A synaptic model of memory: long-term potentiation in the hippocampus. Nature 1993; 361:31–39.
3. Meldrum B, Garthwaite J. Excitatory amino acid neurotoxicity and neurodegenerative disease. Trends Pharmacol Sci 1990; 11:379–387.
4. Olney JW. Neurotoxicity of excitatory amino acids. In: McGeer EG, Olney JW, McGeer PJ, eds. Kainic Acid as a Tool in Neurobiology. New York: Raven Press, 1978:95–171.

5. Benveniste H, Drejer J, Schousboe A, Diemer NH. Elevation of the extracellular concentrations of glutamate and aspartate in rat hippocampus during transient cerebral ischemia monitored by intracerebral microdialysis. J Neurochem 1984; 43:1369–1374.

6. Hagberg H, Lehmann A, Sandberg M, Nystrom B, Jacobson I, Hamberger A. Ischemia-induced shift of inhibitory and excitatory amino acids from intra- to extracellular compartments. J Cereb Blood Flow Metab 1985; 5: 413–419.

7. Watkins JC. The NMDA receptor concept: origins and development. In: Collingridge GL, Watkins JC, eds. The NMDA Receptor. New York: Oxford University Press, 1994:1–30.

8. Simon RP, Swan JH, Griffiths T, Meldrum BS. Blockade of N-methyl-D-aspartate receptors may protect against ischemic damage in the brain. Science 1984; 226:850–852.

9. Meldrum BS, Evans MC, Swan JH, Simon RP. Protection against hypoxic/ischaemic brain damage with excitatory amino acid antagonists. Med Biol 1987; 65:153–157.

10. Wieloch T. Hypoglycemia-induced neuronal damage prevented by an N-methyl-D-aspartate antagonist. Science 1985; 230:681–683.

11. Wong EHF, Kemp JA, Priestley T, Knight AR, Woodroff GN, Iversen LL. The anticonvulsant MK-801 is a potent N-methyl-D-aspartate antagonist. Proc Natl Acad Sci USA 1986; 83:7104–7108.

12. Iversen LL, Kemp JA. Non-competitive NMDA antagonists as drugs. In: Collingridge GL, Watkins JC, eds. The NMDA Receptor. New York: Oxford University Press, 1994:469–486.

13. MacDermott AB, Mayer ML, Westbrook GL, Smith SJ, Barker JL. NMDA-receptor activation increases cytoplasmic calcium concentration in cultured spinal cord neurones. Nature 1986; 321:519–522.

14. Schanne FA, Kane AB, Young EE, Farber JL. Calcium dependence of toxic cell death: a final common pathway. Science 1979; 206:700–702.

15. Patneau DK, Mayer ML. Structure-activity relationships for amino acid transmitter candidates acting at N-methyl-D-aspartate and quisqualate receptors. J Neurosci 1990; 10:2385–2399.

16. Muir KW, Lees KR. Clinical experience with excitatory amino acid antagonist drugs. Stroke 1995; 26:503–513.

17. Anis NA, Berry SC, Burton NR, Lodge D. The dissociative anaesthetics, ketamine and phencyclidine, selectively reduce excitation of central mammalian neurones by N-methyl-D-aspartate. Br J Pharmacol 1983; 79: 565–575.

18. Lodge D, Johnson KM. Noncompetitive excitatory amino acid receptor antagonists. Trends Pharmacol Sci 1990; 11:81–86.

19. Ferrer-Montiel AV, Montal M. Pentameric subunit stoichiometry of a neuronal glutamate receptor. Proc Natl Acad Sci USA 1996; 93:2741–2744.

20. Moriyoshi K, Masu M, Ishii T, Shigemoto R, Mizuno N, Nakanishi S. Molecular cloning and characterisation of the rat NMDA receptor. Nature 1991; 354:31–37.

21. Yamazaki M, Mori H, Araki K, Mori KJ, Mishina M. Cloning, expression and modulation of a mouse NMDA receptor subunit. FEBS Lett 1992; 300: 39–45.

22. Planells-Cases R, Sun W, Ferrer-Montiel A, Montal M. Molecular cloning, functional expression, and pharmacological characterization of an N-methyl-D-aspartate receptor subunit from the human brain. 1993; 90:5057–5061.

23. Foldes RL, Rampersad V, Kamboj RK. Cloning and sequence analysis of cDNAs encoding human hippocampus N-methyl-D-aspartate receptor subunits: evidence for alternative RNA splicing. Gene 1993; 131:293–298.

24. Karp SJ, Masu M, Eki T, Ozawa K, Nakanishi S. Molecular cloning of the key subunit of the human N-methyl-D-aspartate receptor. J Biol Chem 1993; 268:3728–3733.

25. McBain CJ, Mayer ML. N-Methyl-D-aspartic acid receptor structure and function. Physiol Rev 1994; 74:723–760.

26. Laurie DG, Seeburg PH. Regional and developmental heterogeneity in splicing of the rat brain NMDAR1 mRNA. J Neurosci 1994; 14:3180–3194.

27. Laurie DJ, Putzke J, Zieglgansberger W, Seeburg PH, Tolle TR. The distribution of splice variants of the NMDAR1 subunit mRNA in the adult rat brain. Mol Brain Res 1995; 32:94–108.

28. Zukin RS, Bennett MVL. Alternatively spliced isoforms of the NMDAR1 receptor subunit. Trends Neurosci 1995; 18:306–313.

29. Traynelis SF, Hartley M, Heinemann SF. Control of proton sensitivity of the NMDA receptor by RNA splicing and polyamines. Science 1995; 268: 873–876.

30. Monyer H, Sprengel R, Schoepfer R, Herb A, Higuchi M, Lomeli H, Burnashev N, Sakmann B, Seeburg PH. Heteromeric NMDA receptors: molecular and functional distinction of subtypes. Science 1992; 256:1217–1221.

31. Ikeda K, Nagasawa M, Mori H, Araki K, Sakimura K, Watanabe M, Inoue Y, Mishina M. Cloning and expression of the ε4 subunit of the NMDA receptor channel. FEBS Lett 1992; 313:34–38.

32. Kutsuwada T, Kashiwabuchi N, Mori H, Sakimura K, Kushiya E, Araki K, Meguro H, Masaki H, Kumanishi T, Arakawa M, Mishina M. Molecular diversity of the NMDA receptor channel. Nature 1992; 358:36–41.

33. Meguro H, Mori H, Araki K, Kushiya E, Kutsuwada T, Yamazaki M, Kumanishi T, Arakawa M, Sakimura K, Mishina M. Functional characterization of a heteromeric NMDA receptor channel expressed from cloned cDNAs. Nature 1992; 357:70–74.

34. Ishii T, Moriyoshi K, Sugihara H, Sakurada K, Kadotani H, Yokoi M, Akazawa C, Shigemoto R, Mizuno N, Masu M, Nakanishi S. Molecular characterization of the family of the N-methyl-D-aspartate receptor subunits. J Biol Chem 1993; 268:2836–2843.

35. Gallo V, Upson LM, Hayes WP, Vyklicky L, Winters CA, Buonanno A. Molecular cloning and developmental analysis of a new glutamate receptor isoform in cerebellum. J Neurosci 1992; 12:1010–1023.

36. Kumar KN, Tilakaratne N, Johnson PS, Allen AE, Michaelis EK. Cloning of cDNA for the glutamate-binding subunit of an NMDA receptor complex. Nature 1991; 354:70–73.

37. Smirnova T, Stinnakre J, Mallet J. Characterization of a presynaptic glutamate receptor. Science 1993; 262:430–433.

38. Sucher NJ, Akbarian S, Chi CL, Leclerc CL, Awobuluyi M, Deitcher DL, Wu MK, Yuan JP, Jones EG, Lipton SA. Developmental and regional expression pattern of a novel NMDA receptor-like subunit (NMDAR-L) in the rodent brain. J Neurosci 1995; 15:6509–6520.

39. Ciabarra AM, Sullivan JM, Gahn LG, Pecht G, Heinemann S, Sevarino KA. Cloning and characterization of $\chi 1$: a developmentally regulated member of a novel class of the ionotropic glutamate receptor family. J Neurosci 1995; 15:6498–6508.

40. Hollmann M, Heinemann S. Cloned glutamate receptors. Annu Rev Neurosci 1994; 17:31–108.

41. Petralia RS, Wenthold RJ. Light and electron immunocytochemical localization of AMPA-selective glutamate receptors in the rat brain. J Comp Neurol 1992; 318:329–354.

42. Tingley WG, Roche KW, Thompson AK, Huganir RL. Regulation of NMDA receptor phosphorylation by alternative splicing of the C-terminal domain. Nature 1993; 364:70–73.

43. Molnar E, Baude A, Richmond SA, Patel PB, Somogyi P, McIlhinney RAJ. Biochemical and immunocytochemical characterization of antipeptide antibodies to a cloned GluR1 glutamate receptor subunit: cellular and subcellular distribution in the rat forebrain. Neuroscience 1993; 53:307–326.

44. Molnar E, McIlhinney RAJ, Baude A, Nusser Z, Somogyi P. Membrane topology of the GluR1 glutamate receptor subunit: epitope mapping by site-directed antipeptide antibodies. J Neurochem 1994; 63:683–693.

45. Baude A, Molnar E, Latawiec D, McIlhinney RAJ, Somogyi P. Synaptic and nonsynaptic localization of the GluR1 subunit of the AMPA-type excitatory amino acid receptor in the rat cerebellum. J Neurosci 1994; 14:2830–2843.

46. Roche KW, Raymond LA, Blackstone C, Huganir RL. Transmembrane topology of the glutamate receptor subunit GluR6. J Biol Chem 1994; 269:11679–11682.

47. Taverna FA, Wang LY, Macdonald JF, Hampson DR. A transmembrane model for an ionotropic glutamate receptor predicted on the basis of the location of asparagine-linked oligosaccharides. J Biol Chem 1994; 269:14159–14164.

48. Stern-Bach Y, Bettler B, Hartley M, Sheppard PO, O'Hara PJ, Heinemann SF. Agonist selectivity of glutamate receptors is specified by two domains structurally related to bacterial amino acid-binding proteins. Neuron 1994; 13:1345–1357.

49. Wo ZG, Oswald RE. Transmembrane topology of two kainate receptor subunits revealed by N-glycosylation. Proc Natl Acad Sci USA 1994; 91:7154–7158.

50. Hollmann M, Maron C, Heinemann S. N-Glycosylation site tagging suggests a three transmembrane domain topology for the glutamate receptor GluR1. Neuron 1994; 13:1331–1343.

51. Bennett JA, Dingledine R. Topology profile for a glutamate receptor: three transmembrane domains and a channel-lining reentrant membrane loop. Neuron 1995; 14:373–384.

52. Hirai H, Kirsch J, Laube B, Betz H, Kuhse J. The glycine binding site of the N-methyl-D-aspartate receptor subunit NR1: identification of novel determinants of coagonist potentiation in the extracellular M3-M4 loop region. Proc Natl Acad Sci USA 1996; 93:6031–6036.

53. Hume RI, Dingledine R, Heinemann SF. Identification of a site in glutamate receptor subunits that controls calcium permeability. Science 1991; 253:1028–1031.

54. Burnashev N, Schoepfer R, Monyer H, Ruppersberg JP, Gunther W, Seeburg PH, Sakmann B. Control by asparagine residues of calcium permeability and magnesium blockade in the NMDA receptor. Science 1992; 257:1415–1419.

55. Burnashev N, Monyer H, Seeburg PH, Sakmann B. Divalent ion permeability of AMPA receptor channels is dominated by the edited form of a single subunit. Neuron 1992; 8:189–198.

56. Mori H, Masaki H, Yamakura T, Mishina M. Identification by mutagenesis of a Mg^{2+}-block site of the NMDA receptor channel. Nature 1992; 358:673–675.

57. Raymond LA, Blackstone CD, Huganir RL. Phosphorylation and modulation of recombinant GluR6 glutamate receptors by cAMP-dependent protein kinase. Nature 1993; 361:637–641.

58. Wang LY, Taverna FA, Huang XP, MacDonald JF, Hampson DR. Phosphorylation and modulation of a kainate receptor (GluR6) by cAMP-dependent protein kinase. Science 1993; 259:1173–1175.

59. Foster AC, Wong, EHF. The novel anticonvulsant MK-801 binds to the activated state of the N-methyl-D-aspartate receptor in rat brain. Br J Pharmacol 1987; 91:403–409.

60. Ransom RW, Stec NL. Cooperative modulation of MK-801 binding to the N-methyl-D-aspartate receptor-ion channel complex by L-glutamate, glycine, and polyamines. J Neurochem 1988; 51:830–836.

61. Dingledine R, ed. Brain Slices. New York: Plenum Press, 1984.

62. Evans RH, Evans SJ, Pook PC, Sunter DC. A comparison of excitatory amino acid antagonists acting at primary afferent C fibres and motoneurones of the isolated spinal cord of the rat. Br J Pharmacol 1987; 91:531–537.

63. Harrison NL, Simmonds MA. Quantitative studies on some antagonists of N-methyl D-aspartate in slices of rat cerebral cortex. Br J Pharmacol 1985; 84:381–391.

64. Wheatley PL, Collins KJ. Quantitative studies of N-methyl-D-aspartate, 2-amino-5-phosphonovalerate and cis-2,3-piperidine dicarboxylate interactions on the neonatal rat spinal cord in vitro. Eur J Pharmacol 1986; 121:257–263.

65. Leach MJ, Marden CM, Canning HM. ($\pm$)-cis-2,3-Piperidine dicarboxylic acid is a partial N-methyl-D-aspartate agonist in the in vitro rat cerebellar cGMP model. Eur J Pharmacol 1986; 121:173–179.

66. Grimwood S, Foster AC, Kemp JA. The pharmacological specificity of *N*-methyl-D-aspartate receptors in rat cerebral cortex: correspondence between radioligand binding and electrophysiological measurements. Br J Pharmacol 1991; 103:1385–1392.

67. Hamil OP, Marty A, Neher E, Sakmann B. Improved patch clamp techniques for high-resolution current recordings from cells and cell-free membrane patches. Pflügers Arch 1981; 391:85–100.

68. Clements JD, Westbrook GL. Activation kinetics reveal the number of glutamate and glycine binding sites on the *N*-methyl-D-aspartate receptor. Neuron 1991; 7:605–613.

69. Benveniste M, Mayer ML. Kinetic analysis of antagonist action at *N*-methyl-D-aspartic acid receptors. Two binding sites each for glutamate and glycine. Biophys J 1991; 59:560–573.

70. Benveniste M, Clements J, Vyklicky L, Jr., Mayer ML. A kinetic analysis of the modulation of *N*-methyl-D-aspartic acid receptors by glycine in mouse cultured hippocampal neurones. J Physiol 1990; 428:333–357.

71. Kemp JA, Priestley T. Effects of (+)-HA-966 and 7-chlorokynurenic acid on the kinetics of *N*-methyl-D-aspartate receptor agonist responses in rat cultured cortical neurons. Mol Pharmacol 1991; 39:666–670.

72. Lester RAJ, Tong G, Jahr CE. Interactions between the glycine and glutamate binding sites of the NMDA receptor. J Neurosci 1993; 13:1088–1096.

73. Priestley T, Kemp JA. Kinetic study of the interactions between the glutamate and glycine recognition sites on the *N*-methyl-D-aspartic acid receptor complex. Mol Pharmacol 1994; 46:1191–1196.

74. Lester RAJ, Clements JD, Westbrook GL, Jahr CE. Channel kinetics determine the time course of NMDA receptor-mediated synaptic currents. Nature 1990; 346:565–567.

75. Lester RAJ, Jahr CE. NMDA channel behaviour depends on agonist affinity. J Neurosci 1992; 12:635–643.

76. Benveniste M, Mayer ML. Kinetic analysis of antagonist action at *N*-methyl-D-aspartic acid receptors. Biophys J 1991; 59:560–573.

77. Benveniste M, Mienville JM, Sernagor E, Mayer ML. Concentration-jump experiments with NMDA antagonists in mouse cultured hippocampal neurons. J Neurophysiol 1990; 63:1373–1384.

78. Benveniste M, Mayer ML. Structure activity analysis of binding kinetics for NMDA receptor competitive antagonists: the influence of conformational restriction. Br J Pharmacol 1991; 104:207–221.

79. Johnson JW, Ascher P. Glycine potentiates the NMDA response in cultured mouse brain neurons. Nature 1987; 325:529–531.

80. Kleckner NW, Dingledine R. Requirement for glycine in activation of NMDA-receptors expressed in *Xenopus* oocytes. Science 1988; 241:835–837.

81. Kuryatov A, Laube B, Betz H, Kuhse J. Mutational analysis of the glycine-binding site of the NMDA receptor: structural similarity with bacterial amino acid–binding proteins. Neuron 1994; 12:1291–1300.

82. Wafford KA, Kathoria M, Bain CJ, Marshall G, Le Bourdelles B, Kemp JA, Whiting PJ. Identification of amino acids in the *N*-methyl-D-aspartate receptor

NR1 subunit that contribute to the glycine binding site. Mol Pharmacol 1995; 47:374–380.

83. Mayer ML, Westbrook GL, Guthrie PB. Voltage-dependent block by Mg^{2+} of NMDA responses in spinal cord neurones. Nature 1984; 309:261–263.

84. Nowak L, Bregestovski P, Ascher P, Herbet A, Prochiantz A. Magnesium gates glutamate-activated channels in mouse central neurones. Nature 1984; 307:462–465.

85. Peters S, Koh J, Choi DW. Zinc selectively blocks the action of N-methyl-D-aspartate on cortical neurons. Science 1987; 236:589–593.

86. Westbrook GL, Mayer ML. Micromolar concentrations of Zn^{2+} antagonize NMDA and GABA responses of hippocampal neurons. Nature 1987; 328:640–643.

87. Christine CW, Choi DW. Effect of zinc on NMDA receptor-mediated channel currents in cortical neurons. J Neurosci 1990; 10:108–116.

88. Gozlan H, Ben-Ari Y. NMDA receptor redox sites: are they targets for selective neuronal protection? Trends Pharmacol Sci 1995; 16:368–374.

89. Tang CM, Dichter M, Morad M. Modulation of the N-methyl-D-aspartate channel by extracellular H^+. Proc Natl Acad Sci USA 1990; 87:6445–6449.

90. Traynelis SF, Cull-Candy SG. Proton inhibition of N-methyl-D-aspartate receptors in cerebellar neurons. Nature 1990; 345:347–350.

91. McGurk JF, Bennett MV, Zukin RS. Polyamines potentiate responses of N-methyl-D-aspartate receptors expressed in xenopus oocytes. Proc Natl Acad Sci USA 1990; 87:9971–9974.

92. Lerma J. Spermine regulates N-methyl-D-aspartate receptor desensitization. Neuron 1992; 8:343–352.

93. Benveniste M, Mayer ML. Multiple effects of spermine on N-methyl-D-aspartic acid receptor responses of rat cultured hippocampal neurones. J Physiol 1993; 464:131–163.

94. Wang L-Y, MacDonald JF. Modulation by magnesium of the affinity of NMDA receptors for glycine in murine hippocampal neurones. J Physiol 1995; 486.1:83–95.

95. Paoletti P, Neyton J, Ascher P. Glycine-independent and subunit-specific potentiation of NMDA responses by extracellular Mg^{2+}. Neuron 1995; 15:1109–1120.

96. Jane DE, Olverman HJ, Watkins JC. Agonists and competitive antagonists: structure-activity and molecular modelling studies. In: Collingridge GL, Watkins JC, eds. The NMDA Receptor. New York: Oxford University Press, 1994:31–104.

97. Grotta J, Clark W, Coull B, Pettigrew LC, Mackay B, Goldstein LB, Meissner I, Murphy D, LaRue L. Safety and tolerability of the glutamate antagonist CGS 19755 (Selfotel) in patients with acute ischemic stroke. Results of a phase IIa randomized trial. Stroke 1995; 26:602–605.

98. Lowe DA, Emre M, Frey P, Kelly PH, Malanowski J, McAllister KH, Neijt HC, Rudeberg CD, Urwyler S, White TG, Herring PL. The pharmacology of SDZ EAA 494. A competitive NMDA antagonist. Neurochem Int 1994; 25:583–600.

99. Huettner JE, Bean BP. Block of N-methyl-D-aspartate-activated current by the anticonvulsant MK-801: selective binding to open channels. Proc Natl Acad Sci USA 1988; 85:1307–1311.

100. Kemp JA, Marshall GR, Priestley T. A comparison of the agonist-dependency of the block produced by uncompetitive NMDA receptor antagonists on rat cortical slices. Mol Pharmacol 1991; 1:65–70.

101. Tricklebank MD, Singh L, Oles RJ, Preston C, Iversen SD. The behavioural effects of MK-801: a comparison with antagonists acting non-competitively and competitively at the NMDA receptor. Eur J Pharmacol 1989; 167:127–135.

102. Lewis SJ, Barres C, Jacob HJ, Ohta H, Brody MJ. Cardiovascular effects of the N-methyl-D-aspartate receptor antagonist MK-801 in conscious rats. Hypertension 1989; 13:759–765.

103. Olney JW, Labruyere J, Price MT. Pathological changes induced in cerebrocortical neurons by phencyclidine and related drugs. Science 1989; 244:1360–1362.

104. Allen HL, Iversen LL. Phencyclidine, dizocilpine, and cerebrocortical neurons. Science 1990; 247:221.

105. Fischer G, Bourson A, Kemp JA, Lorez H-P, Mutel V, Trube G. Characterization of morphanins highly protective in permanent middle cerebral artery occlusion (MCAO) in rats. Soc Neurosci Abstr 1995; 392.7.

106. Subramaniam S, Donevan SD, Rogawski MA. Block of the N-methyl-D-aspartate receptor by remacemide and its des-glycine metabolite. J Pharmacol Exp Ther 1996; 276:161–168.

107. Bannan PE, Graham DI, Lees KR, Mcculloch J. Neuroprotective effect of remacemide hydrochloride in focal cerebral ischemia in the cat. Brain Res 1994; 664:271–275.

108. Fletcher EJ, Lodge D. Glycine reverses antagonism of N-methyl-D-aspartate (NMDA) by 1-hydroxy-3-aminopyrrolidone-2 (HA-966) but not by D-2-amino-5-phosphonovalerate (D-AP5) on rat cortical slices. Eur J Pharmacol 1988; 151:161–162.

109. Foster AC, Kemp JA. HA-966 antagonizes N-methyl-D-aspartate receptors through a selective interaction with the glycine modulatory site. J Neurosci 1989; 9:2191–2196.

110. Kemp JA, Leeson PD. The glycine site of the NMDA receptor–five years on. Trends Pharmacol Sci 1993; 14:20–25.

111. Leeson PD, Iversen LL. The glycine site on the NMDA receptor: structure-activity relationships and therapeutic potential. J Med Chem 1994; 37:4053–4067.

112. Gill R, Hargreaves RJ, Kemp JA. The neuroprotective effect of the glycine site antagonist 3R-(+)-cis-4-methyl-HA966 (L-687,414) in a rat model of focal ischaemia. J Cereb Blood Flow Metab 1995; 15:197–204.

113. Hargreaves RJ, Rigby M, Smith D, Hill RG. Lack of effect of L-687,414 ((+)-cis-4-methyl-HA-966), an NMDA receptor antagonist acting at the glycine site, on cerebral glucose metabolism and cortical neuronal morphology. Br J Pharmacol 1993; 110:36–42.

114. Tricklebank MD, Bristow LJ, Hutson PH, Leeson PD, Rowley M, Saywell K, Singh L, Tattersall FG, Thorn L, Williams BJ. The anticonvulsant and behavioural profile of L-687,414, a partial agonist acting at the glycine modulatory site on the N-methyl-D-aspartate (NMDA) receptor complex. Br J Pharmacol 1994; 113:729–736.

115. Kemp JA, Foster AC, Leeson PD, Priestley T, Tridgett R, Iversen LL, Woodruff GN. 7-Chlorokynurenic acid is a selective antagonist at the glycine modulatory site of the N-methyl-D-aspartate receptor complex. Proc Natl Acad Sci USA 1988; 85:6547–6550.

116. Leeson PD, Baker R, Carling RW, Curtis NR, Moore KW, Williams BJ, Foster AC, Donald AE, Kemp JA, Marshall GR. Kynurenic acid derivatives. Structure-activity relationships for excitatory amino acid antagonism and identification of potent and selective antagonists at the glycine site on the N-methyl-D-aspartate receptor. J Med Chem 1991; 34:1243–1252.

117. Leeson PD, Carling RW, Moore KW, Moseley AM, Smith JD, Stevenson G, Chan T, Baker R, Foster AC, Grimwood S, Kemp JA, Marshall GR, Hogsteen K. 4-Amido-2-carboxytetrahydroquinolines. Structure-activity relationships for antagonism at the glycine site of the NMDA receptor. J Med Chem 1992; 35:1954–1968.

118. Carling RW, Leeson PD, Moseley AM, Baker R, Foster AC, Grimwood S, Kemp JA, Marshall GR. 2-Carboxytetrahydroquinolines. Conformational and stereochemical requirements for antagonism of the glycine site on the NMDA receptor. J Med Chem 1992; 35:1942–1953.

119. Carling RW, Leeson PD, Moore KW, Smith JD, Moyes CR, Mawer IM, Thomas S, Chan T, Baker R, Foster AC. Grimwood S, Kemp JA, Marshall GR, Ticklebank MD, Saywell K. 3-Nitro-3,4-dihydro-2(1H)-quinolones—excitatory amino acid antagonists acting at glycine-site NMDA and (RS)-alpha-amino-3-hydroxy-5-methyl-4-isoxazolepropionic acid receptors. J Med Chem 1993; 36:3397–3408.

120. Rowley M, Leeson PD, Stevenson GI, Moseley AM, Stansfield I, Sanderson I, Robinson L, Baker R, Kemp JA, Marshall GR, Foster AC, Grimwood S, Tricklebank MD, Saywell KL. 3-Acyl-4-hydroxyquinolin-2(1H)-ones—systemically active anticonvulsants acting by antagonism at the glycine site of the N-methyl-D-aspartate receptor complex. J Med Chem 1993; 36:3386–3396.

121. Kulagowski JJ, Baker R, Curtis NR, Leeson PD, Mawer IM, Moseley AM, Ridgill NP, Rowley M, Stansfield I, Foster AC, Grimwood S, Hill RG, Kemp JA, Marshall GR, Saywell KL, Tricklebank MD. 3'-(Arylmethyl)- and 3'-(aryloxy)-3-phenyl-4-hydroxyquinolin-2(1H)-ones: orally active antagonists of the glycine site on the NMDA receptor. J Med Chem 1994; 37:1402–1405.

122. Sullivan JM, Traynelis SF, Chen HS, Escobar W, Heinemann SF, Lipton SA. Identification of two cysteine residues that are required for redox modulation of the NMDA subtype of glutamate receptor. Neuron 1994; 13:929–936.

123. Ginsberg MD, Reivich M, Frinak S, Harbig K. Pyridine nucleotide redox state and blood flow of the cerebral cortex following middle cerebral artery occlusion in the cat. Stroke 1976; 7:125–131.

124. Lipton SA. Prospects for clinically tolerated NMDA antagonists: open-channel blockers and alternative redox states of nitric oxide. Trends Neurosci 1993; 16:527–532.

125. Rigby M, Le Bourdelles B, Heavens RP, Kelly S, Smith D, Butler A, Hammans R, Hills R, Xuereb JH, Hill RG, Whiting PJ, Sirinathsinghji DJS. The messenger RNAs for the N-methyl-D-aspartate receptor subunits show region-specific expression of different subunit composition in the human brain. Neuroscience 1996; 73:429–447.

126. Wafford KA, Bain CJ, Le Bourdelles B, Whiting PJ, Kemp JA. Preferential coassembly of recombinant NMDA receptors composed of three different subunits. Neuroreport 1993; 4:1347–1349.

127. Chazot PL, Coleman SK, Cik M, Stephenson FA. Molecular characterization of N-methyl-D-aspartate receptors expressed in mammalian cells yields evidence for the coexistence of three subunit types within a discrete receptor molecule. J Biol Chem 1994; 269:24403–24409.

128. Sheng M, Cummings J, Roldan LA, Jan YN, Jan LY. Changing subunit composition of heteromeric NMDA receptors during development of rat cortex. Nature 1994; 368:144–147.

129. Williams K. Ifenprodil discriminates subtypes of the N-methyl-D-aspartate receptor: selectivity and mechanisms at recombinant heteromeric receptors. Mol Pharmacol 1993; 44:851–859.

130. Williams K, Russell SL, Shen YM, Molinoff PB. Developmental switch in the expression of NMDA receptors occurs in vivo and in vitro. Neuron 1993; 10:267–278.

131. Priestley T, Ochu E, Kemp JA. Subtypes of NMDA receptor in neurones cultured from rat brain. Neuroreport 1994; 5:1763–1765.

132. MacKenzie ET, Gotti B, Nowicki J-P, Young AR. Adrenergic blockers as cerebral antiischaemic agents. In: Mackenzie ET, Seylaz J, Bes A, eds. Neurotransmitters and the Cerebral Circulation. New York: Raven Press, 1984:219–243.

133. Carter C, Benavides J, Legendre P, Vincent JD, Noel F, Thuret F, Lloyd KG, Arbilla S, Zivkovic B, MacKenzie ET, Scatton B, Langer SZ. Ifenprodil and SL 82.0715 as cerebral anti-ischemic agents. II. Evidence for N-methyl-D-aspartate receptor antagonist properties. J Pharmacol Exp Ther 1988; 247:1222–1232.

134. Carter CJ, Lloyd KG, Zivkovic B, Scatton B. Ifenprodil and SL 82.0715 as cerebral antiischaemic agents. III. Evidence for antagonistic effects at the polyamine modulatory site within the N-methyl-D-aspartate receptor complex. J Pharmacol Exp Ther 1990; 253:475–482.

135. Reynolds IJ, Miller RJ. Ifenprodil is a novel type of N-methyl-D-aspartate receptor antagonist: interaction with polyamines. Mol Pharmacol 1989; 36:758–765.

136. Gallagher MJ, Huang H, Pritchett DB, Lynch DR. Interactions between ifenprodil and the NR2B subunit of the N-methyl-D-aspartate receptor. J Biol Chem 1996; 271:9603–9611.

137. Williams K, Kashiwagi K, Fukuchi J-I, Igarashi K. An acidic amino acid in the N-methyl-D-aspartate receptor that is important for spermine stimulation. Mol Pharmacol 1995; 48:1087–1098.

138. Kashiwagi K, Fukuchi J-I, Chao J, Igarashi K, Williams K. An aspartate residue in the extracellular loop of the N-methyl-D-aspartate receptor controls sensitivity to spermine and protons. Mol Pharmacol 1996; 49:1131–1141.

139. Graham D, Darles G, Langer SZ. The neuroprotective properties of ifenprodil, a novel NMDA receptor antagonist, in neuronal cell culture toxicity studies. Eur J Pharmacol 1992; 226:373–376.

140. Gotti B, Duverger D, Bertin J, Carter C, Dupont R, Frost J, Gaudilliere B, MacKenzie ET, Rousseau J, Scatton B, Wick A. Ifenprodil and SL 82.0715 as cerebral anti-ischemic agents. I. Evidence for efficacy in models of focal cerebral ischemia. J Pharmacol Exp Ther 1988; 247:1211–1221.

141. Jackson A, Sanger DJ. Is the discriminative stimulus produced by phencyclidine due to an interaction with N-methyl-D-aspartate receptors? Psychopharmacology 1988; 96:87–92.

142. Perrault G, Morel E, Sanger DG, Zivkovic B. Comparison of the pharmacological profiles of four NMDA antagonists, ifenprodil, SL 82.0715, MK-801 and CPP, in mice. Br J Pharmacol 1989; 97:580P.

143. Kew JNC, Trube G, Kemp JA. A novel mechanism of activity-dependent NMDA receptor antagonism describes the effect of ifenprodil in rat cultured cortical neurones. J Physiol (1996).

144. Biton BP, Granger A, Carreau H, Depoortere H, Scatton B, Avenet, P. The NMDA receptor anatagonist eliprodil (SL 82.0715) blocks voltage-operated calcium channels in rat cultured cortical neurons. Eur J Pharmacol 1994; 257:297–301.

145. Church J, Fletcher EJ, Baxter K, Macdonald JF. Blockade by ifenprodil of high voltage-activated Ca^{2+} channels in rat and mouse cultured hippocampal pyramidal neurones: comparison with N-methyl-D-aspartate receptor antagonist actions. Br J Pharmacol 1994; 113:499–507.

146. Muir KW, Grosset DG, Gamzu E, Lees KR. Pharmacological effects of the noncompetitive NMDA antagonist CNS 1102 in normal volunteers. Br J Clin Pharmacol 1994; 38:33–38.

147. Fischer G, Mutel V, Trube G, Malherbe P, Kew JNC, Mohacsi E, Heitz MP, Kemp JA. Ro 25-6981, a highly potent and selective blocker of N-methyl-D aspartate receptors containing the NR2B subunit. Characterization in vitro. J Pharmacol Exp Ther 1997; 283:1285–1292.

148. Chenard BL, Bordner J, Butler TW, Chambers LK, Collins MA, De Costa DL, Ducat MF, Dumont ML, Fox CB, Mena EE, Menniti FS, Nielson J, Pagnozzi MJ, Richter KEG, Ronau RT, Shalaby IA, Stemple JZ, White WF. (1S,2S)-1-(4-Hydroxphenyl)-2-(4-hydroxy-4-phenylpiperidino)-1-propanol: a potent new neuroprotectant which blocks N-methyl-D-aspartate responses. J Med Chem 1995; 38:3138–3145.

149. Fischer G, Bourson A, Kemp JA, Lorez HP. The neuroprotective activity of RO 25-6981, a NMDA receptor NR2B subtype selective blocker. Soc Neurosci Abstr 1996; 693.5.

150. Sakimura K, Kutsuwada T, Ito I, Manabe T, Takayama C, Kushiya E, Yagi T, Aizawa S, Inoue Y, Sugiyama H, Mishina M. Reduced hippocampal LTP and spatial learning in mice lacking NMDA receptor e1 subunit. Nature 1995; 373:151–155.
151. Laube B, Hirai H, Sturgess M, Betz H, Kuhse J. Molecular determinants of agonist discrimination by NMDA receptor subunits: analysis of the glutamate binding site on the NR2B subunit. Neuron 1997; 18:493–503.

14

Computational Chemistry in Receptor-Based Drug Design

Ad P. IJzerman and Eleonora M. van der Wenden
Leiden/Amsterdam Center for Drug Research, Leiden, The Netherlands

Wilma Kuipers
Solvay Pharma, Weesp, The Netherlands

I. INTRODUCTION

"Using computers to design drugs" was on the cover of the 1993 December issue of *Scientific American*. Apparently, molecular modeling and computer graphics are appealing to the lay press. How realistic, though, is such enthusiasm? Most of today's marketed medicines were found by chance or by systematic screening of large collections of compounds either from natural sources or man-made. Current emphasis on high-throughout screening and combinatorial chemistry approaches suggest that for the discovery of tomorrow's drugs "educated" chance will remain pivotal. Where, then, does computer-assisted drug design fit in the receptor-based drug discovery process?

Let us first examine some key terms. Computer-assisted molecular modeling (CAMM) is a relatively new and rapidly developing tool in drug design. Computer graphics techniques allow the transformation of complex data sets obtained, e.g., from theoretical chemical calculations or X-ray diffraction patterns, into a picture on a computer screen. Chemical structures, whether small ligand or macromolecule, and their properties may thus be visualized, manipulated, and matched or combined with other relevant molecules. Thus, the perspective of conventional molecular models, such as Dreiding or CPK, has been dramatically expanded. In addition, the ever-increasing rate at which the atomic coordinates of soluble biomacromolecules such as enzymes are being elucidated and subsequently deposited in a publicly accessible form has indeed fueled structure-based drug design. The availability of the three-dimensional architecture of HIV protease, thymidylate synthase, thrombin, carbonic anhydrase, and many more has led to at least an equal number of new chemical entities being evaluated and used in the clinic now. In such cases CAMM has certainly led to a more rational approach toward drug design, in particular when the binding domains for inhibitors and substrates are known from cocrystallization experiments. Knowing the "lock" definitely helps in designing the "key" (1), although the interaction between newly synthesized ligands and the target macromolecule may appear unexpected (Fig. 1).

However, crystallization of the nonsoluble, membrane-bound proteins has proven to be a much more difficult task. No crystal structures of G-protein-coupled receptors (GPCRs), all membrane-bound, are available. As a consequence, experimental data on structural aspects of these proteins is scarce. Recently, Schertler et al. (2) determined a projection map of rhodopsin, the mammalian G-protein-coupled visual pigment, although at low (9 Å) resolution. It confirmed the existence of seven transmembrane domains, as had been evident from the atomic coordinates of bacteriorhodopsin, a similar bacterial protein, although not coupled to a G-protein. The three-dimensional structure of the transmembrane segments of bacteriorhodopsin had been determined before (at a better but still modest resolu-

Figure 1 "Lock and key" model as modified from a painting by René Magritte.

tion), and served as a first indication of the general architecture of G-protein-coupled receptors (3). Both structures have been used as templates to generate receptor models (4–6), despite the debate as to how similar these two protein structures are (6,7).

Which, then, could be the computational strategies to "design" new ligands—agonists, antagonists, inverse agonists—for G-protein-coupled receptors? The purpose of the present chapter is to address this question.

II. COMPUTATIONAL STRATEGIES

Two complementary approaches can be distinguished. First, information embedded in the ligands themselves can be used to obtain a fingerprint of their binding site on the receptor. Biological data, such as affinity values obtained in radioligand-binding studies on wild-type or mutant receptors, are of invaluable help. Equally important are (physico)chemical and structural properties quantifying the spatial, electronic, and lipophilic characteristics of the compounds studied.

A second approach is to use the receptor as an additional source of coded information. This may range from a purely statistical analysis of receptor sequences to the construction of three-dimensional receptor models, as mentioned above. It is the thrust of this chapter that the two approaches should be combined and, if possible, used simultaneously to come to verifiable conclusions. It will be shown that detailed knowledge of structure-activity relationships, either qualitative (SAR) or quantitative (QSAR), is a *conditio sine qua non* for receptor modeling. In Figure 2 important elements in the two interdependent approaches are brought together with the ultimate focus on the ligand-receptor interaction. Mutual equivalents can be discriminated in Figure 2. First, there is the design process. It can be focused on the ligand, either unchanged or modified ("lead" compounds vs. novel chemical structures), or the protein (wild type vs. mutant receptor). A second equivalent is the actual synthesis of ligand (via conventional or combinatorial chemistry) and/or receptor (i.e., site-directed mutagenesis and other protein engineering techniques). A third, and last, is the biological evaluation of unchanged and modified entities. With respect to the latter it has to be kept in mind that the biological readout should be as close as possible to the process of ligand-binding. All too often structure-activity relationships have been based on biological parameters quite distant from initial receptor recognition.

III. THE FIRST APPROACH: THE LIGAND

Ligand-based design is in essence a successor of the combination of (2D!) "pen and paper" chemistry and "traditional" structure-activity relation-

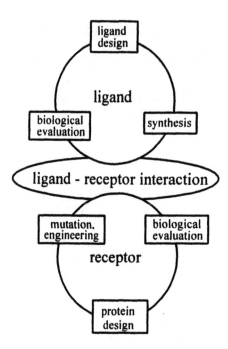

Figure 2 Interdependency of important elements in receptor structure-based drug design.

ships. In general, the binding of a ligand to a receptor is governed by three important factors. First, the ligand should fit sterically to the receptor. This directly implies that size and conformational freedom of both ligand and receptor are important. Second, there should be electrostatic complementarity: parts of the receptor and ligand with opposite charges, or better molecular electrostatic potentials (MEPs), should be in close proximity to each other. Third, to have optimal hydrophobic interaction, lipophilic regions should match. The most important assumption in comparing various ligands by molecular modeling techniques is that the molecules studied bind essentially to the same site on the receptor. A priori this is not self-evident. For example, a competitive antagonist may displace an endogenous or synthetic agonist just by occupying a small, but mutually exclusive part of the agonist binding site. We have observed such a phenomenon when studying the interaction of ligands with the β_2-adrenoceptor (vide infra).

An important concept in ligand-based design is the active analog approach (8). It was developed to derive a pharmacophoric pattern ("pharmacophore") from a set of ligands and, in addition, to obtain an (indirect) view of the receptor binding site. The common volume of the active ligands

(Fig. 3) is the pharmacophoric region, whereas the combined volume of active ligands denotes the space that is available in the receptor including accessory sites for accommodating ligands. The total volume of the inactive ligands that is not present in the set of active compounds represents the so-called "forbidden" areas, i.e., a volume that is occupied by receptor residues and therefore not available for accommodating ligands. It is obvious that from such reasoning a fingerprint of the ligand-binding site can be gleaned.

A. Application of Ligand-Based Design to Adenosine Receptor Antagonists

The active analog approach was applied by Van Galen et al. (9) in a study of antagonists for one of the subtypes of the adenosine receptor, the A_1 receptor. As mentioned, this approach is usually applied for steric comparisons only, but the authors showed that it can also be very useful in studying electrostatic properties of a set of ligands. Three members of chemically different classes of antagonists (xanthines, triazoloquinolinamines, and triazoloquinazolinamines) with more or less equal (nanomolar) affinity for the adenosine A_1 receptor were studied (Fig. 4). This is a good starting point for molecular modeling, since traditional SAR methods fail here owing to the overall dissimilarity of the compounds, whereas the biology points to a good and comparable recognition of the ligands by the receptor. Indeed,

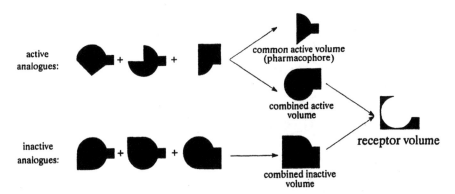

Figure 3 Schematic representation of the active analog approach. The common volume of active analog can be regarded as the full pharmacophore. The combined active volume represents the pharmacophore in combination with substituents that further contribute to activity. Subtracting the combined active volume from the combined inactive volume yields the "forbidden areas," or volume occupied by the receptor. (Modified from Ref. 8.)

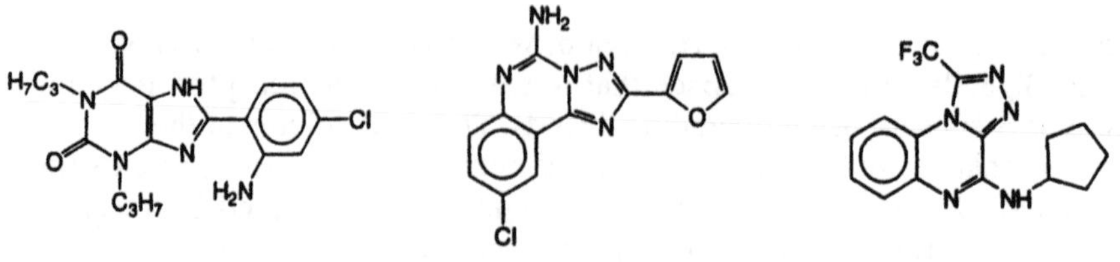

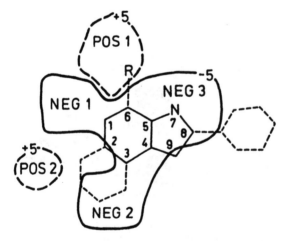

Figure 4 Chemical structures of three high-affinity antagonists used to model the antagonist binding site of the adenosine A_1 receptor; (**1**) PACPX (1,3-dipropyl-8-(2-amino-4-chlorophenyl)xanthine; (**2**) CGS 15943 (9-chloro-2-(2-furyl)[1,2,4] triazolo[1,5-c]quinazolin-5-amine); (**3**) N-cyclopentyl-1-(trifluoromethyl)[1,2,4] triazolo[4,3-a]quinolin-4-amine.

a great advantage over traditional QSAR methods is that the set of ligands, and therefore the model, is not limited to a certain, chemically "narrow" class, thus enabling the development of new scaffolds.

By superimposing the three molecules in various ways it was possible to select one particular mode of superposition that was favorable in both steric and electronic terms. In Figure 5 a schematic model of this fit is

Figure 5 Common steric and electrostatic properties of the adenosine antagonists in Figure 4. Additional regions where substitution may enhance affinity are indicated with dashed lines. [Reprinted with permission from Ref. 9. Copyright (1996) American Chemical Society.]

shown in which a large Y-shaped area of negative molecular electrostatic potential (at a level of −5 kcal/mol) is apparent, resulting from the aromatic system of the 6:5 fused heterocycle. This area extends from the ring system at three points, designated NEG1, NEG2, NEG3. NEG1 and NEG2 have various molecular determinants, whereas NEG3 is invariably caused by a nitrogen lone pair. There are also two areas of positive molecular electrostatic potential (at +5 kcal/mol). Three other areas (dashed lines) can be identified, amenable to further (hydrophobic) substitution. All this information was used to design and synthesize a novel class of adenosine receptor antagonists, the imidazoquinolinamines (10). The core structure complies with the steric and electronic pattern in Figure 5, and it allows for further substitution. In Table 1 affinities for the rat adenosine A_1 and A_{2A} receptor are gathered for a selected number of derivatives. The unsubstituted compound (R_1=R_2=H) had moderate affinity for both subtypes of adenosine receptors, slightly higher than the prototypic antagonist theophylline. A cyclopentyl substituent was most favorable on R_1, whereas position R_2

Table 1 Affinities (K_i values in nM, SEM ≤ 15%) for the Rat A_1 and A_{2A} Adenosine Receptor of 1*H*-Imidazo[4,5-c]quinolin-4-amines

R_1	R_2	$A_1 K_i{}^a$	$A_{2A} K_i{}^b$	$A_{2A}/A_1{}^c$
H	H	1600	1400	0.9
Phenyl	H	2100	10500	5.0
Cyclopentyl	H	43	290	6.7
H	Cyclopentyl	270	740	3.0
Phenyl	Cyclopentyl	230	6%^{d}	>4
Cyclopentyl	Cyclopentyl	39	450	11
H	Phenyl	34	290	8.5
Phenyl	Phenyl	460	9%^{d}	>2
Cyclopentyl	Phenyl	10	450	45

^{a} [³H]DPCPX (1,3-dipropyl-8-cyclopentylxanthine) binding to rat brain cortical membranes.
^{b} [³H]NECA (5′N-ethylcarboxamidoadenosine) binding to rat striatal membranes.
^{c} Ratio A_{2A}-vs. A_1-affinity in the rat.
^{d} Percentage of displacement at 1 μM.
Source: Reprinted with permission from Ref. 10. Copyright (1996) American Chemical Society.

accommodated a phenyl substituent. Combined substitution (R_1=cyclo-pentyl, R_2=phenyl) was synergistic, yielding a compound with nanomolar affinity and some selectivity for adenosine A_1 receptors.

B. Use of Pharmacophore Structure

Implicit in the above example is the definition of the pharmacophore for adenosine receptor antagonism. Apparently the flat aromatic 6:5 heterobi-cyclic structure with its typical charge distribution was such a pharmaco-phore. A search of the 1995 MEDLINE database yielded almost 100 hits for this keyword, and on inspection it appeared that pharmacophore models have been developed for virtually each G-protein-coupled receptor cur-rently known. A pharmacophore is often described only by a set of distance constraints between atoms or centroids (such as the center of a phenyl ring) to which some typical characteristics such as hydrogen-bonding capac-ity (donor or acceptor) and charge may be added. Special computer pro-grams are even available that assist in extracting a pharmacophore model from known structure-activity relationships. Today pharmacophores are also used to search 3D databases of known compounds, either publicly/commercially available (e.g., the Cambridge Structural Database with X-ray structures and others) or privately owned (e.g., in-house databases of pharmaceutical companies). The aim of this "data mining" is, of course, to provide new lead structures to the medicinal chemist. This area is of tremendous commercial interest, but is beyond the scope of this review. The reader is referred to recent reviews on this subject (11,12).

IV. THE SECOND APPROACH: THE RECEPTOR

In the example of the foregoing paragraph the adenosine receptor was nothing more than a black box that remained largely unused in the process of designing new adenosine receptor antagonists. Now we will move step by step from the ligand toward the receptor. As an example we will use the prototypic β_2-adrenoceptor and its ligands. With no experimentally determined receptor structure available, circumstantial evidence becomes the sole source for information. Information embedded in the ligands—agonists such as isoproterenol and antagonists such as propranolol—is again our starting point. We derived quantitative structure-affinity relationships for both agonists and antagonists. Chemically agonists are usually phenyl-ethanolamines, whereas antagonists, by virtue of the oxymethylene bridge between aromatic nucleus and aliphatic side chain, are mainly phenoxypro-panolamines (see Fig. 6). At physiological pH the side-chain amine function in both agonists and antagonists is positively charged. Within each of the

(A)   agonists (phenylethanolamines)

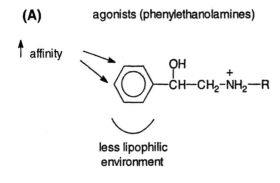

less lipophilic
environment

(B)   antagonists (phenoxypropanolamines)

more lipophilic
environment

Figure 6 Qualitative summary of QSAR analysis of (A) β-adrenoceptor agonists (phenylethanolamines) and (B) antagonists (phenoxypropanolamines). (Reprinted with permission from Ref. 24.)

two classes we selected compounds with identical side chains but different in the substitution patterns of their aromatic ring systems. Their affinities for the β_2-adrenoceptor, determined in radioligand-binding studies, were correlated with the physicochemical characteristics of the aromatic ring systems in the compounds (13). The results of these multiple regression correlations are summarized qualitatively in Figure 6. Apparently, introduction of *meta* and/or *para* substituents to the phenyl rings in both classes has discriminating effects. This substitution pattern is favorable for agonist affinity, whereas, in contrast, antagonist affinity is decreased. Furthermore, from the contribution of lipophilicity to affinity it emerged that its influence is different for agonists compared to antagonists. The receptor region accommodating the phenyl ring of the antagonists is significantly more lipophilic than the surroundings of the aromatic moiety of the agonist. These results question all ligand models in which the aromatic nuclei of agonists and antagonists coincide. The apparent assumption in such models is that for a competitive interaction a maximal degree of overlap between agonists and antagonists is essential. Thus, Jen and Kaiser (14) have proposed a "rigid" bicyclic structure for the phenoxypropanolamines, which can be superimposed on the phenylethanolamines in a way that both phenyl rings are in the same position. More recently, Strosberg and co-workers (15) superimposed the aromatic moieties as well as the protonated amino function of both agonists and antagonists when docking these compounds in their β_2-adrenoceptor model.

As mentioned before, a partial, but mutually exclusive, overlap of agonists and antagonists can also be a molecular explanation for competitive antagonism. From site-directed mutagenesis studies on the β_2-adrenoceptor it became evident that substitution of Asp-113 (somewhere halfway in helix III if we think of GPCRs as seven-helix bundles) with asparagine led to dramatic decreases in both agonist and antagonist affinity. Since other G-protein-coupled receptors that bind biogenic amines such as serotonin and acetylcholine also have this aspartate present, it was concluded that this residue was the counterion for the protonated amino function in the ligands (16). It has also been postulated that two serine residues in helix V of the β_2-adrenoceptor (Ser-204 and Ser-207, respectively) are involved in binding the *meta* and *para* positions on the agonist phenyl ring (e.g., the catechol—two hydroxyl groups—moiety in isoproterenol) without significant effects on antagonist binding (17). This quite hydrophilic interaction via hydrogen bonds could form an explanation for the relatively modest contribution of lipophilicity to agonist affinity apparent from the QSAR analysis. Consequently, the receptor region around the aromatic nucleus of antagonists should contain amino acid residues more lipophilic than the two serines. For antagonists another anchor site was found in an asparagine residue in helix VII (Asn-312). It was proposed that this residue forms a hydrogen bond with the oxygen atom in the oxymethylene bridge. All these findings are incorporated in Figure 7, from which it is suggested that Asp-113 is the crucial element in the competition between isoproterenol (agonist) and propranolol (antagonist), whereas the other residues are relatively specific for either agonists or antagonists.

A. Receptor Models

At this stage the question emerges inevitably whether it would be possible to create 3D models of GPCRs, allowing for a further inspection of putative

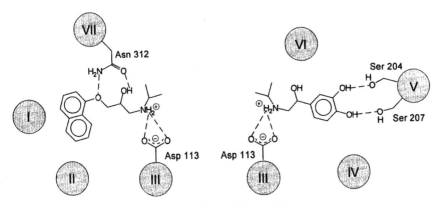

Figure 7 Binding sites for β-adrenoceptor agonists (right: isoproterenol) and antagonists (left: propranolol) as suggested from site-directed mutagenesis studies (shaded circles represent transmembrane α-helices).

binding sites. Since Figure 7 is not more than a visualization of QSAR and mutation data, it would be helpful to have additional information on other amino acid residues that surround the ligands, such as the ones close to the aromatic ring of propranolol, anticipated to be hydrophobic. As no experimental data are available on the intact molecular architecture of GPCRs, receptor modeling based on homology with other proteins is the only, albeit very speculative, option.

The pivotal suggestion that the GPCR family bears structural homology to bacteriorhodopsin, a proton pump present in the cell wall of *Halobacterium halobium*, has been an impetus to our current understanding of receptor structure. Bacteriorhodopsin is of similar size and shares other characteristics with the mammalian G-protein-coupled receptors. It displays the typical seven-transmembrane α-helical architecture that was mentioned before (3), and, hence, it could serve as a template for receptor modeling. Furthermore, retinal, its endogenous ligand, binds also to its G-protein-coupled mammalian equivalent rhodopsin. On the other hand, there is virtually no sequence (i.e., amino acid) homology between the two types of proteins. Another problem is that the structural map of bacteriorhodopsin has a resolution of 3.5 Å at the best. Hibert and co-workers (4,5) were the first to analyze in depth the opportunities and pitfalls associated with this approach. In the meantime the overall structure of rhodopsin has also been published, although not at a resolution good enough for identifying individual amino acid residues (2). Baldwin combined structural information on rhodopsin with a sequence analysis of other GPCRs to suggest a probable arrangement of the seven α-helices (6). Although met with scepticism, the receptor models developed on the basis of either bacteriorhodopsin or rhodopsin have been useful in clarifying the putative molecular basis of receptor-ligand recognition, in particular when combined with and adjusted to available pharmacological and structure-activity relationship data.

B. The Use of 3D Receptor Models

Over the years we have developed 3D receptor models for biogenic amines such as epinephrine (β_2-adrenoceptor) and serotonin (5-HT$_{1A}$ receptor), for adenosine (adenosine A$_1$, A$_{2A}$, and A$_3$ receptors), and for peptides (CCK$_A$ receptor) (18–23). Similarly, many more models on similar and other GPCRs have been published. For a detailed collection and discussion the reader is referred to Ref. 24. In some cases different models exist for one and the same receptor due to different interpretation of existing data, again indicating the high degree of speculation in such models. Therefore, receptor models should be seen as a potential source of novel hypotheses and ideas amenable to validation and verification rather than as accurate descriptions of biochemical reality. We have used receptor models mainly for the selection of amino acid candidates for site-directed mutagenesis to further probe the ligand-binding site. The β_2-adrenoceptor will be discussed below once more as an example. Second, SAR and QSAR data may be visualized and rationalized to some extent in receptor models, sometimes giving rise to the design of new ligands. These aspects will be discussed in the light of 5-HT$_{1A}$/D$_2$ receptor selectivity of arylpiperazine compounds.

C. Application of 3D Receptor Models: The β_2-Adrenoceptor

Based on the atomic coordinates of bacteriorhodopsin and the amino acid sequence of the human β_2-adrenoceptor (25), a homologous 3D model of the membrane-spanning α-helices of the receptor was built, essentially according to a method described by Lewell (26). The overall architecture of our receptor models is virtually identical to that of bacteriorhodopsin because of the procedure followed, i.e., on-screen mutation and subsequent minimization of the side chains only. We used the information from site-directed mutagenesis and QSAR studies to dock agonists and antagonists, along the lines that led to Figure 7. It appeared relatively straightforward to position isoproterenol in such a way that the two phenolic hydroxy groups of the catechol function interact with the two serine residues on helix V. A hydrogen bond is formed between the oxygen atom of Ser-204 and the hydrogen atom of the *meta*-hydroxy group of isoproterenol. A similar hydrogen bond occurs between Ser-207 and isoproterenol's *para*-hydroxy group. An ionic interaction is seen between the protonated amino group of the ligand and the negatively charged aspartate residue 113 in helix III. A similar procedure can be applied to antagonists such as propranolol. Again, it appeared feasible to visualize the mutation data. Because of its positively charged amino group, propranolol also interacts with Asp-113, whereas Asn-312 may form some kind of double hydrogen bond with the β-hydroxy group and the oxygen atom of the ether function. When we consider in more detail all amino acids that surround the two molecules within 4 Å (i.e., the binding pocket), it is readily apparent that the naphthyl nucleus of propranolol is in a hydrophobic region, consisting of tryptophan, methionine, phenylalanine, and tyrosine residues (data not shown). This is unlike the area where the catechol function of isoproterenol binds; the two serine residues (204 and 207) form a more hydrophilic environment. As mentioned before, these findings are in agreement with the QSAR data described above; the different contribution of lipophilicity to affinity can be explained as different receptor environments.

From the receptor modeling studies a clue to understand the ligands' stereoselectivity also emerges. It is well known that the $(-)$-isomers of both agonists (R designation) and antagonists (S designation) have considerably higher affinity for β_2-adrenoceptors than the corresponding $(+)$-isomers. Depending on optical purity and receptor recognition, stereoselectivity ratios can be as high as 1000 (27). This critical property, however, has not yet been attributed to specific receptor residues. A direct consequence of the proposed limited overlap between agonists and antagonists on the β-adrenoceptor is the unequivalence of the two β-hydroxy groups that determine the stereoselectivity. In our model Asn-312 (helix VII) interacts with the β-hydroxy group of propranolol, but not with the same functionality in isoproterenol. This may agree with site-directed mutagenesis data, not only on β-adrenoceptors but also on $5HT_{1A}$ receptors that recognize phenoxypropanolamines too (28,29).

Lewell (26) proposed that a serine residue in helix IV (Ser-161) forms a hydrogen bond with the β-hydroxy group in agonists. However, mutation of this residue into alanine had no effect on agonist affinity (17). Hibert and co-workers in their seminal modeling paper (4) proposed Ser-165 instead, but mutagenesis studies did not yield more definitive proof, since substitution into alanine failed to produce immunoreactive protein (17). The conformational flexibility of isoproterenol while anchored via the catechol and amine functions was still considerable in our modeling efforts. Indeed, residues in helix IV such as the two serines can be brought near the ethanolamine side chain. On the other hand, residues in helix VI are also in close proximity. In particular, an asparagine in position 293 was capable of forming a hydrogen bond with isoproterenol's β-hydroxy group. We mutated Ser-165 into Ala, and Asn-293 into Leu, respectively. We were able to express the alanine mutant transiently in HEK-293 cells, but no effects were observed on the binding of either isoproterenol stereoisomer.

However, the leucine mutant showed substantial loss of stereospecificity of isoproterenol in receptor binding studies, and in the activation of adenylate cyclase after stable expression of the leucine mutant in CHO cells. Intriguingly, the loss of affinity on the leucine mutant in a series of agonists appeared strongly correlated with the intrinsic activity of the compounds. Stereospecific recognition of antagonists such as propranolol was unaltered in this mutant (30).

D. Application of 3D Receptor Models: 5-HT$_{1A}$/D$_2$ Receptor Selectivity

In Tables 2 and 3 radioligand-binding data are gathered for a series of arylpiperazines with high affinity for 5-HT$_{1A}$ receptors but with varying degrees of selectivity with respect to dopamine D$_2$ receptors. From the data in Table 2 it is obvious that changes in the ethylcarboxamido substituent influence the two receptor affinities to a similar, although not identical, extent. The change from a methyl to a phenyl and other aromatic substituents is very favorable in terms of affinity, but not so much for selectivity, suggesting that this side chain is recognized equally well by both receptors. Changes in the N^1-aryl part (Table 3), however, influence selectivity rather than affinity, yielding the relatively selective 5-HT$_{1A}$ receptor agonist flesinoxan (the last compound in Table 3).

A three-dimensional model, similar to the one described for the β_2-adrenoceptor, was developed for the 5-HT$_{1A}$ receptor (19). Flesinoxan and

Table 2 K_i Values (nM ± SEM) of N^4-Substituted 2-methoxy-N^1-phenylpiperazines on 5-HT$_{1A}$[a] and Dopamine D$_2$[b] receptors

R	5-HT$_{1A}$	D$_2$	D$_2$/5-HT$_{1A}$
CH$_3$	800 ± 70	1100 ± 40	1.4
Phenyl	1.3 ± 0.3	13 ± 4	10
4-Fluorophenyl	1.0 ± 0.3	5.0 ± 0.8	5.0
2-Thienyl	1.6 ± 0.3	17 ± 2	11
2-Furanyl	18 ± 4	73 ± 21	4.1
4-Pyridyl	18 ± 1	230 ± 40	13

[a] From displacement studies with [^{3}H]-8-OH-DPAT (2-dipropylamino-8-hydroxy-1,2,3,4-tetrahydronaphthalene) on rat frontal cortex membranes.
[b] From displacement studies with [^{3}]spiperone from rat striatal membranes.

Table 3 K_i Values (nM ± SEM) of N^4-Substituted N^1-Arylpiperazines on 5-$HT_{1A}{}^a$ and Dopamine $D_2{}^b$ Receptors

N^1-Aryl substituent (Ar)	5-HT_{1A}	D_2	D_2/5-HT_{1A}
	4.9 ± 1.5	92 ± 17	19
	1.0 ± 0.3	5.0 ± 0.8	5.0
	0.15 ± 0.01	5.3 ± 0.9	35
	0.30 ± 0.04	14 ± 1	47
	1.7 ± 0.2	140 ± 30	83

[a] From displacement studies with [³H]-8-OH-DPAT (2-dipropylamino-8-hydroxy-1,2,3,4-tetrahydronaphthalene) on rat frontal cortex membranes.
[b] From displacement studies with [³H]spiperone from rat striatal membranes.

related compounds were docked into the 5-HT_{1A} receptor model, very much like the docking of similar nor-compounds, i.e., compounds lacking the N^4-ethylcarboxamido substituent (31). In Figure 8 residues that are part of the putative 5-HT_{1A} receptor binding site, defined as all residues within a sphere of 4 Å from flesinoxan, are in bold. The corresponding residues in the D_2 receptors are partly identical and partly different. The residues that surround the N^4-substituent in the 5-HT_{1A} receptor model (i.e., Asp-116, Cys-120, Ser-123, Ile-24, Leu-127: helix III; Phe-354, Cys-

| 5-HT$_{1A}$ | 109 | C D L F I A L D V L C C T S S I L H L C A I | III |
| D$_2$ | 107 | C D I F V T L D̲ V M M C̲ T A S̲ I̲ L N L̲ C A I | |

| 5-HT$_{1A}$ | 153 | A A A L I S L T W L I G F L I S I P P M L | IV |
| D$_2$ | 152 | V T V M I A I *V* W̲ V L *S* F T I S C P L L F | |

| 5-HT$_{1A}$ | 195 | Y T I Y S T F G A F Y I P L L I M L V L Y | V |
| D$_2$ | 189 | F V V Y S̲ *S* I V S *F* Y V P F I V T L L V Y | |

| 5-HT$_{1A}$ | 347 | L G I I M G T F I L CW L P F F I V A L V L P F | VI |
| D$_2$ | 376 | L A I V L G V *F* I I C̲W̲L̲ P *F* *F* I T *H* *I* L N I H | |

| 5-HT$_{1A}$ | 381 | L G A I I N W L G Y S N S L L N P V I Y A Y F N | VII |
| D$_2$ | 408 | L Y S A F T W L G Y V N *S* A V N P I I Y T T F N | |

Figure 8 Aligned sequences of the putative transmembrane domains III–VII of the rat 5-HT$_{1A}$ and the rat dopamine D$_2$ receptor. Marked residues were found to be part of the binding site (residues within a sphere of 4 Å from flesinoxan) in the 5-HT$_{1A}$ receptor model. There are no such residues in helices I and II. Residues in the D$_2$ receptor sequence that differ from those in the modeled binding site are in italics. Residues in the D$_2$ sequence that correspond with those in the modeled 5-HT$_{1A}$ receptor binding site are underlined.

357, Trp-358, Phe-361: helix VI; and Ser-393: helix VII) are identical to the ones present in the D$_2$ receptor. This may indeed explain the similar SARs of the N^4-substituent at 5-HT$_{1A}$ and D$_2$ receptors. In contrast, a number of residues in the model that are in contact with the N^1-heteroaryl moiety (i.e., Thr-160, Trp-161, Gly-164: helix IV; Ser-199, Thr-200, Ala-203, Phe-204: helix V; Phe-362, Ala-365, Leu-366: helix VI) differ from those in D$_2$ receptors. Observed differences are Thr-160↔Val, Gly-164↔Ser, Thr-200↔Ser, Ala-203↔Ser, Ala-365↔His, and Leu-366↔Ile, respectively. In particular, the dioxane ring of flesinoxan is in close proximity to the small and lipophilic Ala-365 in the 5-HT$_{1A}$ receptor model. This residue is replaced with a more bulky and polar histidine in D$_2$ receptors, which may explain the relative selectivity of flesinoxan for the 5-HT$_{1A}$ receptor.

In conclusion, 3D receptor models can only be rough approximations of "biochemical reality" and are best seen as visualizations of available experimental data. Nevertheless, from them intriguing suggestions of how ligands may bind emerge, they provide clues for further experimentation, and help us in thinking at an almost atomic level about important pharmacological parameters such as affinity, selectivity, and intrinsic efficacy.

E. Advanced Sequence Analysis

The triad sequence-structure-function is one of the central paradigms in protein research. Today, over 800 members of G-protein-coupled receptors have been cloned and sequenced. Second, much functional data is available, ranging from glycosylation status, selectivity for other proteins, such as G-protein subunits and subtypes thereof, and for ligands, such as agonists, antagonists, and inverse agonists, to desensitization and down-regulation. As discussed above, 3D structural data is scarce, and, hence, 3D receptor models can only be preliminary, albeit informative. From a more principal point of view it would be useful to bypass the structural element and link sequence with function directly. In that case, for instance, there would be little need any more to rely on the atomic coordinates of bacteriorhodopsin. The class of GPCRs in itself then includes ample information on, e.g., conserved residues within subclasses, likely boundaries of transmembrane domains, and so forth. It will take some time, however, to process all intrinsic, "hidden" information from these primary sequences statistically to derive more definitive clues concerning receptor structure and function.

A first approach may be found in multiple sequence alignments, effectively yielding a matrix of amino acid residues within one sequence versus residues corresponding positionally over an entire range of sequences available. We used a "profile"-based alignment procedure developed by Sander and Schneider (32). This profile is directly related to the frequency of occurrence of residue types in the sequences, and avoids potential bias when a single arbitrary sequence is used as starting point. The eventual matrix can be used for any conceivable search strategy.

Such a strategy is the analysis of correlated mutations. It has been observed that there is a tendency of residues to either stay conserved or mutate in tandem between (sets of) sequences. As an example, Goebel et al. (33) found that the correlation coefficient is related to the chance that the residue pair is in contact in the structure. This is a *structural* correlate, but one could wonder whether *functional* correlates (for instance, between ligand-binding and one or more residues) can also be retrieved from a matrix of aligned sequences. In Figure 9 this concept is illustrated for a hypothetical set of 10 sequences.

We applied this strategy to the aryloxypropanolamine binding site as described above for the β-adrenoceptor (34). We gathered affinity data for both propranolol and pindolol from the literature, and learned that these two compounds are also active on some subtypes of the 5-HT receptor (35,36). It appeared that 27 and 36 GPCR sequences (including species homologs) were available for which the affinities of pindolol and propranolol, respectively, had been determined. All receptors with high affinity

	sequence position	compound
	1 2 3 4 5 6 7 8 9 10	1 2 3
sequence number		
1	P I A V H F A A A G	+ + -
2	P I V V H G S A A G	+ + +
3	P V I D H S S G A G	+ + +
4	G I L R H T S G A G	- - -
5	P I Y I H V T S A F	+ + -
6	P I G L W V A S A F	+ - -
7	P A T C W W A T A F	+ - -
8	G I S C W W S V A L	- - -
9	P I C T W L S I A L	+ - -
10	P I C T W L S L L I	+ - -

Figure 9 A hypothetical set of 10 sequences of 10 residues each. The affinities of three hypothetical compounds for each receptor are coded with pseudo-residue "+" for high affinity, and "−" for low affinity. Binding of compound 1 is correlated only with the presence of a proline (**P**) at position 1. Compound 2 correlates with the presence of the same proline (**P**) at position 1 *and* a histidine (**H**) at position 5. Compound 3 depends on the combined presence of proline (**P**) 1, histidine (**H**) 5, and serine (**S**) 7. (Slightly adapted and reprinted with permission from Ref. 24.)

(pK$_i$ ≥ 7.0) were coded "+", the others "−". These codes are regarded as "pseudo-residues" (see also Fig. 9). It appeared that only the asparagine residue in helix VII (Asn-312 in the β-adrenoceptor) displayed a 100% correlation with high affinity for both compounds (the pseudo-residues, coded +), despite the fact that two different data sets were used. Since many mutation studies (also reviewed in Ref. 34) had already shown this asparagine residue to be highly relevant for binding these compounds, the analysis showed the potential of this method correlating sequence elements (Asn-312) with function (pseudo-residues).

Similarly, we analyzed the binding of 5-carboxamidotryptamine to serotonin receptor subtypes. This analog of serotonin displays higher affinity for 5-HT$_{1A,B,D}$, 5-HT$_5$, and 5-HT$_7$ receptors than serotonin itself (coding +), whereas the affinity is remarkably lower (coding −) on 5-HT$_{1E,F}$, 5-HT$_2$, and 5-HT$_6$ receptors (35). For 23 sequences available it appeared that only one residue, a proline in helix VI, was positively correlated with high affinity for 5-carboxamidotryptramine. So far this proline has not been mutated, suggesting that this analysis of correlated mutations may indeed yield directions for further, mandatory, experimentation.

V. CONCLUSIONS

Figure 1 is based on a painting by the Belgian surrealist René Magritte. The title given by the artist is *Le sourire du diable—The smile of the devil.* Although some scientists doubt the existence of this malicious creature, it may be that the artist foreknew the complexities inherent to the "lock and key" concept. It was the thrust of the present chapter to show and discuss these complexities in receptor-based drug design, how to possibly deal with or even bypass them, to derive meaningful clues for drug design. It is hoped that combined efforts in computational and organic chemistry as well as molecular biology may lead to validation and optimization of the proposed models and strategies, contributing to the more rational design of new chemical entities.

Note Added in Proof

Further structural analysis of both bacteriorhodopsin and mammalian rhodopsin has corroborated the concept of a similar overall architecture of the two proteins. Current status is that the 3D structure of bacteriorhodopsin including its extra- and intracellular domains has been largely resolved (37,38). Rhodopsin's structure is far from complete, however, but the relative orientation of the seven helices appears quite well established now (39), putting emphasis on the important, albeit subtle, differences between the mammalian and bacterial form of the protein.

ACKNOWLEDGMENTS

The authors thank Gert Vriend (EMBL, Heidelberg, Germany) and Martin Lohse (Univ. Würzburg, Germany) for helpful discussions and collaboration.

REFERENCES

1. Fischer E. Einfluss der Configuration auf die Wirkung der Enzyme. Berl Dtsch Chem Ges 1894; 27:2985–2993.
2. Schertler GFX, Villa C, Henderson R. Projection structure of rhodopsin. Nature 1993; 362:770–772.
3. Henderson R, Baldwin JM, Ceska TA, Zemlin F, Beckmann E. Downing KH. Model for the structure of bacteriorhodopsin based on high-resolution electron cryo-microscopy. J Mol Biol 1990; 213:899–929.
4. Hibert MF, Trumpp-Kallmeyer S, Bruinvels A, Hoflack J. Three-dimensional models of neurotransmitter G-binding protein-coupled receptors. Mol Pharmacol 1991; 40:8–15.

5. Trumpp-Kallmeyer S, Hoflack J, Bruinvels A, Hibert MF. Modeling of G protein-coupled receptors: application to dopamine, adrenaline, serotonin, acetylcholine, and mammalian opsin receptors. J Med Chem 1992; 35:3448–3462.

6. Baldwin JM. The probable arrangement of the helices in G protein-coupled receptors. EMBO J 1993; 12:1693–1703.

7. Hoflack J, Trumpp-Kallmeyer S, Hibert M. Re-evaluation of bacteriorhodopsin as a model for G protein-coupled receptors. TIPS 1994; 15:7–9.

8. Marshall GR, Motoc I. Approaches to the conformation of the drug bound to the receptor. In: Burgen ASV, Roberts GCK, Tute MS, eds. Molecular Graphics and Drug Design. Amsterdam: Elsevier, 1986:115–156.

9. Van Galen PJM, Van Vlijmen HWT, IJzerman AP, Soudijn W. A model for the antagonist binding site on the adenosine A_1 receptor based on steric, electrostatic and hydrophobic properties. J Med Chem 1990; 33:1708–1713.

10. Van Galen PJM, Nissen P, Van Wijngaarden I, IJzerman AP, Soudijn W. 1H-Imidazol[4,5-c]quinolin-4-amines: novel non-xanthine adenosine antagonists. J Med Chem 1991; 34:1202–1206.

11. Humblet C, Dunbar JB Jr. 3D database searching and docking strategies. Annu Rep Med Chem 1993; 28:275–284.

12. Martin YC. 3D Database searching in drug design. J Med Chem 1992; 35:2145–2154.

13. IJzerman AP, Aué GHJ, Bultsma T, Linschoten MR, Timmerman H. Quantitative evaluation of the β_2-adrenoceptor affinity of phenoxypropanolamines and phenylethanolamines. J Med Chem 1985; 28:1328–1334.

14. Jen T, Kaiser JC. Adrenergic agents 5. Conformational analysis of 1-alkylamino-3-aryloxy-2-propanols by proton magnetic resonance studies. Implications relating to the steric requirements of adrenoceptors. J Med Chem 1977; 20:693–698.

15. Strosberg AD, Camoin L, Blin N, Maigret B. In receptors coupled to GTP-binding proteins, ligand-binding and G-protein activation is a multistep dynamic process. Drug Design Disc 1993; 9:199–211.

16. Strader CD, Sigal IS, Register RB, Candelore MR, Rands E, Dixon RAF. Identification of residues required for ligand-binding to the β-adrenergic receptor. Proc Natl Acad Sci USA 1987; 84:4384–4388.

17. Strader CD, Candelore MR, Hill WS, Sigal IS, Dixon RAF. Identification of two serine residues involved in agonist activation of the β-adrenergic receptor. J Biol Chem 1989; 264:13572–13578.

18. IJzerman AP, Zuurmond HM. Molecular modelling of β-adrenoceptors. In: Findlay J, ed. Membrane Protein Models: Experiment, Theory and Speculation Oxford: BIOS Scientific Publishers, 1996; 133–144.

19. Kuipers W, Van Wijngaarden I, IJzerman AP. A model of the serotonin 5-HT_{1A} receptor: agonist and antagonist binding sites. Drug Design Disc 1994; 11:231–249.

20. IJzerman AP, Van Galen PJM, Jacobson KA. Molecular modeling of adenosine receptors. I. The ligand-binding site on the A_1 receptor. Drug Design Disc 1992; 9:49–67.

21. IJzerman AP, Van der Wenden EM, Van Galen PJM, Jacobson KA. Molecular modeling of adenosine receptors. II. The ligand-binding site on the rat A_{2a} receptor. Eur J Pharmacol Mol Pharmacol Sect 1994; 268:95–104.

22. Van Galen PJM, Van Bergen AH, Gallo-Rodriguez C, Melman N, Olah ME, IJzerman AP, Stiles GL, Jacobson KA. A binding site model and structure-activity relationships for the rat A_3 adenosine receptor. Mol Pharmacol 1994; 45:1101–1111.

23. Van der Bent A, IJzerman AP, Soudijn W. Molecular modelling of CCK-A receptors. Drug Design Disc 1994; 12:129–148.

24. Findlay J. Membrane Protein Models: Experiment, Theory and Speculation. Oxford: BIOS Scientific Publishers, 1996.

25. Kobilka B. Adrenergic receptors as models for G protein-coupled receptors. Annu Rev Neurosci 1992; 15:87–114.

26. Lewell XQ. A model of the adrenergic beta-2 receptor and binding sites for agonist and antagonist. Drug Design Disc 1992; 9:29–48.

27. Morris TH, Kaumann AJ. Different steric characteristics of β_1- and β_2-adrenoceptors. Naunyn-Schmiedeberg's Arch Pharmacol 1984; 327:176–179.

28. Suryanarayana S, Kobilka BK. Amino acid substitutions at position 312 in the seventh hydrophobic segment of the β_2-adrenergic receptor modify ligand-binding specificity. Mol Pharmacol 1993; 44:111–114.

29. Guan XM, Peroutka SJ, Kobilka BK. Identification of a single amino acid residue for the binding of a class of β-adrenergic receptor antagonists to 5-hydroxytryptamine$_{1A}$ receptors. Mol Pharmacol 1992; 41:695–698.

30. Wieland K, Zuurmond HM, Krasel C, IJzerman AP, Lohse MJ. Involvement of Asn-293 in stereospecific agonist recognition and in activation of the β_2-adrenergic receptor. Proc Natl Acad Sci USA 1996; 93:9276–9281.

31. Kuipers W, Van Wijngaarden I, Kruse CG, Ter Horst–Van Amstel M, Tulp MTM, IJzerman AP. N^4-Unsubstituted N^1-arylpiperazines as high-affinity 5-HT$_{1A}$ receptor ligands. J Med Chem 1995; 38:1942–1954.

32. Sander C, Schneider R. Database of homology-derived protein structures and the structural meaning of sequence alignment. Proteins 1991; 9:56–68.

33. Goebel U, Sander C, Schneider R, Valencia A. Correlated mutations and residue contacts in proteins. Proteins 1994; 18:309–317.

34. Kuipers W, Oliveira L, Paiva ACM, Rippmann F, Sander C, Vriend G, IJzerman AP. Analysis of G protein-coupled receptor function. In: Findlay J, ed. Membrane Protein Models: Experiment, Theory and Speculation. Oxford: BIOS Scientific Publishers, 1996:27–45 (The method described in this reference is part of the molecular modeling program WHAT IF. For further information contact the author: email: vriend@embl-heidelberg.de.)

35. Boess FG, Martin IL. Molecular biology of 5-HT receptors. Neuropharmacology 1993; 33:275–317.

36. Seeman PMD. Receptor Tables, Vol 2. Drug Dissociation Constants for Neuroreceptors and Transporters. Toronto: SZ Research, 1993.

37. Grigorieff N, Ceska TA, Downing KH, Baldwin JM, Henderson R. Electron-crystallographic refinement of the structure of bacteriorhodopsin. J Mol Biol 1996; 259:393–421.

38. Kimura Y, Vassylyev DG, Miyazawa A, Kidera A, Matsushima M, Mitsuoka K, Murata K, Hirai T, Fujiyoshi Y. Surface of bacteriorhodopsin revealed by high-resolution electron crystallography. Nature 1997; 389:206–211.
39. Unger VM, Hargrave PA, Baldwin JM, Schertler GFX. Arrangement of rhodopsin transmembrane α-helices. Nature 1997; 389:203–206.

Recombinant G-Protein-Coupled Receptors as Screening Tools for Drug Discovery

A. Donny Strosberg
Institut Cochin de Génétique Moléculaire (ICGM), Paris, France

Prabhavathi B. Fernandes
Small Molecule Therapeutics, Inc., Monmouth Junction, New Jersey

I. INTRODUCTION

Membrane receptors may be studied both by ligand-binding and by effector triggering. Assays to identify agonists or antagonists make use of both properties. Initial, high-throughput screening may be used to screen large combinatorial libraries of synthetic or natural compounds to identify "lead" candidates. Once purified, these leads may be assayed in second-messenger assays to differentiate agonists from antagonists.

Receptor-based screening assays have largely progressed through advances in genetic engineering, cell biology, and bioelectronics. The molecular identification of multiple receptor subtypes, and their expression as single recombinant proteins in model systems, the coupling of these receptors to natural or contrived signal transmission proteins, and the automated and sensitive analysis of binding and signaling constitute major characteristics of the new assay systems now used by pharmaceutical laboratories keen to delay as long as possible costly and complex screening based on animal tissues or live animals.

The power of these new in vitro multiarray assays translates into impressive numbers of analyses performed simultaneously on the same compounds, thus eliminating, at an early stage, potential side effects due to multiple specificities. We will summarize here some of these new assays as they are being applied to G-protein-coupled receptors (GPCR).

II. THE G-PROTEIN-COUPLED RECEPTOR FAMILY

A. General Structure

Since the cloning of the first β_2-adrenoceptor only a decade ago, over 300 members of the GPCR family of a variety of species have been analyzed at the molecular level (see Refs. 1,2).

All these proteins display the same general structure: a single polypeptide chain characterized by an extracellular N-terminal region of variable length, seven hydrophobic, probably transmembrane regions, linked by extracellular and intracellular mostly short loops, and an intracellular C-terminal region also of variable length. These proteins display a ligand-binding site, which may be composed solely of residues from the transmembrane regions, as is the case for the biogenic amines and for other small ligands including peptides, or which may also involve residues from the extracellular regions, as was shown for other peptides (e.g., melanocortins, chemokines) or larger hormones. In some cases, the recognition site for synthetic antagonists was reported to be physically distinct from that for the natural agonist.

The regions involved in coupling to the G-proteins are composed of the membrane proximal parts of the intracellular loops and the C-terminus. All seven transmembrane regions are indispensable for ligand-binding to any of the members of the family, with the possible exception of receptors for the large glycohormones (TSH, FSH, LH) for which initial studies reported that the binding site is extracellular. However, even in this subfamily, signaling also requires the integrity of all seven transmembrane regions as is true for all other members of the family.

B. Subtypes Encoded by Intronless Genes

The exponential expansion of the GPCR family over the last 10 years has considerably benefitted from the similarity in structure not so much of the proteins, but more practically, from the corresponding genes. Indeed, after the initial affinity purification, peptide sequencing, and cloning of the first few members of the family, mostly adrenoceptors, it soon became obvious that these were encoded by intronless genes. This finding opened the way for rapid cloning of homologous genes by the use of hybridization or polymerase chain reaction triggered by "primers" based on the very conserved transmembrane region encoding oligonucleotides (3). These methodological breakthroughs resulted in the identification not only of the genes encoding receptors predicted to exist on the basis of pharmacology, but also of a number of additional unexpected homologous subtypes.

C. Subtypes Encoded by Genes Containing Introns

After the initial identification of a number of receptor subtypes encoded by intronless genes, it became obvious that such a property was not general for all GPCR: in fact, quite a number of members are now known to be encoded by genes containing introns. Interestingly, however, it appears that the exon-intron organization is quite well conserved in subfamilies. Most adrenoceptors (whether α_1, α_2, or β) are thus devoid of introns, except for the β_3AR (4), whereas the seven dopamine receptors are mostly encoded by genes that contain introns situated in homologous positions. Despite the complications due to the existence of introns, the power of the PCR-based or high-throughput sequencing techniques soon helped identify many receptors belonging to the GPCR family.

D. Orphan Receptors

In addition to new subtypes homologous to those predicted from the pharmacology, a number of receptor genes were identified at the molecular level, but not at the functional level (5). These "orphan" receptors, for which the natural ligand remains unknown, constitute a considerable challenge for today's molecular pharmacologists. Indeed, these receptors could serve to identify new natural ligands (unknown hormones, neurotransmitters) and thus provide new targets for drugs. Others could be the hitherto unidentified receptors for amyline, CGRP, Gaba-B, or the leukotriene receptor subtypes LTB4, LTC4, and LTD4, or the histamine H3 receptor. The discovery of the orphan receptors has contributed to the impetus to develop new "generic" signal transmission assays for screening combinatorial libraries of compounds (6–8).

Hundreds of orphan receptor sequences are being identified from the efforts to sequence the human genome. The function of these receptors could be identified by using screening to identify ligands for the orphan receptors and then using the ligands in vivo to determine function.

III. EXPRESSION SYSTEMS

The need to individually express, in similar cellular environments, large numbers of different receptors has led to the development of a number of systems, which include cells of a variety of origins: microbial, mammalian, insect, amphibian. Each of these systems has its own advantages.

A. Microbial Expression Systems

Microbial expression systems were initially discarded because functional expression attempts with bacteriorhodopsin, the model protein for seven-transmembrane receptors, failed, due to faulty folding. Another reason why expression in microorganisms was rejected was either lack of protein glycosylation (*Escherichia coli* bacteria) or altered glycosylation (*Saccharomyces cerevisiae* yeasts). When it was shown that the polysaccharides on GPCR play no role in either binding or signal transmission, microbial expression systems became again a valuable alternative to the more expensive and complex mammalian systems.

After the initial functional expression of the human β_1- and β_2-adrenoceptors in *E. coli* (9), a number of other GPCR have been successfully expressed in *E. coli* (see Refs. 10,11a,b). Ease to develop, low cost, excellent reproductibility in the binding studies, low background, wide applicability, and automation through the use of robots combine to make expression in *E. coli* very good cellular support to design primary screens for lead compounds.

Unfortunately, bacteria lack endogenous G-proteins and corresponding effector systems. While Gα subunits can be expressed in bacteria (12), and/or recombined with GPCR (13,14) to reconstitute receptor–G-protein complexes, the absence of compatible resident adenylyl cyclase still prevents demonstration of agonist activity.

Yeasts, on the other hand, contain the pheromone mating signaling system composed of G-protein-like components and a responsive effector. Genetic engineering led King et al. (15) to develop a hybrid system by which a mammalian β_2-adrenoceptor was coupled to the yeast effector, and signaling could be triggered by epinephrine. This system, which has reportedly been difficult to reproduce by several other laboratories, has now apparently been optimized by at least two groups, one of which has recently started to report data, suggesting that yeast functional expression systems for GPCR have now become more widely applicable (16–18). Expression of GPCR in yeast has been coupled with expression of peptide ligands of GPCR to develop an autocrine system for screening for receptor antagonists. This simple screening system, may in part eliminate the need for costly cell cultures as well as for labeled ligands.

B. Mammalian Systems

Mammalian expression systems have the major advantage of combining a variety of endogenous G-proteins and effectors. High expression levels attained in transient or stable cellular clones considerably facilitate receptors and/or ligand characterization.

1. Transient Expression

The COS system has been developed to obtain high-level expression in some individual cells, to rapidly determine whether receptors are active. This system is extremely easy to use, but highly variable, and therefore difficult to standardize.

2. Stable Expression

A variety of other cell types have been used for stable expression of GPCR. Regardless of the problem of endogenous subtypes and clone variability, recent reports have documented variations in receptor ligand-binding properties linked to levels of receptor expression. It has thus been shown that dissociation constants and effector activation constants may vary in function of number of receptors per cell.

3. Endogenous GPCR

Chinese ovary cells (CHO), human kidney cells (HEK 293), monkey kidney cells (COS), rat glioma cells (CG), and mouse fibroblasts (3T3), which are among the most popular cells used for transfection with GPCR genes, also often contain endogenous receptors, thus causing high background or complicating interpretations linked to simultaneous activation of distinct subtypes. An effort is underway, on the Internet, to list all endogenous GPCR, G-proteins, and effectors identified in the most common mammalian expression systems. Caution should, however, be used in interpreting these data, as important variations may be linked to particular cellular clones or variants used in different laboratories. Some groups thus reported the simultaneous presence of both AT1 and AT2 subtypes of angiotensin II receptors in NG115 murine neuroblastoma cells whereas others found only the AT2 subtype (19).

C. Insect Systems

Transfection of *Spodoptera frugiperda* (Sf9 or Sf21) insect cells by a recombinant baculovirus containing a GPCR gene may result in high levels of receptor expressed in the cells. This possibility has initially raised considerable hope for the mass production of purified, functional GPCR. However, several characteristics of the currently available systems severely limit their general application. First, and foremost, a high proportion of GPCR produced in the Sf9 or Sf21 insect cells are not functional, probably because the receptors are not or only partially glycosylated, and as result, or for other reasons, are not folded correctly. For example, β_2-adrenoreceptor preparations expressed in baculovirus-infected Sf9 cells contain only 10–20% of molecules that can be affinity-purified in an alprenolol-sepharose

Table 1 GPCR Functionally Expressed in Microorganisms

Receptor	Type	Species	Host	Expression Level[a]	Ref.[b]
Adrenoreceptor	β_1	Human	E. coli	50/cell (1 pmol/mg)[c]	Marullo et al. (31)
	β_1	Turkey	E. coli	30/cell (0.6 pmol/mg)[c]	Chapot et al. (32)
	β_2	Human	E. coli	220/cell (4 pmol/mg)[c]	Marullo et al. (9)
			S. cerevisiae	115 pmol/mg	King et al. (15)
			S. pombe		Ficca et al. (33)
Muscarinic acetylcholine receptor	m1	Human	S. cerevisiae	0.02 pmol/mg	Payette et al. (34)
	m5	Rat	S. cerevisiae	0.12 pmol/mg	Huang et al. (35)
Serotonin receptor	5HT$_{1A}$	Human	E. coli	120/cell (2 pmol/mg)[c]	Bertin et al. (13)
	5HT$_{5A}$	Mouse	S. cerevisiae	16 pmol/mg	Bach et al. (16)
			P. pastoris	22 pmol/mg	Weiss et al. (36)
Endothelin receptor	ET$_B$	Human	E. coli	41/cell (0.8 pmol/mg)[c]	Haendler et al. (37)
Neurotensin receptor		Rat	E. coli	15 pmol/mg	Grisshammer et al. (38)
				800/cell	Tucker & Grisshammer (39)
				(26 pmol/mg)[c]	

Opsin	oRh	Octopus	*E. coli*	1–10 mg/L	Harada et al. (40)
Neurokinin A receptor	NK2	Rat	*E. coli*	2.5–7 pmol/mg	Grisshammer et al. (41)
Neuropeptide Y1 receptor	NPY Y1	Human	*E. coli*	9 pmol/mg	Herzog et al. (42)
Adenosine receptor	A$_1$	Human	*E. coli*	0.2–0.4 pmol/mg	Jockers et al. (43)
	A$_{2a}$	Rat	*S. cerevisiae*	0.4 pmol/mg	Price et al. (18)
Dopamine receptor	D$_{2S}$	Human	*S. cerevisiae*	1–2 pmol/mg	Sander et al. (44–46)
			S. pombe	15 pmol/mg	Sander et al. (44–46)
	D$_3$	Human	*E. coli*	0.5–1 pmol/10^9 cells	Vanhauwe et al. (47)
Somatostatin receptor	SST2	Rat	*S. cerevisiae*	0.1–0.2 pmol/mg	Price et al. (17)
Opioid receptor	μ	Human	*P. pastoris*	0.4 pmol/mg	Talmont et al. (48)

E. coli, Escherichia coli; H. salinarium, Halobacterium salinarium; P. pastoris, Pichia pastoris; S. cerevisiae, Saccharomyces cerevisiae; S. pombe, Schizosaccharomyces pombe.

[a] Expression level of receptors; L, liter of culture medium.

[b] For clarity only, the first reported reference has been listed (for more complete references see text). More than one reference has been mentioned if considerable improvement of expression levels has been obtained.

[c] For uniformity, this value has been estimated from the original value given "sites/cell" using the approximation that 15 pmol/mg crude membrane proteins corresponds to approximately 800 sites/*E. coli* cell.

Source: Modified from Ref. 11b.

gel, or shown otherwise to be functional. The baculoviral particles contain fully functional receptors, but in smaller amounts (20).

Second, and possibly related to the first point, infected Sf9 cells die quickly after producing recombinant virus, and optimal time for harvesting receptors varies between 24 and 72 hr. The exact peak has to be determined for every recombinant virus, as high levels of proteolytic activity may result in drastic reduction in levels of active receptor.

Several groups have attempted to bypass cell culture by directly infecting *Spodoptera* caterpillars with the GPCR recombinant virus. Although extremely high levels of virus could be obtained, purification of active receptor yielded disappointing results, most likely again because of the high levels of proteolytic enzymes present in the animals.

One alternative strategy recently developed to avoid this rapid loss of the infected cells consists in directly transfecting the GPCR gene into the Sf9 cells, rather than relying on infection by the virus. While the levels of receptors were considerably lower, the cells were much more stable, and allowed for a much better control of the cultures and receptor production.

D. Amphibian Systems

Frog *Xenopus* oocytes have been among the first systems to be used to identify mRNA encoding GPCR. Numa's laboratory thus cloned the genes encoding several subtypes of muscarinic acetylcholine receptors by injecting fractions of RNA into the eggs and then measuring receptor activity after treatment with the ligand. Nakanishi's laboratory systemically pursued this strategy by measuring electric currents in response to treatment of oocytes with a variety of ligands, after injecting mRNA fractions encoding a variety of GPCR including all the tachykinin receptors (substance K, substance P, etc.; see Ref. 21).

For a while, *Xenopus* oocytes thus appeared as a good expression system for GPCR. Unfortunately, the difficulty in setting up the system in laboratories removed from the frog producers and the seasonal variability in the properties of the eggs have discouraged most investigators from pursuing the use of this system as a means of expression receptors for ligand screening.

E. Need for Several Expression Systems

These various expression systems are not mutually exclusive and indeed the use of combinations may present considerable advantages to weed out potential lead compounds that bind to receptors when these are expressed in one system but not in another.

It should also be noted that various levels of expression of the same receptor in the same system may be required to define compounds merely as "ligands" or as "agonists" or "antagonists." In the first case, one prefers to use cells with high levels of receptors; in the second case, one prefers levels compatible with coupling to the normal signal transmission system. Overexpression may actually result in overriding genuine receptor-effector association and lead to what has been described as "promiscuous" interactions.

IV. RECEPTOR-BASED ASSAYS

A. Specific Assays

Different binding affinities of ligands for the human versus the animal receptors have been noted in several cases. The use of receptors from rodent, dog, bovine, and other animals for identifying ligands for human receptors has thus led to the development of compounds that work well in animals but not in humans. The most potent β_3-adrenoceptor lipolytic ligand known today, f.i. BRL 37344, is 38 times more active in stimulating adenylyl cyclase modulated by mouse or rat than human β_3-adrenoceptors. Not surprisingly, the strong and specific lipolytic action induced by this and similar compounds in rodents or dogs was not seen in humans involved in various clinical trials. Significant differences in the binding of ligands have also been reported between the rat and human endothelin ET_A receptors. Elshourbaby et al. (22) noted that BQ123 is 100-fold more potent than ET-3 when tested in rat aortic smooth muscle cells, but these compounds have very similar potencies when tested against the human ET_A receptor.

In several instances, single amino acid residue substitutions may be sufficient to explain striking differences in binding affinities of ligands for animal or human receptors. The replacement of asparagine for threonine residue at position 355 in the human 5-HT$_{1B}$ receptor renders the binding characteristics of the human receptor similar to those of the rat receptor (23). The human neurokinin-1 (NK-1) receptor, which is the receptor for substance P, differs from the rat NK-1 receptor in only 22 of 407 amino acids. Of these, residue 290 in the seventh transmembrane domain, which is serine in rat and isoleucine in human, is responsible for the 20-fold greater affinity of CP-96,345 for the human receptor than for the rat receptor (24).

It is of interest that in some instances, species differences may not affect agonist binding while significantly modifying affinity for antagonists. For example, the human and canine cholecystokinin-B (CCK-B) receptors that share 90% amino acid identity show no difference in agonist binding, but for the antagonist, L365260, a benzodiazepine analog, the binding affinities for the human and canine receptors are quite different (24,25). In competition, binding assays using ^{125}I-labeled CCK-8 as the radioligand, L365260, has an affinity of 7.0 nM for the human receptor and 139 nM for the canine receptor. These few examples stress the importance of using human rather than animal receptors for screening purposes.

In addition to radioligand-binding assays, Scintillation Proximity assays, from Amersham Corporation, are used for identifying ligands for GPCR. This technology is based upon the coating of scintillant. This method has the advantage of being a homogenous assay and no filtration or separation step is involved. A further advance has been with the development of the Delfia fluorescence labeling methods of Wallac International, whereby the ligand is labeled with a fluorescent label. This eliminates the use of radioactive label and is also homogeneous.

B. Generic Assays

Radioligand-binding assays are useful for finding antagonists. Functional assays in mammalian cells or in yeast are useful in finding agonists as well as in finding noncompetitive antagonists.

1. Melanophore Response to cAMP

The natural phenomenon of cell surface receptors coupling to G proteins in *Xenopus laevis* melanocytes has been used to develop an innovative system that can be used for assaying ligand interaction with cloned receptors

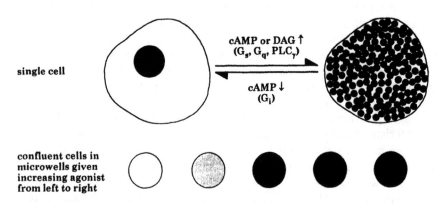

Figure 1 Melanophore assay. The cell at the top contains thousands of darkly pigmented melanosomes. Generation of either cAMP or DAG following stimulation of a G-protein-coupled or a tyrosine kinase receptor causes pigment dispersion. Alternatively, diminution of cAMP following stimulation of a G_i-coupled receptor induces pigment aggregation. The middle of the figure shows an example of agonist-mediated pigment dispersion of all cells which is dose-dependent and provides concentration-response curves. (Obtained from Michael R. Lerner, M.D., Ph.D., Yale University School of Medicine, Department of Internal Medicine and Pharmacology.)

(26–29). A melanophore cell line was obtained from *X. laevis* skin. Aggregation of melanosomes filled with melanin in the center of the cell and/or their movement throughout the cytoplasm results in white-black color transformation of the cells. The distribution of the melanosomes is dependent upon intracellular levels of cAMP. In nature, activation of the MSH receptor on the cell surface couples the receptor to its G-protein leading to signal transduction and elevation in intracellular cAMP levels. This results in the dispersal of melanosomes producing a camouflaging pigment for the frog. Melatonin has the opposite effect of MSH in that it inhibits cAMP accumulation and causes melanosome aggregation. The dispersal-aggregation of the melanosomes is related to the cAMP dependent phosphorylation-dephosphorylation of a protein associated with the melanosomes. Messenger systems other than cAMP, such as the phospholipase C pathway and PIP_2 hydrolysis, can also cause dispersal of the pigment. Activation of kinases and phosphatases is involved in pigment dispersal also. These cells have the flexibility of coupling to different G_α-proteins, and coupling of heterologous receptors to $G_{\alpha i}$ and $G_{\alpha s}$ has been demonstrated. Functional expression of the human β_2 adrenoceptors (22) and the bombesin receptor (29) has been reported. Transient or stable expression of GPCR is achieved by transfection with expression plasmids such as pcDNA1/NEO that are the same as those used in many laboratories for expression in mammalian cells.

In addition to its advantage of flexibility provided by a variety of endogenous G-proteins in melanocytes, suitability of mammalian expression vectors, and the easily visible black-white response to ligand, *Xenopus* cells have the further advantage of growing at room temperature and not needing elevated levels of CO_2. The assay is easily adaptable to microtiter format and the cells can be fixed with 70% ethanol before or after to quantification by phototransmission readers. The pigment dispersal also can be automated to suit image analysis systems. As with any system, the *Xenopus* melanocyte system does have a few disadvantages. Frog melanocytes express, for example, an endogenous functional receptor that is β_1-adrenoceptor-like. However, activity related to the heterologously expressed β_2-adrenoceptor could still be detected because of the low response of the endogenous receptor to β_2-adrenoceptor ligands. Therefore, when new receptor types are expressed in the *Xenopus* melanocytes, it is advisable to test these cells for endogenous expression of related receptors to avoid confounding results. Another disadvantage is the long generation time of the melanocyte cell line (approximately 3–4 days). The melanocytes must be grown in conditioned medium in which fibroblasts have been grown. It is possible that with more experience and further growth medium modifications, more suitable growth conditions can be achieved.

2. Microphysiometer for Change of pH

The ligand can be a functional agonist or antagonist or a partial antagonist/agonist. Functional effects of receptor-ligand interaction can be determined by measuring intracellular messengers such as adenylyl cyclase levels, phosphatidyl inositol accumulation, calcium efflux, etc., which are fairly tedious assays. The metabolic effect of ligand interaction with a cell receptor also can be measured by using a Cytosensor microphysiometer made by Molecular Devices Inc. In this case, the rate of proton excretion by the cell, a measure of increased metabolic activity, results in increased acidification of the medium and can be quantified. The method and instrumentation is described in detail by McConnell et al. (30). The instrument uses silicon technology and a light-sensitive potentiometric sensor to measure the rate of acidification of the medium. The cells are grown in a flow chamber on a membrane that is in contact with a pH-sensitive surface of the potentiometric sensor through an aqueous environment. The potentiometric sensor is illuminated from the bottom by an infrared-emitting device. Culture medium is pumped in pulses over the cells, and the pH of the medium leaving the cell at each pulse is measured. Acid metabolites that accumulate in the medium are measured as a decrease in pH. The degree of acidification is the same at each pulse as long as the cells are left undisturbed. When an agonist or antagonist is introduced into the medium, the metabolism of the cell changes. In the case of an agonist, increased metabolism is reflected by an increased rate of acidification and the rate of change relative to medium without agonist is recorded by the instrument. The decrease in the rate of acidification is measured for antagonists. Strong buffers are avoided in the medium to allow detection of small pH changes. The microphysiometer has been useful in identifying the agonistic or antagonistic nature of ligands in binding assays, and in the future, could be used as a very sensitive primary screening tool for novel ligands.

3. Luciferase Response to cAMP or Other Messengers

At least two mammalian cellular screening systems have been described that measure biological activity of ligands acting on receptors. The first system (6) relies on a CHO cell line stably transformed with a reporter plasmid containing the firefly luciferase gene under the transcriptional control of multiple cAMP-responsive elements (CRE). Stimulation of these cells with forskolin, an adenylyl cyclase activator, led to a 20–30-fold induction of luciferase activity. The reporter cells were transfected with the human dopamine D1 or D5 receptor genes. Treatment of these cells with dopamine receptor agonists modulated the luciferase expression in a dose-dependent manner (6).

A second system was based on the transformation of the human adenocarcinoma cell line A549 with the luciferase gene under the control of a phorbol esters–responsive regulatory region belonging to the ICAM-1 gene promoter. These cells were then transfected with the human neurokinin 2 (NK2) or the serotonin 2 (5-HT2) receptor genes and were shown to display luciferase responses after treatment with either the various neurokinins (neurokinin A, substance P, neuromedin K) or 5-HT2 specific agonists or antagonists (7). Analogous test cell lines for the other adenylyl cyclase or phospholipase C–coupled receptors could thus serve as convenient cellular screening systems.

Aequorin

A wide range of GPCR may be coupled in a nonselective fashion to the $G_{\alpha16}$ subunit, which modulates phospholipase Cb activity. Stimulation of the receptor by an agonist results in production of inositol phosphates leading to increase in intracellular calcium. Aequorin, a phosphoprotein from *Aequorea victoria*, emits luminescence that is directly related to the concentration of calcium. Rees et al. (8) have developed a generic agonist screening system by coexpressing receptors $G_{\alpha16}$ and aequorin. These authors have demonstrated that the normally $G_{\alpha i}$-coupled adenosine A_1 and melatonin $Rmel_{1A}$ receptors and the normally $G_{\alpha s}$-coupled A_2 and β_2 adrenoreceptor all may be coupled to $G_{\alpha16}$ and generate an increase in calcium detected using aequorin, in response to the appropriate receptor agonist (8).

V. PERSPECTIVES

Small-molecule drug discovery currently relies on good part in high-throughput screening of combinatorial as well as natural libraries of compounds. Screening of these large numbers of molecules on GPCR is still considered to be a major challenge mainly because these receptor proteins have remained difficult to purify in an active form in amounts sufficient for the design of screening assays. The need today of keeping these GPCR in plasma cell membrane constitutes a major source of background that has to be solved in the future. This first step only concerns binding of lead compounds. The second step is identification of agonists or antagonists, and this requires functional signal transmission screening, thus adding another level of complexity. Surprisingly, this step seems to have progressed faster, probably because it uses enzymatic amplification procedures, which render the signal detection quite sensitive. We can be confident that biophysical methods coupled with robotics should soon solve the remaining difficulties of high-throughput screening of ligands on GPCR embedded in synthetic materials.

ACKNOWLEDGMENTS

Our work is supported by the Centre National de la Recherche Scientifique, the Institut National de la Santé et de la Recherche Médicale Française, the University of Paris VII, the French Ministry for National Education and Research, the Ligue Nationale contre le Cancer, the Fondation pour la Recherche Médicale Française, the European Union MIEC CHRX-CT 94-0490, and the ENBST CHRX-CT 94-0689 Human Capital and Mobility contracts.

REFERENCES

1. Sealfon SC. Receptor Molecular Biology. San Diego: Academic Press, 1993.
2. Watson S, Girdlestone D. Receptor and ion channel nomenclature supplement. Trends Pharmacol Sci 1996; 7th Edition, pp. 10–11.
3. Libert F, Parmentier M, Lefort A, Dinsart C, Van Sande J, Maenhaut C, Simons MJ, Dumont JE, Vassart G. Selective amplification and cloning of four new members of the G protein-coupled receptor family. Science 1989; 244:569–572.
4. Strosberg AD. Structure, function and regulation of adrenoceptors. Protein Sci 1993; 12:1198–1209.
5. Parmentier M, Libert F, Vassart G. La famille de récepteurs couplés aux protéines G et ses orphelins. Med Sci 1995; 11:222–231.
6. Himmler A, Strowa C, Czernilofsky AP. Functional testing of human dopamine D1 and D2 expressed in stable cAMP-responsive luciferase reporter cell lines. J Recept Channels 1993; 13:79–84.
7. Weyer U, Schäfer R, Himmler A, Mayer SK, Bürger E, Czernilofsky AP, Stratowa, C. Establishment of a cellular assay system for G-protein linked receptors: coupling of human NK2 and 5-HT2 receptors to phospholipase C activates a luciferase reporter gene. Receptors Channels 1993; 1:193–200.
8. Stables J, Green A, Marshall F, Fraser N, Knight E, Sautel M, Milligan G, Lee M, Rees S. Shedding light on G-protein-coupled receptors. Anal Biochem; 1997; 252:115–126.
9. Marullo S, Delavier-Klutcho C, Eshdat Y, Strosberg AD, Emorine LJ. Human β2-adrenoceptors expressed in *Escherichia coli* membranes retain their pharmacological properties. Proc Natl Acad Sci USA 1988; 85:7551–7555.
10. Strosberg AD, Marullo S. Functional expression of G protein–coupled receptors in microorganisms. Trends Pharmacol Sci 1992; 13:95–98.
11a. Jockers R, Petit L, Lacroix I Marullo S, Strosberg AD. Novel isoforms of Mel1c melatonin receptors modulating intracellular cGMP levels. Mol Endocrinol 1997; 11:1070–1081.
11b. Jockers R, Strosberg AD. Expression of G-protein-coupled receptors in microorganisms. In: Davies JE, ed. Manual of Industrial Microbiology and Biotechnology, 2nd ed. (in press).

12. Freissmuth M, Selzer E, Marullo S, Schütz W, Strosberg AD. Expression of two human β-adrenoceptors in *E. coli*: functional interaction, with two forms of Gs. Proc Natl Acad Sci USA 1991; 88:8548–8552.

13. Bertin B, Freissmuth M, Breyer RM, Schütz W, Strosberg AD, Marullo S. Functional expression of the human serotonin 5HT1A receptor in *Escherichia coli*. J Biol Chem 1992; 267:8200–8206.

14. Bertin B, Freissmuth M, Jockers R, Strosberg AD, Marullo S. Cellular signaling by an agonist-activated receptor/Gsα fusion protein. Proc Natl Acad Sci USA 1994; 91:8827–8831.

15. King K, Dohlman HG, Thorner J, Caron MG, Lefkowitz RJ. Control of yeast mating signal transduction by a mammalian beta 2-adrenoceptor and Gs alpha subunit. Science 1990; 250:121–123.

16. Bach M, Sander P, Haase W, Reiländer H. Pharmacological and biochemical characterization of the mouse 5HT(5A) serotonin receptor heterologously produced in the yeast *Saccharomyces cerevisiae*. Receptors and Channels, 1996; 4:129–139.

17. Price LA, Kajkowski EM, Hadcock JR, Ozenberger BA, Pausch MH. Functional coupling of a mammalian somatostatin receptor to the yeast pheromone response pathway. Mol Cell Biol 1995; 15:6188–6195.

18. Price LA, Strand J, Pausch MH, Hadcock JR. Pharmacological characterization of the rat A2a adenosine receptor functionally coupled to the yeast pheromone response pathway. Mol Pharmacol 1996; 50:829–837.

19. Nahmias C, Cazaubon S, Briend-Sutren MM, Lazard D, Villageois P, Strosberg AD. Angiotensin II AT2 receptors are functionally coupled to protein tyrosine dephosphorylation in N1E-115 neuroblastoma cells. Biochem J 1995; 306:87–92.

20. Loisel T, Ansanay H, Saint-Onge S, Gay B, Boulanger P, Strosberg AD, Marullo S, Bouvier M. Functional expression of properly folded human β2-adrenoceptor at the surface of extracellular baculovirus viral particles; a new tool for large scale production, biochemical and structural characterization. Nature Biotech 1997; 15:1300–1304.

21. Masu Y, Nakayama K, Tamaki H, Harada Y, Nakanishi S. cDNA cloning of substance K receptor through oocyte expression system. Nature 1987; 329:836–838.

22. Elshourbagy NA, Korman DR, Wu HL, Sylvester DR, Lee JA, Nuthalgant P, Bergsma DJ, Kumar CS, Nambi P. Molecular characterization and regulation of the human endothelin receptors. J Biol Chem 1993; 268:3873–3879.

23. Parker EM, Grisel DA, Iben LG, Shapiro RA. A single amino acid difference accounts for the pharmacological distinctions between the rat and human 5-hydroxytryptamine$_{1B}$ receptors. J Neurochem 1993; 60:380–383.

24. Sachais BR, Snider RM, Lowe III JA, Krause JE. Molecular basis for the species selectivity of the substance P antagonist CP-96,345. J Biol Chem 1993; 268:2319–2323.

25. Beinborn M, Lee YM, McBride EW, Quinn SM, Kopin AS. A single amino acid of the cholecystokinin-B/gastrin receptor determines specificity for non-peptide antagonists. Nature 1993; 362:348–353.

26. Potenza MN, Graminski GF, Lerner MR. A method for evaluating the effects of ligands upon G_s protein-coupled receptors using a recombinant melanophore-based bioassay. Anal Biochem 1992; 206:315–322.

27. McClintock TS, Graminiski GF, Potenza MM, Jayawickreme CK, Roby-Shemkovitz A, Lerner MR. Functional expression of recombinant G-protein coupled receptors monitored by video imaging of pigment movement in melanophores. Anal Biochem 1993; 209:298–305.

28. Graminski GF, Jayawickreme CK, Potenza MN, Lerner MR. Pigment dispersion in frog melanophores can be induced by a phorbol ester or stimulation of a recombinant receptor that activates phospholipase C. J Biol Chem 1993; 268:1–8.

29. Karne S. Jayawickreme CK, Lerner MR. Cloning and characterization of an ET-3 receptor (ET_c receptor) from Xenopus laevis dermal melanophores. J Biol Chem 1993; 268:19126–19133.

30. McConnell HL, Owicki JC, Parce JW, Miller DL. Baxter GT, Wada HG, Pitchford S. The Cytosensor microphysiometer: biological applications of silicon technology. Science 1992; 257:1906–1912.

31. Marullo S, Delavier-Klutchko C, Guillet JG, Charbit A, Strosberg AD, Emorine LJ. Expression of human beta1 and beta2 adrenoceptors in E. coli as a new tool for ligand screening. Bio/Technology 1989; 7:923–927.

32. Chapot MP, Eshdat Y, Marullo S, Guillet JG, Charbit A, Strosberg AD, Delavier KC. Localization and characterization of three different beta-adrenoceptors expressed in Escherichia coli. Eur J Biochem 1990; 187:137–144.

33. Ficca AG, Testa L, Valentini GPT. The human beta(2) adrenoceptor expressed in Schizossacharomyces pombe retains its pharmacological properties. FEBS Lett 1995; 377:140–144.

34. Payette P, Gossard F, Whiteway M, Denis M. Expression and pharmacological characterization of the human M1 muscarinic receptor in Saccharomyces cerevisiae. FEBS Lett 1990; 266:21–25.

35. Huang HJ, Liao CF, Yang BC, Kuo TT. Functional expression of rat M5 muscarinic acetylcholine receptor in yeast. Biochem Biophys Res Commun 1992; 182:1180–1186.

36. Weiss HM, Haase W, Michel Hand Reiländer H. Expression of functional mouse 5-HT5A serotonin receptor in the methylotrophic yeast Pichia pastoris: pharmacological characterization and localization. FEBS Lett 1995; 377:451–456.

37. Haendler B, Hechler U, Becker A, Schleuning WD. Expression of human endothelin receptor ETB by Escherichia coli transformants. Biochem Biophys Res Commun 1993; 191:633–638.

38. Grisshammer R, Duckworth R, Henderson R. Expression of a rat neurotensin receptor in Escherichia coli. Biochem J 1993; 295:571–576.

39. Tucker J, Grisshammer R. Purification of a rat neurotensin receptor expressed in Escherichia coli. Biochem J 1996; 317:891–899.

40. Harada Y, Senda T, Sakamoto T. Takamoto K, Ishibashi T. Expression of octopus rhodopsin in Escherichia coli. J Biochem (Tokyo) 1994; 115:66–75.

41. Grisshammer R, Little J, Aharony D. Expression of rat NK-2 (neurokinin A) receptor in *E. coli*. Receptors Channels 1994; 2:295–302.
42. Herzog H, Münch G, Shine J. Human neuropeptide Y1 receptor expressed in *Escherichia coli* retains its pharmacological properties. DNA Cell Biol 1994; 13:1221–1225.
43. Jockers R, Linder RE, Hohenegger M, Nanoff C, Bertin B, Strosberg AD, Marullo S, Freissmuth M. Species difference in the G protein selectivity of the human and bovine A(1)-adenosine receptor. J Biol Chem 1994; 269:32077–32084.
44. Sander P, Grünewald S, Bach M, Haase W, Reiländer H, Michel H. Heterologous expression of the human D2S dopamine receptor in protease-deficient *Saccharomyces cerevisiae* strains. Eur J Biochem 1994; 226:697–705.
45. Sander P, Grünewald S, Maul G, Reiländeer H, Michel H. Constitutive expression of the human D2S-dopamine receptor in the unicellular yeast *Saccharomyces cerevisiae*. Biochim Biophys Acta 1994; 1193:255–262.
46. Sander P, Grünewald S, Reiländer H, Michel H. Expression of the human D2S dopamine receptor in the yeasts *Saccharomyces cerevisiae* and *Schizosaccharomyces pombe*: a comparative study. FEBS Lett 1994; 344:41–46.
47. Vanhauwe J, Luyten, W HLM, Josson K, Fraeyman N, Leysen JE. Expression of the human dopamine D3 receptor in *Escherichia coli*: investigation of the receptor properties in the absence and presence of G proteins. Abstract, p. 206 Proceedings of the 8th International Catecholamine Symposium Oct 12–18, 1996. Asilomar Conference Center, Pacific Grove, CA, 1996.
48. Talmont F, Sidobre S, Demange P, Milon A. Expression and pharmacological characterization of the human μ-opioid receptor in the methylotrophic yeast *Pichia pastoris*. FEBS Lett 1996, 394:268–272.

Inverse Agonists and G-Protein-Coupled Receptors

Richard A. Bond
University of Houston, Houston, Texas

Michel Bouvier
University of Montreal, Montreal, Quebec, Canada

I. INTRODUCTION

G-protein-coupled receptors represent the single largest family of cell surface receptors involved in signal transduction. It is estimated that several hundred distinct members of this receptor family in humans direct responses to a wide variety of chemical transmitters, including biogenic amines, amino acids, peptides, lipids, nucleosides, and large polypeptides. They therefore represent major targets for the development of new drug candidates with potential application in all clinical fields. Many currently used therapeutics act by either activating (agonists) or blocking (antagonists) these receptors; widely used examples are β-adrenoceptor agonists for asthma and antagonists for hypertension, histamine H_1- and H_2-receptor antagonists for allergies and duodenal ulcers, respectively, opioid receptor agonists (e.g., morphine) as analgesics, dopamine receptor antagonists as antipsychotics, and 5HT receptor agonists (e.g., sumitriptan) for migraine. The concept of efficacy is a fundamental parameter in the analysis of drug action on receptors, and the ranking of drugs based on the amplitude of the response they induce can be a determining factor in their therapeutic utility. Classical receptor theory describes receptors (R) as quiescent until activated by the binding of an agonist (A), with the resulting binary complex (AR) having affinity for its cognate G-protein (1–3). Thus, the concept of ligand efficacy in traditional theory may be viewed as the ligand *inducing* a conformational change in the receptor. In such an induced-fit model, antagonists are traditionally considered to have zero efficacy; i.e., they occupy the site but do not affect either coupling activity or the response level. Their physiological effects were thus usually attributed to their ability to prevent activation of the receptors by endogenous neurotransmitters or hormones (competitive antagonism), by inducing either no conformational change or a change that did not affect the affinity of the receptor for the G-protein.

Recently, however, evidence has been accumulating that suggests that some unliganded G-protein-coupled receptors exhibit spontaneous activity and are signaling cellular responses in the absence of any agonist stimulation and that ligands previously considered as competitive antagonists can inhibit such spontaneous activity. These ligands, which have now been termed inverse agonists, therefore appear to possess negative intrinsic activities. The simplest way to rationalize these data is to propose a model consisting of the receptor in equilibrium between at least two conformational states, the classic, inactive conformation, which lacks affinity for G-proteins (R), and a conformation exhibiting affinity for G-proteins in the absence of ligand (R*). Also, central to drug development is the possible redefinition of ligand efficacy. The most parsimonious explanation of ligand efficacy in a multistate model is simply the differential affinity of the ligand for the conformational states (4–7). In the two-state model, ligands exhibiting higher affinity for the active conformation (R*) would stabilize that conformation and reset the equilibrium to increase the total amount of R* and thus function as agonists, ligands with higher affinity for R would result in a decrease in R* and be inverse agonists, and ligands with equal affinity for R and R* would not alter the equilibrium but would be antagonists of both agonists and inverse agonists. Thus, ligand efficacy would perhaps be viewed as conformational *selection* of already preexisting states. Although this selected-fit model offers an easy explanation for the mode of action of inverse agonists, various lines of evidence including molecular-dynamic simulation (8) still suggest that different ligands may promote ligand-specific conformational changes. In that respect, the evidence suggesting that different agonists may stabilize and/or promote distinct active conformations has recently been reviewed and termed agonist trafficking (9). Conformational induction and conformational selection may therefore be two inseparable consequences of receptor binding systems.

This chapter will review the functional evidence for the existence of spontaneously active G-protein-coupled receptors and inverse agonists. We will also briefly discuss multistate models that can simulate the data and their possible impact on drug development. Finally, we will describe some recently identified diseases where spontaneous activity of receptors may be contributing to the pathology of the disease.

II. EVIDENCE OF SPONTANEOUS RECEPTOR ACTIVITY AND INVERSE AGONISTS

A. Evidence from GTPase Activity Assays

In 1989, Costa and Herz provided the first evidence of spontaneous activity and inverse agonism for G-protein-coupled receptors (10). Using NG108-15 cells, which endogenously express δ-opioid receptors, they showed three classes of effects by ligands on GTPase activity. The opioid agonist, DADLE ([D-Ala^2,D-Leu^5]enkephalin), produced increases in GTPase activity, while ICI 174864 ([N,N'-diallyl-Tyr^1,a-aminoisobutyric acid^{2,3}]Leu^{5}enke-phalin), a compound previously classified as an antagonist, decreased basal, unstimulated GTPase activity (Fig. 1). Another antagonist, MR 2266 did not produce any effects on basal GTPase activity, but competitively and stereospecifically antagonized the positive and inhibitory effects of DADLE and ICI 174864, respectively (Fig. 1). A determining factor in the ability to detect inverse agonism in this study resided in the elevation of the baseline GTPase activity by the substitution of Na$^+$ by K$^+$. Indeed, this ion substitution increased baseline GTPase activity by about 50% thus allowing detection of the inverse agonist activity of some ligands (Fig. 1). As will be reviewed further below, an increase in the baseline of the parameter being measured greatly facilitated the detection of inverse agonism by improving the signal/noise ratio. Consistent with this notion, decreasing the concentration of Na$^+$ in a [35S]GTPγS binding assay, which increases basal binding levels, also allowed the detection of an α_{2D}-adrenergic receptor spontaneous activity that could be inhibited by the inverse agonist rauwolscine (11).

B. Evidence from Second-Messenger Assays

For spontaneous activity and inverse agonism to have physiological significance, the effects observed at the level of the G-proteins need to be productively transfered to signaling effectors. Studies using indices of adenylate cyclase or phosphoinositide hydrolysis as a readout of cellular signaling have demonstrated that spontaneous activity of unliganded receptors and inverse agonist effects can indeed be transduced into cellular responses. In Sf9 cells infected with recombinant baculovirus encoding the human β_2-adrenoceptor, baseline adenylate cyclase activity and receptor density were shown to increase proportionately with time of infection, and several compounds previously classified as β-adrenoceptor antagonists were shown to be inverse agonists and to inhibit baseline adenylyl cyclase activity in the

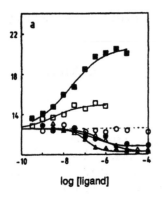

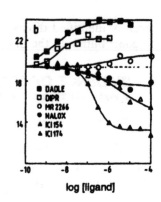

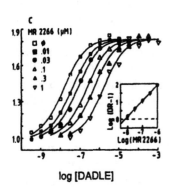

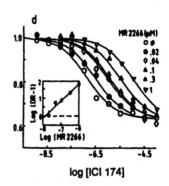

Figure 1 Effect of opioid ligands on GTPase activity in membranes of NG108-15 cells. (a and b) High-affinity GTPase was assayed in a reaction mixture containing 150 mM NaCl (a) or KCL (b) and various concentrations of opioid agonists and antagonists (■ = DADLE; □ = diprenorphine; O = MR 2266; ● = naloxone; Δ = ICI154129; ▲ = ICI 174864). Basal activity is indicated by a broken line; points are means of triplicate determinations. Data are representative of three additional experiments performed with different batches of membranes (c and d). Concentration–response curves for the stimulatory effect of the agonist DADLE, studied in 150 mM NaCl (c), and the inhibitory effect of ICI 174864, studies in the 150 mM Kcl (d), in the presence of 0-1 μM MR 2266. Fractional response is the ratio of GTPase activity in the presence of ligand to that in its absence. Insets: Schild plots of the data. (DR = dose ratio.)

absence of agonist stimulation (12). Interestingly, not all antagonists showed the same efficacy to inhibit spontaneous activity. Indeed, similarly to the phenomenon of partial agonism, various level of "inverse efficacy" were found, and a continuum going from full inverse agonist to neutral antagonist was observed (Fig. 2). The fact that the rank order of efficacy of the inverse agonists could be distinguished from their binding affinities excluded the contribution of competitive antagonism to the effects observed. Moreover, quasi-neutral antagonist could competitively block the actions of both agonist and inverse agonists illustrating that the inhibitory actions of inverse agonist represented intrinsic activities of these compounds. Although overexpression obtained in Sf9 cells facilitated the detection of spontaneous activity and inverse agonism, it could also be readily observed in CHW cells expressing as little as 200 fmol of human β_2-adrenoceptor per milligram of proteins (12).

Similar results were obtained for the 5-hydroxytryptamine 5-HT$_{2C}$ receptor expressed in NIH 3T3 fibroblasts, using phosphoinositide hydrolysis as the functional endpoint. Again, some ligands previously classified as 5-HT antagonists functioned as inverse agonists and decreased baseline phosphoinositide hydrolysis formation (13). The authors attempted to correlate the

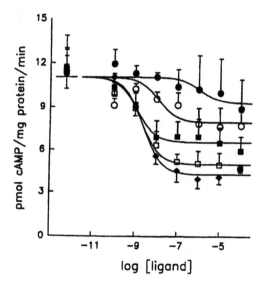

Figure 2 Inhibition of agonist-independent adenylyl cyclase activity in membranes from CHW cells expressing β_2AR. Production of cAMP was measured in the presence of labetalol (●), pindolol (○), alprenolol (■), propanolol (□), or timolol (♦), at the concentrations indicated on the *abscissa*. The data presented represent the mean values of three experiments carried out in duplicate. The averaged data sets were analyzed simultaneously, assuming a common value of $Y_{x=0}$. The fitted values of EC$_{50}$ are as follows: labetalol, 1.1 ± 1.7 μm; pindolol, 13 ± 10 nM; alprenolol, 1.3 ± 0.7 nM; propranolol, 2.0 ± 0.8 nM; timolol, 2.5 ± 0.9 nM.

inverse agonist activity of the compounds upon IP formation, measured in 5-HT_{2C} expressing NIH 3T3 cells, with their ability to produce down-regulation of 5-HT_{2C} receptors in choroid plexus epithelial cells. They showed that mianserin, which was found to be an efficacious inverse agonist in NIH 3T3 cells, but not the neutral antagonist BOL produced a decrease in brain receptor density (13). However, when the inverse efficacy and ligand-promoted down-regulation of nine 5-HT ligands were compared in Sf9 cells expressing the 5-HT_{2C} receptor, no correlation between the inverse agonist effect on PI formation and the down-regulation could be found (14), suggesting that these two processes may not involve the same mechanisms.

Spontaneous activity of receptors expressed in cell lines and the occurrence of inverse agonism, detected at the level of the second messengers generated, have now been shown for a number of G-protein-coupled receptors and their respective antagonists. These include: The β_2-adrenoceptor (12), the D_1- and D_5-dopamine (15), the 5HT_{2C}-serotonin (13,14), the δ opioid (16), and the H_2-histamine receptor (17). In regard to the D_1- and D_5-dopamine receptors, the authors found that despite displaying a similar pharmacology, as determined by the affinities of various compounds for both receptor subtypes, the receptors were markedly different in their ability to increase baseline signaling in the absence of hormone. The D_5-receptor produced greater increases in basal adenylyl cyclase activity at comparable levels of expression (5), which suggests that one reason receptor subtypes exist may be to set the ligand-independent versus ligand-dependent signaling levels ("signal to noise") in different tissues.

In all the cases discussed above, heterologous overexpression systems were used. However, inverse agonism toward native receptors expressed in untransfected cells has also been demonstrated. These include the β-adrenoceptor of turkey erythrocytes, using adenylyl cyclase activity as the endpoint (18), the muscarinic receptors of cardiac myocytes from various species in which electrophysiological measurements of Ca^{2+} and K^+ currents were assessed (19,20), and the bradykinin B2-receptor stimulating IPs production in rat myometrial cells (21). This suggests that in some tissues, the endogenous level of receptor expression or the extent of spontaneous activity may be sufficient for the inverse agonists to have measurable effect under physiological conditions.

Overall, these findings are difficult to reconcile with the classic model of G-protein-coupled receptor signaling, which envisions that the ternary complex formed by the agonist, the receptor, and the G-protein is the only active form of the receptor. However, they are entirely consistent with the "two-state" receptor model whereby the receptor exists in at least two states, R (inactive) and R* (active) even in the absence of ligands. This spontaneous isomerization to an active conformer would be responsible for the spontaneous activity observed in systems expressing high levels of receptor. Under basal conditions, the inactive state would largely predominate and spontaneous activity could be detected only if the absolute number of receptor in the active conformation is sufficient to promote a sizable signal. In such a model, agonists and inverse agonists would then be characterized by their preferential affinities for a given state, with inverse agonists and agonists preferentially stabilizing the inactive and the active state, respectively. A neutral antagonist would not discriminate between the two conformers. Accordingly, differences in drug efficacy can be rationalized by assuming that different compounds will have distinct relative preferences for the active and inactive conformers. In a first effort to formalize this model, Samama et al. (22,23) proposed the extended ternary complex model. One implication of this model, is that G-proteins bind only to R* but not R. However, there is no a priori reason to rule out the process by which GR* relaxes to GR or the binding of R to G (albeit with a lower affinity than R*). A more general model, termed the cubic model, has thus been proposed (24) and simulation using both models can mimic the behavior of agonists and inverse agonists. A number of recent papers reviewed the theoretical implications of inverse agonism on our understanding of the thermodynamics of G-protein-coupled receptor activation (6,25–28).

Inhibition of the agonist-independent activity of G-protein-coupled receptors has also been demonstrated using receptors rendered constitutively active by mutation. Indeed, site-directed mutagenesis of residues located in the third cytoplasmic loop of the β_2- and α_1-adrenoceptors has been shown to significantly increase the level of their spontaneous activity (23,29). The activity of these receptors refered to as "constitutively active" can also be inhibited by inverse agonists (22,30). Interestingly, the order of efficacy of inverse agonists toward the constitutively active mutant β_2-adrenoceptor was found to be virtually indistinguishable from that observed for the overexpressed receptor. Therefore, in the context of the two-state model, mutations leading to constitutive activation would be assumed to stabilize or favor the transition toward the active isomer. However, the conformational changes imposed by the mutations cannot be seen as rigid or irreversible since, as mentioned above, inverse agonists can inhibit their spontaneous activity and thus can still bind to and stabilize the inactive conformer.

In contrast to the considerable details available for ligand-binding domains, few studies have addressed the issue of identifying structural elements within the receptor that determine drug efficacy. On the basis of the two-state model described above, it would be predicted that agonists would promote receptor activation through interactions with residues forming a high-affinity binding site in the active conformation (R*) that is able to interact with G-proteins, whereas inverse agonists would populate the inactive state through binding to residues appropriately positioned in the R conformation. Experiments to identify the specific residues forming the agonist and inverse agonist binding sites are complicated by the fact that R and R* exist in a dynamic equilibrium, with the relative proportions of the two states under "resting conditions" (i.e., no ligand) being determined by the rate of spontaneous isomerization. Receptor ligands, as well as other factors such as G-proteins, would be expected to displace this equilibrium, and experiments based on equilibrium binding analyses yield little information on this dynamic aspect.

C. Evidence from Transgenic Mice Overexpressing the Human β_2-Adrenoceptor

The data reviewed so far convey the notion that cellular manifestation of spontaneous activity and inverse agonism at G-protein-coupled receptors can be observed in overexpression systems, in cells expressing constitutively activated mutant receptors, and in a few cases in native tissues. However, the strongest available data supporting a possible role for inverse agonism and spontaneous activity in animal physiology come from work carried out in transgenic mice specifically overexpressing the normal human β_2-adrenoceptor (31) in their cardiomyocytes.

1. Biochemical Evidence

In one of the transgenic lines studied, the α-myosin heavy-chain promoter directed the cardiospecific expression of the human β_2AR to levels reaching 200 times that of the endogenous receptor. In cardiac membrane preparations derived from these animals, the basal adenylyl cyclase activity was comparable to the maximal agonist-stimulated (isoproterenol) response observed in control mice (31) and could be only marginally stimulated further by isoproterenol thus suggesting a significant level of spontaneous ligand-independent activity. Consistent with this notion, the inverse agonist ICI-118,551 was found to inhibit the basal adenylyl cyclase activity. Further evidence supporting the idea that overexpression of the β_2AR led to significant spontaneous signaling activity came from co-immunoprecipitation experiments. Indeed, the extent of coupling between the receptor and its endogeneous G-protein can be assessed by determining the amount of receptor that can be immunoprecipitated using an anti-Gαs antibody. Under basal conditions, 50 times more receptor could be co-immunoprecipitated in membranes derived from transgenic mice overexpressing the receptor as compared with their control littermate, suggesting that receptor-G-protein coupling occurred even in the absence of hormonal stimuli.

2. Isolated Tissue Evidence

Evidence that spontaneous signaling activity detected at the level of the second-messenger production is accompanied by functional changes was first obtained by measuring isometric contractility in the isolated, paced, left atria of control and transgenic mice. In a manner similar to the results obtained for adenylyl cyclase activity, baseline contractility in mice overexpressing the β_2AR was significantly elevated compared to controls and was comparable to the contractility observed in controls following maximal β-adrenoceptor stimulation with isoproterenol (10 nM) (31). However, unlike the adenylyl cyclase data in which further increases in enzyme activity were observed after isoproterenol administration, no agonist-dependent contractile response could be observed in the transgenic mice, suggesting that maximal contractile response was achieved by the spontaneously active receptors.

As was the case in the in vitro assays, the inverse agonist ICI-118,551 produced a concentration-dependent inhibition of left atrial tension, and the maximal inhibition correlated with the receptor density, suggesting a receptor-mediated event (7). These data therefore suggest that inverse agonists can inhibit physiological responses evoked as a result of spontaneous receptor activity. Three lines of evidence exclude the possibility that the observed effects could result from competitive antagonism with the endogenous catecholamines most likely present in the isolated atria. First, if the inhibition produced by the antagonist were due to displacement of endogenous catecholamines, then the inhibition produced by various antagonists should be equal if the amount of such displacement were equal. However, at variance with expectations of the competition of endogenous hormone explanation, four different β_2-adrenoceptor antagonists all produced varying amounts of inhibition when used at concentrations that were 300 times their respective K_B values for the β_2-adrenoceptor. Second, animals were treated with an intraperitoneal (i.p.) dose of reserpine (0.3 mg/kg 24 hr prior to sacrifice), which produced a greater than 98% depletion of cardiac catecholamines as measured by HPLC coupled to electrochemical detection. This treatment with reserpine failed to alter the concentration-response curve (CRC) to ICI-118,551, again indicating that displacement of endogenous catecholamines was not the mechanism of inhibition by ICI-118,551 (7). Third, a quasi-neutral antagonist, alprenolol, was used to block the inhibitory effects of ICI-118,551. Alprenolol (100 nM) produced an approximately 30-fold shift to the right of the ICI-118,551 CRC. Thus, the ICI-118,551 effect was exerted via the β-adrenoceptor but not by competition with endogenous agonists.

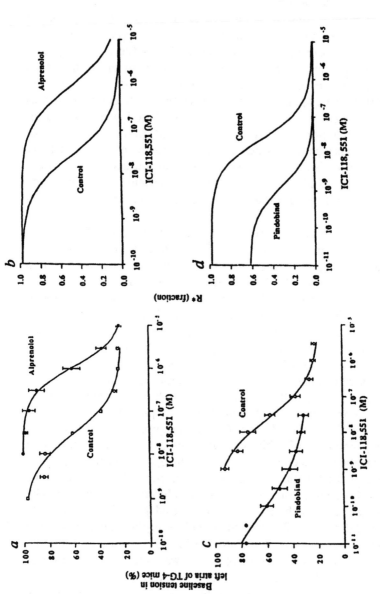

Figure 3 *a*, The effect of alprenolol on the inhibitory response of ICI-118,551 on the baseline left atrial tension in TG-4 mice; control, ○; alprenolol (1 × 10⁻⁷), ●. The results are expressed as the mean percent inhibition of baseline ± s.e.m. *b*, simulation based on the simple form of the two-state model using the following parameters; K_R (basal, R/R*) = 0.02; for ICI-118,551, K_L = 5 × 10⁻¹⁰ M, and K_{L^*} = 3 × 10⁻⁶ M: alprenolol was defined as a neutral competitive antagonist, K_L = K_{L^*} = 3 × 10⁻⁹ M (based on a pA_2 (−logK_R) value of 8.5.

3. In Vivo Evidence

The functional changes resulting from spontaneous activity of the overexpressed $\beta_2 AR$ that were observed in the isolated organ assay could also be detected in vivo. Indeed, cardiac catheterization and hemodynamic measurements performed in anesthetized transgenic and control mice revealed that the maximum first derivative of left ventricular pressure (LV dP/dt_{max}; which is used as an index of in vivo cardiac contractility) was significantly higher in mice overexpressing the $\beta_2 AR$. Systemic administration of the inverse agonist ICI-118,551 caused a significant reduction of the LV dP/dt_{max}, which resulted from the inhibition of the receptor spontaneous activity since it could be blocked by prior administration of the quasi-neutral antagonist alprenolol (7).

III. CONSTITUTIVE ACTIVITY AND PATHOLOGICAL CONDITIONS

As illustrated by the studies above, most demonstrations of compounds possessing inverse agonist activity are derived from systems that have been manipulated in some way to increase basal activity (usually by overexpressing the receptor). Demonstration of inverse agonist activity in native systems is quite rare, and thus the functional importance of spontaneous activity in normal physiology is still an open question. In systems expressing low levels of receptors with intrinsically low levels of spontaneous activity, inverse agonists and neutral antagonists will most likely behave in clinically indistinguishable manner. On the other hand, if the local concentration of a given receptor is sufficient or if the intrinsic spontaneous activity of a receptor subtype is high enough, one could predict that at least theoretically, the two classes of compounds could have different physiological responses. Whether receptor concentrations can be sufficiently high in normal tissues (e.g., in postsynaptic specialization in the brain) or whether the spontaneous activity of receptor subtypes such as the D5-dopamine receptor is sufficient to detect functional responses in humans that would be specific to inverse agonists remains to be seen. However, mounting evidence suggests that pathological conditions may exacerbate spontaneous activity and thus that inverse agonists may have advantageous therapeutic value in certain diseases.

The best indication suggesting that inverse agonists may be prefered therapeutic agents over neutral antagonist in some diseases is provided by the recent observation that several human diseases result from mutations of G-protein-coupled receptors that promote their constitutive activity. For example, in a type of male precocious puberty, a mutation of the luteinizing hormone receptor in the testes produces spontaneous function of the receptor, in the absence of circulating LH, thus leading to an exaggerated testosterone secretion and a precocious onset of puberty (32). Additional examples of human diseases resulting from constitutively activating mutations of G-protein-coupled receptors include: the Jansen-type metaphyseal chondrodysplasia, which results from a constitutively activating mutation (CAM) of the PTH receptor (33), congenital and somatic hyperthyroidism caused by CAMs of the thyrotropin receptor (34,35), and retinitis pigmentosa and stationary night blindness as a consequence of CAMs of rhodopsin (36). Based on the observation previously reviewed that inverse adrenergic agonists can inhibit the spontaneous activity of constitutively activated β_2AR and α_1AR mutants, it could be argued that inverse agonists would be of greater therapeutic utility than neutral antagonists in the diseases cited above since the problem is spontaneous activity of the receptor, not excess hormonal stimulation.

Although more speculative, it could be argued that inverse agonists could also be useful in pathologies resulting from the overexpression of a given receptor subtype. Schizophrenia may represent an example of such a pathology. Indeed, a study comparing the densities of dopamine receptors in brains of schizophrenic and nonschizophrenic patients showed the schizophrenic tissues had four times the level of the D4-dopamine receptor (37). Although several dopamine antagonists have been used clinically as effective neuroleptics, it has been difficult to correlate their therapeutic efficacy with known parameters such as affinity for this receptor subtype; perhaps a component of the therapeutic benefit lies in inverse agonist property of some of these antagonists at the dopamine D4-receptor subtype. Obviously, other possibilities, including their action on the 5-HT receptors, exist to explain the antipsychotic actions of some dopaminergic antagonists, but their potential role as inverse agonists deserves further investigation.

An even more speculative case could be made for possible differences between inverse agonists and neutral antagonists in the beneficial effects of β-adrenoceptor antagonists in congestive heart failure (CHF). The use of β-blockers in heart failure is increasing as clinical results show that some are beneficial in increasing the quality of life and decreasing mortality. However, not all drugs classified as β-adrenoceptor antagonists are effective in CHF. Clinical effectiveness in CHF is not correlated to known drug characteristics such as cardioselectivity (preferential β_1-adrenoceptor antagonist) or lipophilicity. A recent study using a constitutively active mutant of the human β_2-adrenoceptor has shown that inverse agonists, but not neutral antagonists, produce up-regulation of the receptor in transfected NG108-15 cells (38). It may be that the β-blockers demonstrating clinical improvement are really inverse agonists that are producing receptor up-regulation to improve heart function. Clearly, this is a very speculative hypothesis and more data concerning the inverse efficacy of clinically effective cardiotonic β-blockers toward β_1AR and β_2AR are required before it can be formerly proposed as a potential mode of action of the β-antagonists in heart failure.

It may also be possible that in some circumstances a neutral antagonist may display better clinical efficacy than an inverse agonist. It has recently been demonstrated that histamine H_2 receptors can be up-regulated by cimetidine and ranitidine, both of which displayed inverse agonist activity in transfected CHO cells, but not by the neutral antagonist burimamide (17). The authors suggest that these findings may yield a plausible explanation for the observed development of tolerance with the use of cimetidine and ranitidine. Their chronic use may produce up-regulation of the H2 receptor, which would oppose its therapeutic value of preventing histamine activation of the H2 receptor and subsequent release of hydrochloric acid.

Although, the physiological importance of GPCRs' spontaneous activity and of inverse agonism in normal conditions still remains to be demonstrated, data obtained to date clearly indicate that agonist-independent receptor activity, at least in the case of CAMs, contributes to various pathophysiological processes. In these cases, agents that can inhibit the spontaneous signaling activity such as inverse agonists will certainly prove to be therapeutically useful. Therefore, further investigation aimed at defining the molecular determinants of inverse agonism will undoubtedly lead to the development of a new arsenal of clinically useful agents with distinct indications to that of neutral competitive antagonists.

REFERENCES

1. Kenakin TP. Are receptors promiscuous? Intrinsic efficacy as a transduction phenomenon. Life Sci 1988; 43:1095–1101.
2. Leff P, Harper D. Do pharmacological methods for the quantification of agonists work when the ternary complex mechanism operates? J Theor Biol 1989; 140:381–397.
3. Mackay D. Continuous variation of agonist affinity constants. Trends Pharmacol Sci 1988; 9:156–157.
4. Monod J, Wyman J, Changeux J-P. On the nature of allosteric transitions: a plausible model. J Mol Biol 1965; 12:88–118.
5. Colquhoun D. Drug Receptors. London: Macmillan, 1973.
6. Leff P. The two-state model of receptor activation. Trends Pharmacol Sci 1995; 16:89–97.
7. Bond RA, Johnson TD, Milano CA, Leff P, Rockman HA, Mcminn T, Apparsunduram S, Hyek MF, Kenakin TP, Allen LF, Lefkowitz RJ. Physiologic effects of inverse agonists in transgenic mice with myocardial overexpression of the beta$_2$-adrenoceptor. Nature 1995; 374:272–276.

8. Luo X, Zhanq D, Weinstein H. Ligand-induced domain motion in the activation mechanism of a G-protein-coupled receptor. Protein Eng 1994; 7:1441–1448.

9. Kenakin T. Agonist-receptor efficacy. II. Agonist trafficking of receptor signals. Trends Pharmacol Sci 1995; 16:232–238.

10. Costa T, Herz A. Antagonists with negative intrinsic activity of δ-opioid receptors coupled to GTP-binding proteins. Proc Natl Acad Sci USA 1989; 86:7321–7325.

11. Wang-ni T, Duzic E, Lanier SM, Deth RC. Determinants of α_2-adrenergic receptor activation of G proteins: evidence for a precoupled receptor/G protein state. Mol Pharmacol 1994; 45:524–531.

12. Chidiac P, Hebert TE, Valiquette M, Dennis M, Bouvier M. Inverse agonist activity of β-adrenergic antagonists. Mol Pharmacol 1994; 45:490–499.

13. Barker EL, Westphal RS, Schmidt D, Sanders-Bush E. Constitutively active 5-hydroxytryptamine$_{2c}$ receptors reveal novel inverse agonist activity of receptor ligands. J Biol Chem 1994; 269:11687–11690.

14. LaBrecque J, Fargin A, Bouvier M, Chidiac P, Dennis M. Serotonergic antagonists differentially inhibit spontaneous activity and decrease ligand binding capacity of the rat 5-hydroxytryptamine type 2C receptor in Sf9 cells. Mol Pharmacol 1995; 48:150–159.

15. Tiberi M, Caron MG. High agonist-independent activity is a distinguishing feature of the dopamine D_{1B} receptor subtype. J Biol Chem 1994; 269:27925–27931.

16. Mullaney I, Carr IC, Milligan G. Analysis of inverse agonism at the delta opioid receptor after expression in rat 1 fibroblasts. Biochem J 1996; 315:227–234.

17. Smit MJ, Leurs R, Alewijnse AE, Blauw J, Van Nieuw Amerongen PG, Van De Vrede Y, Roovers E, Timmerman H. Inverse agonism of histamine H2 antagonists accounts for upregulation of spontaneously active histamine H2 receptors. Proc Natl Acad Sci USA 1996; 93:6802–6807.

18. Gotze K, Jakobs KH. Unoccupied β-adrenoceptor-induced adenylyl cyclase stimulation in turkey erythrocyte membranes. Eur J Pharmacol 1994; 268:151–158.

19. Mewes T, Dutz S, Ravens U, Jakobs KH. Activation of calcium currents in cardiac myocyte by empty β-adrenergic receptors. Circulation 1993; 88:2916–2922.

20. Hanf R, Li Y, Szabo G, Fischmeister R. Agonist-independent effects of muscarinic antagonists on Ca^{2+} and K^+ currents in frog and rat cardiac cells. J Physiol (Lond) 1993; 461:743–765.

21. Leeb-Lundberg LMF, Mathis SA, Herzig MCS. Antagonists of bradykinin that stabilize a G-protein-uncoupled state of the B2 receptor act as inverse agonists in rat myometrial cells. J Biol Chem 1994; 269:25970–25973.

22. Samama P, Pei G, Costa T, Cotecchia S, Lefkowitz RJ. Negative antagonists promote an inactive conformation of the β_2-adrenergic receptor. Mol Pharmacol 1994; 45:390–394.

23. Samama P, Cotecchia S, Costa T, Lefkowitz RJ. A mutation-induced activated state of the β_2-adrenergic receptor. J Biol Chem 1993; 268:4625–4636.

24. Weiss JM, Morgan PH, Lutz MW, Kenakin TP. The cubic ternary complex receptor-occupancy model. I. Model description. J Theor Biol 1996; 178:151–167.
25. Milligan G, Bond RA, Lee M. Inverse agonism: pharmacological curiosity or potential therapeutic strategy? Trends Pharmacol Sci 1995; 16:10–13.
26. Kenakin T. Agonists, partial agonists, antagonists, inverse agonists and agonist/antagonists? Trends Pharmacol Sci 1987; 8:423–426.
27. Black JW, Shankley NP. Inverse agonists exposed. Nature 1995; 374:214–215.
28. Leff P. Inverse agonism: theory and practice. Trends Pharmacol Sci 1995; 16:356–358.
29. Cotecchia S, Exum S, Caron MG, Lekkowitz RJ. Regions of the α_{1A}-adrenergic receptor involved in coupling to phosphatidylinositol hydrolysis and enhanced sensitivity of biological function. Proc Natl Acad Sci USA 1990; 87:2896–2900.
30. Kjelsberg MA, Cotecchia S, Ostrowski J, Caron MG, Lefkowitz RJ. Constitutive activation of the α_{1B} adrenergic receptor by all amino acid substitution at a single site. J Biol Chem 1992; 267:1430–1433.
31. Milano CA, Allen LF, Dolber P, Rockman H, Chien KR, Johnson TD, Bond RA, Lefkowitz RJ. Enhanced myocardial function in transgenic mice with cardiac overexpression of the human β_2-adrenergic receptor. Science 1994; 264:582–586.
32. Shenker A, Laue L, Kosug S, Merendino JJJ, Minegishi T, Cutler GBJ. A constitutively activating mutation of the luteinizing hormone receptor in familial male precocious puberty. Nature 1993; 365:652–654.
33. Schipani E, Kruse K, Juppner H. A constitutively active mutant PTH-PTHrP receptor in Jansen-type metaphyseal chondrodysplasia. Science 1995; 268: 98–100.
34. Kopp P, van Sande J, Parma J, Duprez L, Gerber H, Joss E, Jameson JL, Dumont JE, Vassart G. Congenital hyperthyroidism caused by a mutation in the thyrotropin receptor gene. N Engl J Med 1995; 332:150–154.
35. Parma J, Duprez L, van Sande J, Cochaux P, Gervy C, Mockel J, Dumont J, Vassart G. Somatic mutations in the thyrotropin receptor gene cause hyperfunctioning thyroid adenomas. Nature 1993; 365:649–651.
36. Rim J, Oprian DD. Constitutive activation of opsin: interaction of mutants with rhodopsin kinase and arrestin. Biochemistry 1995; 34:11938–11945.
37. Seeman P, Guan HC, Van Tol HHM. Dopamine D_4 receptors elevated in schizophrenia. Nature 1993; 365:441–445.
38. Macewan DJ, Milligan G. Inverse agonist-promoted up-regulation of the human β_2-adrenergic receptor in transfected neuroblastoma X glioma hybrid cells. Mol Pharmacol 1996; 50:1479–1486.

Index

Adrenoceptors
α_1-Adrenoceptor agonists
as nasal decongestants, 235–236
α_2-Adrenoceptor agonists
as antihypertensive drugs,
233–235
β_2-Adrenoceptor agonists
combined therapy with steroids
in asthma, 166–168
design of agonist drugs, 141–147
long-acting agonists, 144–148,
152–158
mechanism of action, 149–158
pharmacology of salmeterol,
146–148, 153–157
therapeutic use in asthma,
135–158, 162–168
α_1-Adrenoceptor antagonists
as antihypertensive drugs,
232–233
α_2-Adrenoceptor antagonists
as treatment for benign prostatic
hyperplasia, 237–241
α-Adrenoceptors
drugs in development 232–244
existing drugs 231–238
subtypes and role in disease
231–243
β_2-Adrenoceptors
in asthma, 131–138, 162–168
cloning and expression, 81,
369–372

[β_2-Adrenoceptors]
computer-aided ligand design,
330–332
constitutive activity, 369–372
structure, 81, 334–336, 339–340,
346
3–D receptor models, 334–336
classification, 8–9, 138–141
Agonists
α_1-adrenoceptor agonists, 235–236
α_2-adrenoceptor agonists, 233–235
β_2-adrenoceptor agonists, 135–158
analysis of affinity and efficacy,
28–37
computer-aided design, 325–332
5–$HT_{1B/1D}$ receptor agonists,
176–190
inverse agonists, 71–74, 98–100,
364–375
P2U (P2Y$_2$) purinoceptor agonists,
256–266
Angiotensin receptor antagonists
design of candesartan, 216–217
design of losartan, 215
development of AT$_1$-receptor
antagonists, 214–217, 219–221
nonselective antagonists, 217–219
peptide antagonists, 209–210
synthetic compounds, 210–212
Angiotensin receptors, 209–214
molecular biology, 213–214
subtypes in the cardiovascular
system, 209–213

Angiotensin
  role in cardiovascular system,
    208–209
Antagonists
  adenosine receptor antagonists,
    327–330
  α_1-adrenoceptor antagonists,
    232–233, 241–243
  α_2-adrenoceptor antagonists,
    237–241
  analysis of affinity, 39–44
  angiotensin AT_1-receptor
    antagonists, 209–212
  computer-aided design, 325–332
  histamine H_2-receptor antagonists,
    196–204
  muscarinic antagonists, 274–289
  NMDA-receptor antagonists,
    300–309
Assays of receptor function
  binding assays, 50–58, 353–354
  biochemical assays, 92–98,
    354–357, 365–368
  electrophysiological analysis,
    110–113, 115–125
  high throughput screening, 93–94,
    353–357
Asthma
  combined use of steroids and
    β_2-adrenoceptor agonists,
    166–168
  concern over use of β_2-adreno-
    ceptor agonists, 162–166
  control of airway smooth muscle
    tone, 136–138
  design of β_2-adrenoceptor agonists
    for, 141–147
  epinephrine and cortisol in,
    135–136
  long-acting β_2-adrenoceptor
    agonists in, 144–148, 152–158
  muscarinic antagonists in, 285–288
  synthetic steroids in, 159–162
ATP (adenosine triphosphate)
  actions on purinoceptors, 253–264
  role in epithelial function, 256–259

Benign prostatic hypertrophy
  design of subtype specific
    α-adrenoceptor antagonists
    for, 238–241
  treatment using α_2-adrenoceptor
    antagonists, 237–238

Computer-aided drug design
  amino acid sequence analysis,
    339–340
  design of receptor ligands,
    325–332
  receptor models, 332–338
Cystic fibrosis
  role of P2U ($P2Y_2$) receptors in,
    256–259, 263–266

Electrophysiology
  analysis of drug-receptor
    interactions, 127
  analysis of receptor structure and
    function, 115–125
  role in understanding disease,
    125–126
Epilepsy
  NMDA receptor antagonists in,
    303–309

Glaucoma
  treatment using α-adrenoceptor
    agonists and antagonists,
    236–237
G Proteins
  downstream signaling, 85–92,
    354–357, 365–368
  role of receptor coupling, 12–19,
    69–74, 83–85, 348–349
  structure, 83–84
  subtypes, 84–85

H_2-receptor antagonists
  design and discovery, 196–200
  development of cimetidine,
    201
  therapeutic use of antagonists,
    202–204

Histamine
  role in gastric acid secretion,
    200–202
Histamine receptors
  classification, 196–200
Hypertension
  clinical use of AT_1-receptor
    antagonists, 214–217, 219–220
  design of sub-type specific α_1-
    adrenoceptor antagonists for,
    241–243
  role of renin-angiotensin system,
    207–209
  therapeutic use of α_1-adrenoceptor
    antagonists, 232–233
  therapeutic use of α_2-adrenoceptor
    antagonists, 233–235

Ligands
  determination of binding affinity,
    50–68
  influence of G-protein coupling on
    binding, 68–74

Migraine
  5-$HT_{1B/D}$ receptor agonists in,
    176–190
  control of cranial blood flow,
    176–177, 179–180
  role of sensory neurons, 180–181,
    186–189
  sumatriptan, 177–180
  zolmitriptan, 180–188
Molecular biology
  chimeric receptors, 117–118
  impact on drug design, 3, 213–214
  receptor cloning and expression,
    15, 81–83, 115–116, 120–122,
    261, 347–353
  receptor structure, 14–15, 18,
    81–83, 108–110, 116–120,
    122–124, 173–176, 259–261,
    298–300, 333–336, 339–340, 346
  site-directed mutagenesis, 118–120,
    260–261, 332
  subunit swapping, 116–117

Muscarinic receptor antagonists
  allosteric antagonists, 281
  drugs in development, 285–288
  irreversible antagonists, 281–284
  in peptic ulcer disease, 284
  use in receptor classification,
    274–281
Muscarinic receptors
  classification and subtypes, 274–281

Nasal congestion
  treatment using α_1-adrenoceptor
    agonists, 235–236
NMDA (N-Methyl-D-aspartate)
  role in neuronal damage, 297–298
NMDA receptor antagonists
  analysis of activity, 300–301
  clinical experience in stroke and
    epilepsy, 303–309
  design of channel blockers,
    303–304
  design of competitive antagonists,
    303
  design of glycine site antagonists,
    304–305
  redox site blockers, 306
  subunit selective antagonists,
    306–308
NMDA receptors
  molecular structure, 110, 119,
    298–300
  sites of antagonist action, 301–309

Peptic ulcer disease
  role of histamine, 200–202
  therapeutic use of H_2-antagonists,
    202–204
  therapeutic use of muscarinic
    antagonists, 284
P_2-purinoceptors
  classification and function,
    253–256, 266–267
P2U ($P2Y_2$) receptor agonists
  in cystic fibrosis, 256–259,
    263–266
  UTP and analogues, 263–266

P2U (P2Y$_2$) receptors
  cloning and expression, 259–262
  role in epithelial function,
    256–259
  structure, 259–261
  targets for cystic fibrosis therapy,
    256–259

Quantitative analysis of drug action
  Clark analysis for antagonists,
    43–44
  ligand association analysis, 61–63
  ligand displacement analysis, 65–67
  ligand dissociation analysis, 59–61
  ligand saturation analysis, 63–65
  null methods for agonists, 34–38
  operational analysis of agonists,
    29–34
  Schild analysis for antagonists,
    39–43

Receptor coupling
  downstream coupling, 86–92,
    365–368
  G-protein coupling, 12–19, 69–74,
    83–85, 348–349
  in recombinant expression systems,
    354–357
  transduction information in
    classification, 11–12
Receptors
  3–D models, 332–338
  cloning and expression, 15, 81–83,
    115–116, 120–122, 261,
    347–353
  conformational states, 19, 71–74,
    98–100, 364, 368–369
  constitutive activity, 71–74, 100,
    364–375

[Receptors]
  fundamental concepts, 1–2, 7–10
  orphan receptors, 95, 347
  principles of classification, 8–19
  structure, 14–15, 18, 81–83,
    108–110, 116–120, 122–124,
    173–176, 259–261, 298–300,
    333–336, 339–340, 346
Receptor theory
  Law of Mass Action, 28–29, 50–53
  occupancy theory, 25–26
  operational model of agonism,
    29–31
  ternary complex model, 69–74
  two-state model of agonism,
    363–365

Serotonin (5–HT$_{1B/D}$) receptor
    agonists
  development of sumatriptan,
    177–180
  development of zolmitriptan,
    180–188
Serotonin (5–HT) receptor
    antagonists
  drugs in development, 173–175
Serotonin (5–HT) receptors
  agonists in migraine, 176–190
  classification and subtypes,
    173–176
  structure, 173–176
Stroke
  NMDA receptor antagonists in,
    303–309

UTP (uridine triphosphate)
  actions on purinoceptors, 253–264
  hydrolysis resistant analogs,
    264–266
  use in cystic fibrosis, 263–265